# Satellite Personal Communications for Future-generation Systems

Springer-Verlag London Ltd.

http://www.springer.de/engine/

Enrico Del Re and Laura Pierucci (Eds.)

# Satellite Personal Communications for Future-generation Systems

## Final Report: COST 252 Action

With 222 Figures

 Springer

Enrico Del Re, Professor
Laura Pierucci, Assistant Professor

Department of Electronics and Telecommunications,
University of Florence, Via S. Marta 3, 50139 Firenze, Italy

British Library Cataloguing in Publication Data
Satellite personal communications for future-generation
  systems : final report - COST 252 Action
  1. Personal communication service systems 2. Artificial
  satellites in telecommunication - System engineering
  I. Del Re, Enrico, 1947- II. Pierucci, Laura
  621. 3' 8456
ISBN 978-1-85233-537-3

Library of Congress Cataloging-in-Publication Data
A catalog record for this book is available from the Library of Congress

ISBN 978-1-85233-537-3          ISBN 978-1-4471-0131-4 (eBook)
DOI 10.1007/978-1-4471-0131-4
http://www.springer.co.uk

© Springer-Verlag London 2002
Originally published by Springer-Verlag London Berlin Heidelberg in 2002

Typeset by Florence Production Ltd, Stoodleigh, Devon
69/3830-543210   Printed on acid-free paper   SPIN 10830774

# Preface

From December 1995 to April 2000, European COST Action 252 "Evolution of Satellite Personal Communications from Second to Future Generation Systems" promoted cooperation on this subject among many European research groups, including a significant number from central and eastern European countries.

It is recognised internationally that satellite systems are necessary to provide the required global coverage of future mobile and personal communications. Within the European Union, it was therefore of strategic importance that efforts pursued at a national level be complemented and enhanced on a much larger scale by an European concerted research action aiming at the definition and development not only of an European system but also of a worldwide global system. Features expected to be implemented by the new generation systems include: personalised services with a capability to respond to new services, facilities and applications (e.g. multimedia and personal communications); mobile terminals to offer the same services, facilities and applications as a fixed terminal in a common-feel way; freedom to roam on a worldwide basis; parity of quality, performance, privacy and cost between fixed and mobile access. These systems are intended to realise true personal mobile radio-communications from anywhere and to allow people to communicate freely and in any form with each other from homes or offices, cities or rural areas, fixed locations or moving vehicles (land, sea, air).

The two main objectives of the COST 252 Action have been:

- a mid-term activity focused on the adaptability of the GSM standard (second generation mobile terrestrial system) to mobile satellite systems (MSS) considering the future requirements of an integrated third generation system Universal Mobile Telecommunications System (UMTS);
- a longer-term activity dealing with the exploitation of the satellite component for the MBS (Mobile Broadband System) able to provide services at higher rates (> 2Mb/s) than presently assumed in the UMTS.

Based on selected technical contributions of the European Action COST 252 this book provides many innovative results which can be the basis for new global (mobile /terrestrial/satellite) telecommunications systems providing efficient multimedia services at high rates.

The latest research results and new perspectives on communications problems are presented in areas such as:

- Satellite systems.
- Management signalling and resource allocation.
- CDMA system and receivers.
- Protocols.
- Coding.
- Satellite-ATM and Satellite-UMTS.

The book describes the research work carried out within the COST 252 Action in an exaustive and user-friendly way. After the Introduction, each of the following three chapters report the activities in Working Groups 1, 2 and 3 respectively. Chapter 5 summarises the conclusions and suggests some possible research topics for future Actions.

This book would not be possible without the enthusiastic effort and friendly cooperation among all the COST 252 participants.

We would like to express our sincere gratitude to them.

Enrico Del Re
Laura Pierucci

# Contents

# List of Acronyms

| | |
|---|---|
| ABRS | Adaptive Bandwidth Reservation Scheme |
| ACTS | Advanced Communications Technologies and Services |
| ATM | Asynchronous Transfer Mode |
| BER | Bit Error Rate |
| B-ISDN | Broadband-ISDN |
| CDMA | Code Division Multiple Access |
| COST | European Cooperation in the field of Scientific and Technical Research |
| CTF | Channel Transfer Function |
| DCA | Dynamic Channel Allocation |
| DCM | Discrete Channel Model |
| DECT | Digital Enhanced Cordless Telecommunications |
| ETSI | European Telecommunications Standards Institute |
| FCA | Fixed Channel Allocation |
| FES | Fixed Earth Station |
| FIFO | First Input, First Output |
| GEO | Geostationary Orbit |
| GoS | Grade of Service |
| GRAN | Genetic Radio Access Network |
| GSM | Global System for Mobile communications |
| HEO | Highly Elliptical Orbit |
| IC | Interference Cancellation |
| ICR | Initial Cell Rate |
| IMT-2000 | International Mobile Telecommunications-2000 |
| IN | Intelligent Network |
| IP | Internet Protocol |
| ISDN | Integrated Service Digital Network |
| ISL | Inter-Satellite Link |
| ITU | International Telecommunication Union |
| ITU-R | International Telecommunication Union-Radiocommunication Sector |
| ITU-T | International Telecommunication Union Standardisation Sector |
| LA | Location Area |
| LEO | Low Earth Orbit |
| LMN | Local Mobile Network |
| LOS | Line of Sight |
| LUI | Last Useful Instant |
| MAC | Medium Access Control |
| MAI | Multiple Access |
| MBS | Mobile Broadband System |
| MEO | Medium Earth Orbit |
| MMP | Modulated Markov Process |
| MONET | Mobile Network project |
| MPEG | Moving Pictures Expert Group |
| MS | Mobile Subscriber |
| MSA | Mobile Switching Area |
| MSS | Mobile Satellite Systems |
| MTCM | Multilevel Trellis Coded Modulations |
| MUD | Multiuser Detection |
| OFDM | Onthogonal Frequency Division Multiplexing |

| | |
|---|---|
| OVSF | Orthogonal Variable Spreading Factor |
| PES | Primary Earth Station |
| PIC | Parallel Interference Cancellation |
| PPADCA | Persistent Polite Aggressive DCA |
| PPD | Partial Packet Discard |
| PRMA | Packet Reservation Multiple Access |
| PSTN | Public Switched Telephone Network |
| QH | Queuing of Handover |
| QoS | Quality of Service |
| RACE | Research and Development in Advanced Communications Technologies in Europe |
| RRM | Radio Resource Network |
| SAINT | Satellite Integration in the Future Mobile Network |
| S-ATM | Satellite-ATM |
| S-CDMA | Satellite–CDMA |
| SEECOMS | Satellite EHF Communications for Mobile Multimedia Services |
| SIC | Successive Interference Cancellation |
| S-IP | Satellite-IP |
| S-PCN | Satellite-Personal Communication Network |
| S-PCS | Satellite-Personal Communication Service |
| S-UMTS | Satellite-UMTS |
| SWA | Sliding Window Algorithm |
| TCH | Tomlinson Cercas Hughes Codes |
| TCM | Trellis Coded Modulations |
| TDMA | Time Division Multiple Access |
| TES | Traffic Earth Station |
| UMTS | Universal Mobile Telecommunications System |
| UT | User Terminal |
| VSAT | Very Small Aperture Terminal |
| W-CDMA | Wideband–CDMA |
| WG1 | Working Group 1 |
| WG2 | Working Group 2 |
| WG3 | Working Group 3 |

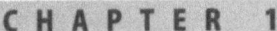

# COST 252
# Framework

**E. Del Re and L. Pierucci** University of Florence, Italy

## 1.1 Introduction

The two main objectives of the European Cooperation in the field of Scientific and Technical Research (COST) 252 Action "Evolution of Satellite Personal Communications from Second-to-Future-generation Systems" have been:

- a mid-term activity focused on the adaptability of the GSM standard (second generation mobile terrestrial system) to mobile satellite systems (MSS) considering the future requirements of an integrated third generation system (UMTS);
- a longer-term activity dealing with the exploitation of the satellite component for the MBS able to provide services at higher rates (> 2 Mb/s) than presently assumed in the UMTS.

The main objective of the International Telecommunications Union (ITU) in the International Mobile Telecommunications after the year 2000 (IMT-2000) is to extend the service provided by the current second-generation systems with high-rate data capabilities to support wireless Internet access but also simultaneous voice and data, multimedia, e-mail and broadband integrated services.

It is recognised internationally that satellite systems are necessary to provide the required global coverage of future mobile and personal communications. Within the European Union, it was therefore of strategic importance that efforts pursued at a national level be complemented and enhanced on a much larger scale by an European concerted research action aiming at the definition and development, not only of a Pan-European system but also, of a worldwide global system. Features expected to be implemented by the new generation systems include: personalised services with a capability to respond to new services, facilities and applications (e.g. multimedia and personal communications); mobile terminals to offer the same services, facilities and applications as a fixed terminal in a common-feel way; freedom to roam on a worldwide basis; parity of quality, performance, privacy and cost between fixed and mobile access. These systems are intended to realise true personal mobile radio-communications from anywhere and to allow people to communicate freely with each other from homes or offices, cities or rural areas, fixed locations or moving vehicles (land, sea, air).

The satellite component of the future systems offers, in particular, an effective means for providing services to areas where terrestrial telecommunication infrastructures are not yet well advanced. Research undertaken in COST 252 has been aimed to consider geostationary and non-geostationary constellations of satellites to provide advanced mobile wideband multimedia services to users thruoughout the world at rates up to 2 Mb/s. For non-geostationary satellites, mobility management signalling, call set-up and resource allocation procedures have been considered in detail and their impact on the protocols of the terrestrial component have been carefully investigated. Thus COST 252 has dealt with the satellite parts for third generation mobile UMTS/IMT-2000 as well as the portable aspects of the mobile broadband systems (MBS).

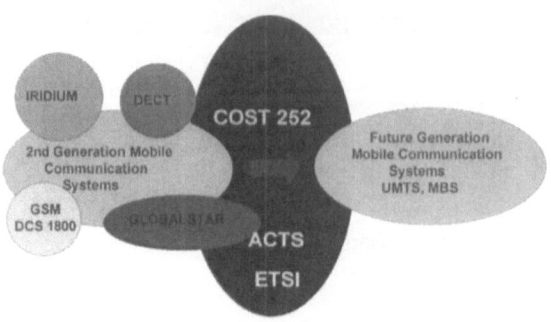

**Figure 1.1** Role of COST 252

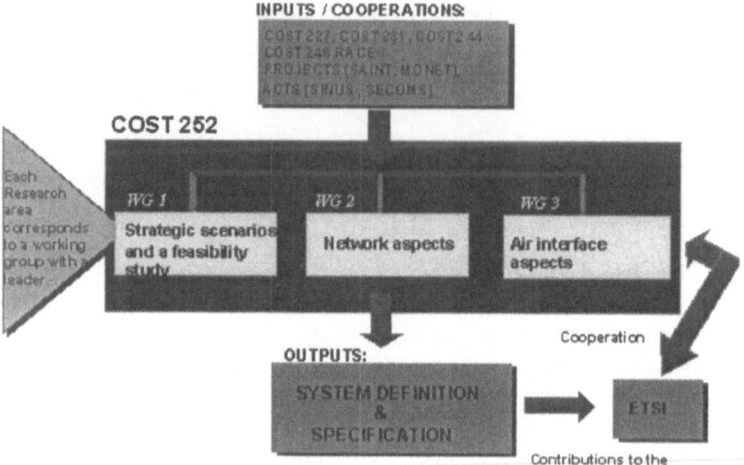

**Figure 1.2** Organisation of COST 252

## 1.2 Technical Description and Implementation

The role of COST252 is depicted in Fig. 1.1, where the cooperation with ACTS and ETSI is also indicated.

The organisation of COST 252, together with the foreseen inputs and outputs, is shown in Fig. 1.2.

### 1.2.1 WG1 – Strategic Scenarios and Feasibility Study

A first important step towards the achievement of future satellite-terrestrial integrated systems is the study of a smooth migration from second generation terrestrial systems (e.g. GSM) to include a satellite component. This is a key point, because it can meet the interests of both users and operators: on one hand we have the interest of operators to save the large investments for the implementation of second generation terrestrial systems; on the other hand, we have the interests of the users to be able to reuse as much as possible their terminals even for future networks.

Inputs from the COST 231 Action ("Evolution of Land Mobile Radio (including personal) Communications") have been used for the research within COST 252 Action. The strategic elements for the deployment of future integrated networks are:

- The ease of implementation of new multimedia and personalised services.
- Services integration into a network.
- Based on the recommendations of ITU, ETSI and of the outcome of RACE, service and security requirements have to be studied.
- The use of non-geostationary orbits for the satellite component of the systems (in particular, low earth orbits (LEO) and medium earth orbits (MEO)) have to be considered.

It has been considered essential to cooperate with COST 248 Action, titled "The Future European Telecommunications User", which studies the user's attitude on future telecommunication services

**Table 1.1.** WG2 tasks

| | |
|---|---|
| WG2100 Mobility management (SUR, UFI, TOR, DLR, IJS) | Multiple Access schemes<br>Multiple Access scheme for ATM via satellite |
| WG2200 Resource allocation (TOR, DLR, UOA, SUR, EPF, UFI, IJS) | Capacity issues<br>Network dimensioning<br>Channel allocation schemes<br>Source traffic modelling and traffic flow identification |
| WG2300 ATM traffic and modelling (UFI, SUR, EPF, DLR) | IP over ATM / protocol encapsulation<br>Wireless ATM protocol efficiency considerations<br>Broadband/multimedia services impact on networking procedures<br>Routing considerations<br>Wireless call admission control and QoS provisioning |

and products and related inputs on regulatory issues and system requirements from other bodies and groups (e.g. ETSI, ITU-T, ITU-R, SAINT, ACTS).

## 1.2.2 WG2 – Network Aspects

It is a matter of fact that many future global coverage MSSs will use a constellation of non-geostationary satellites. Starting from the outputs of COST 227 Action and related RACE projects (e.g. SAINT, MBS, MONET) it is necessary to proceed with the studies to improve or modify the network architecture and add new system functionalities. Then, a further sub-tasks division has been considered within this working group according to the following topics:

- The study of the network architecture.
- The study of procedures and protocols.
- Resources management in the integrated terrestrial-satellite scenario.

## 1.2.3 WG3 – Air Interface Aspects

The interest here focuses on transmissions aspects including also the characterisation of the satellite channel (especially in the case of non-geostationary satellites) and technology issues in order to cope with the link budget requirements that are particularly critical in the case of hand-held mobile terminals. Moreover, specific requirements on the technology are expected to fulfill the requirements related to RF emissions of handsets (health hazard). A tight cooperation with other related COST projects has been deemed as essential (e.g. COST 244 Action, titled "Biomedical Effects of Electromagnetic Fields"). The work within this group has been divided among the following topics:

- Characterisation of non-GEO channels and propagation effects.
- Multiple access techniques and inter-system interference.
- Identification of GSM and UMTS commonalities for a smooth transition.
- Handover strategies.
- Receivers: modulation, coding and equalisation with variable data rates.
- Gateway station.
- Terminal technology including biological constraints.

## 1.2.4 Contacts

COST 252 Action website: http://lenst.det.unifi.it/cost252
E-mail reflector for the COST 252 participants at cost252@bradford.ac.uk

*Chairperson*: Prof. Enrico del Re, Department of Electronics and Telecommunications, Via S. Marta 3, 50139 Firenze- Italy, Telephone: +39 055 4796285, Fax: +39 055 4796485. E.mail:delre@lenst.det.unifi.it

*Scientific Secretary*: Dott. Laura Pierucci, Department of Electronics and Telecommunications, Via S. Marta 3, 50139 Firenze- Italy, Telephone: +39 055 4796271, Fax: +39 055 4796485. E.mail: pierucci@lenst.det.unifi.it

**Table 1.2.** WG3 tasks

| | |
|---|---|
| WG3100 Channel characteristic, propagation, non-GEO satellites | Channel characteristics for measurements, modelling (TOR, SUR, TZD, UOA, FTC)<br>Propagation<br>Measurements<br>Modelling |
| WG3200 Multiple Access techniques (UFI, UOA, FTC, CSE, TOR, IST) | Multiple access and intersystem interference<br>Comparison of TDMA CDMA PRMA via satellite<br>Multimedia applications |
| WG3300 Receivers (IJS, AUT, UCL, CSE, SUR, IST, TZD, UFI) | Modulation<br>Coding<br>Equalisation with variable data rates<br>Smart antennas for space diversity |

**Table 1.3.** Participating institutions

| | |
|---|---|
| BRA Bradford University UK | IJS Institut Jozef Stefan SL |
| CSE CSELT I | SIN SINTEF N CTS Corporate Technology Swiss Telecom CH<br>SUR University of Surrey UK |
| DLR Institut fuer Nachrichtentechnik D | TOR University Rome Tor Vergata I |
| EPF Ecole Polytechnic. Federal de Lausanne CH | UCL University Catholic Louvain B |
| EST Ecole Superior Telecom F | UFI University of Florence I |
| FTC France Telcom – CNET F | WUT Warsaw University of Technology PL |
| IST Institute Superior Tecnico P | AUT Aristotle University of Thessaloniki GR |
| UAU University of Aveiro | UOA University of Athens GR<br>TZD Deuthsche Telecom D |

# CHAPTER 2

# Strategic Scenarios and Feasibility Studies

**R.E. Sheriff** University of Bradford, UK
**M. Mohorcic** Institut Jozef Stefan, Slovenia

## 2.1 Review of Objectives and Chapter Organisation

This chapter is concerned with strategic scenarios and feasibility studies of the next generation satellite personal communication systems. These topics were addressed within working group WG1 of the COST 252 Action with the aim to select a limited number of future satellite service scenarios for in-depth study in working groups specialised on network aspects (WG2) and air interface aspects (WG3). To achieve this aim the activities of WG1 were divided into four working areas: service requirements definition; user requirements definition; system requirements definition; and scenario definition.

- *Service requirements definition* aims to characterise multimedia services according to their service characteristics (i.e. QoS, BER, data rate, etc.). This area of work has mainly been covered in (1) and (5).
- *User requirements definition* considers service provision from a user perspective. Requirements to be categorised include: terminal type, user profiles, security requirements, etc. User requirements have been addressed in (1) and (5).
- *System requirements definition* considers the overall system, including operating environments, potential satellite orbits, visibility aspects, frequency issues, etc. These system requirements have been discussed (1, 2, 3, 4, 5, 6) and (7).
- *Scenario definition* activity is responsible for selecting scenarios for detailed study in WG2 and WG3. Scenarios for satellite personal communications to be studied more in-depth within the COST 252 Action have been outlined in (3) and (5).

In the following, the activities performed and the contributions produced within COST 252 Action are briefly described for each working area, highlighting the relevant results and the main problems that require further investigation.

## 2.2 Service Requirements Definition

### 2.2.1 Introduction

An analysis of future satellite service requirements has been performed in (1), where end-user services were categorised and some potential teleservices were described. In order to facilitate a smooth migration from current mobile satellite services (MSS) to future satellite personal communication services (S-PCS), the next generation of satellite communication networks will have to provide teleservices available in current second-generation mobile systems, as well as providing services compatible with those of the fixed networks.

Third generation mobile systems will be capable of delivering services and applications at data rates up to 2 Mbit/s and will integrate cordless, cellular, satellite, paging and wireless local loop technology. The third generation mobile system that will be deployed in Europe will be known as the universal mobile telecommunications system (UMTS), and will be a member of the global IMT-2000 (International Mobile Telecommunications-2000) system family. Thus, the next generation of satellite communications is expected to serve as the satellite component of the IMT-2000 system family (UMTS in Europe).

Focusing on the satellite-UMTS (S-UMTS) environment, several services will need to be restricted to lower data rates, when compared to the corresponding terrestrial-UMTS (T-UMTS) environment. On the other hand, S-UMTS can offer a range of satellite specific services not feasible in the terrestrial-UMTS component.

### 2.2.2 Classification of End-user Services

The most common approach for service classification is division into bearer services, teleservices, and supplementary services. Furthermore, in the age of multimedia services it is also suitable to define service components, which can be perceived as basic building elements of both, mono- and multimedia services.

In order to provide a smooth migration from the present to future generation personal communications, it is expected that the latter will accommodate a number of existing service networks, such as GSM 900, GSM 1800, GSM 1900, DECT, etc. Therefore, the future generation system will have to provide teleservices available in existing mobile networks and besides that, it is expected also to provide services compatible with those of the ISDN.

In order to achieve this aim, this is likely to require the use of ISDN bearer services (circuit and packet- mode services), B-ISDN bearer services (defined in ATM adaptation layer) and the Internet service using protocol IPv6. As for the supplementary services, these only add value to teleservices and cannot be offered independently.

For the following identification of services and applications, considered feasible to be offered in next generation mobile systems, it is convenient to define the basic service components, which compose mono- and multimedia services, according to the type of information source:

- *Speech*: optimised for voice communication services with low-quality demands utilising vocoders for achieving lower required bit rates.
- *Audio*: used for the transmission of radio programme and CD quality audio service.
- *Image*: used for transmission of compressed high-quality images (e.g. in MPEG 4 format).
- *Video*: for transmission of compressed video information at different quality levels.
- *Data*: used for data flow transmission in circuit or packet-mode.

A typical set of S-UMTS services should embrace telephony, teleconferencing, paging, messaging high quality audio broadcasting, video telephony, video conferencing, remote lecturing, database access, remote medical consultation, group travel services, vehicle navigation, freight management, etc.

In Fig. 2.1, the trade-off between achievable services bit-rate and mobility of users is depicted.

Figure 2.2 summarises the above in the context of UMTS services and illustrates the effect of mobility on service bit-rates.

In addition to traditional services (e.g. telephony, short messaging, low bit rate data transmission in voice channel) currently available in mobile communication systems of the second-generation, S-UMTS services will be able to support a wide range of new applications, such as:

- multimedia mailboxes;
- transfer of documents and files containing text, images and voice;

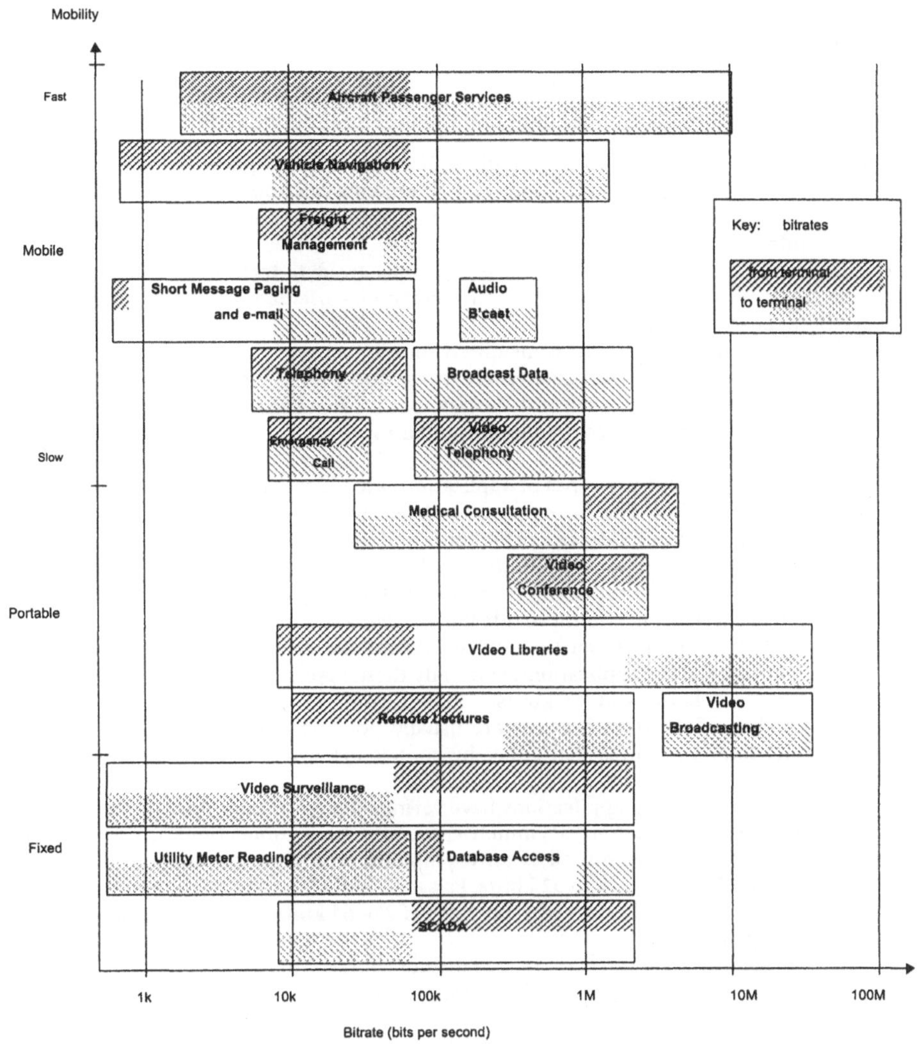

Mobility

Fast

Aircraft Passenger Services

Vehicle Navigation

Mobile

Freight Management

Short Message Paging and e-mail

Audio B'cast

Key: bitrates

from terminal

to terminal

Telephony

Broadcast Data

Slow

Emergency Call

Video Telephony

Medical Consultation

Video Conference

Portable

Video Libraries

Remote Lectures

Video Broadcasting

Video Surveillance

Fixed

Utility Meter Reading

Database Access

SCADA

1k     10k     100k     1M     10M     100M

Bitrate (bits per second)

**Figure 2.1.** Mobility vs. bit-rate for services.

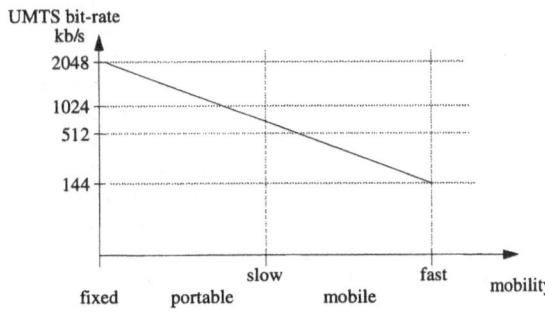

UMTS bit-rate kb/s

2048

1024

512

144

fixed     portable     slow     mobile     fast     mobility

**Figure 2.2.** Bit-rate of UMTS services vs. user mobility.

- messaging services;
- directory services;
- access to different databases;
- transactional applications;
- video information transfer;
- advanced traffic telematics applications; etc.

## 2.2.3 Expected Types of S-UMTS Services

*Telephony* – Examples of the telephony type of applications are two-party (telephone service) and multiparty (teleconference) voice applications and emergency call applications. They are symmetric real-time connection-oriented services. Typical user bit-rates are between 2.4–32 kbit/s, depending on the speech codec used.

*Voice-band data* – Typical voice-band data applications are data transmission by use of modems and facsimile. They are symmetric non-real-time connection-oriented services. Typical bit-rates are in the range between 2.4–28.8 kbit/s.

*Sound* – According to the quality of transmitted sound, two types of applications can be distinguished; program sound services (AM quality) are transmitted in the range 0–7 kHz, and high quality audio services (CD quality) in the range 0–20 kHz. Both, program sound and high quality audio services are asymmetric real-time connection-oriented services. They can be realised at 128 kbit/s and 940 kbit/s, respectively.

*Videotelephony* – Videotelephony types of applications are videotelephone, videoconferencing and video surveillance/monitoring. The former two applications are symmetric real-time connection-oriented services, with minimum bit-rates of 64 kbit/s and 384 kbit/s, respectively. However, for good quality of service, bit-rates up to 1920 kbit/s can be implemented. Remote video surveillance/ monitoring is asymmetric real-time connection-oriented service with typical throughput equal to videotelephony. All of these applications provide the user with voice, video and data communications.

*Teleaction* – Teleaction services can be divided into remote control, remote terminal and user/ network profile editing types of applications. Normally these applications have low bit-rate demands in the range between 1.2 kbit/s and 9.6 kbit/s, although some of them may have requirements for bit-rates up to 64 kbit/s. Teleaction services are usually non-real-time services, they can be symmetric or asymmetric and are provided either by connection-oriented or connectionless network service.

*Message handling* – Messaging applications have various demands regarding the network service. Short message service and paging are asymmetric non-real-time connectionless services with very low bit-rates (0.1–9.6 kbit/s). Voice mail is an asymmetric real-time connection-oriented service with throughputs from 2.4 kbit/s up to 32 kbit/s. Fax mail and electronic mail are asymmetric non-real-time connection- oriented services with bit-rates of 2.4–64 kbit/s and 1.2–64 kbit/s, respectively, depending on the type of terminal used.

*Telefax G4* – Telefax is an asymmetric non-real-time connection-oriented service, aimed for the exchange of documents containing coded information. Group 4 telefax requires a bit-rate of 32 or 64 kbit/s.

*Teletex* – Teletex is an asymmetric non-real-time connection-oriented service, used for the exchange of documents containing teletex coded information.

*Videotex* – Videotex is an asymmetric non-real-time connection-oriented service. It is used to retrieve text and image information from a videotex database.

*Directory* – Directory services are mainly used to provide an interaction with databases. A typical application is telephone directory, which can provide information on telephone numbers via voice or data connection. Typical bit-rates are between 2.4 and 9.6 kbit/s.

*Database access* – According to the source of information contained, databases are divided into simple (text) and special (text, voice, image) databases. Hence, different types of access applications must be developed for different types of databases. Simple database access can be implemented as an asymmetric non-real-time connection-oriented service. Bit-rates between 2.4 and 9.6 kbit/s should be sufficient for transmission of textual and data information. Special database access applications are also asymmetrical non-real-time connection-oriented services, but since they are designed for transmission of text, voice and image type of data, higher throughputs from 64 kbit/s and up to 1920 kbit/s are proposed.

*Teleshopping* – Teleshopping is an asymmetric non-real-time connection-oriented service, which should provide a transmission of text or voice in the direction from user to service provider and a transmission of text, voice and/or image in the opposite direction. Consequently, the bit-rate required

**Table 2.1.** Telephone service requirements

| | Mean | 95% | Target | Least acceptable value |
|---|---|---|---|---|
| Establishment delay | < 2 s | < 4 s | < 2 s | < 4 s |
| Establishment failure probability | $< 10^{-4}$ | $< 10^{-3}$ | $< 10^{-4}$ | $< 10^{-3}$ |
| Release delay | < 2 s | < 2 s | < 2 s | < 4 s |
| Release failure probability | $< 10^{-5}$ | $< 10^{-5}$ | $< 10^{-4}$ | $< 10^{-3}$ |
| Probability of premature release | $< 10^{-5}$ | $< 10^{-5}$ | $< 10^{-6}$ | $< 10^{-5}$ |
| Probability of unacceptable transmission | $< 10^{-6}$ | $< 10^{-5}$ | $< 10^{-6}$ | $< 10^{-5}$ |

for one direction is between 1.2 and 16 kbit/s and in the other direction in the range 128 kbit/s–768 kbit/s.

*Broadcasting* – Broadcasting applications include message broadcast/multicast (1.2–9.6 kbit/s), data broadcast/multicast (1.2–64 kbit/s) and emergency/public announcement (1.2–9.6 kbit/s textual, 2.4–32 kbit/s voice). Broadcast applications are asymmetric non-real-time connectionless services, while multicast applications are connection-oriented services. Textual emergency/public announcement is an asymmetric non-real-time connectionless service, while voice emergency/public announcement is an asymmetric real-time connection-oriented service.

*Electronic news* – Electronic news is an asymmetric non-real-time connection-oriented service, which allows a user to search for information in the accessed database. The information can be presented only as text or as text, image and voice. Bit-rate requirements for text, voice and image are 1.2–16 kbit/s, 2.4–64 kbit/s and 128–768 kbit/s, respectively.

*News distribution/agency* – News distribution is an asymmetric non-real-time connection-oriented service, which will provide textual or multimedia information transfer from a news agency to a number of users within a certain area. Textual type of service requires a bit-rate in the range of between 1.2 and 16 kbit/s, while multimedia (text, voice, image) type of service needs a bit-rate up to 768 kbit/s for the image part of the information.

*Unrestricted digital information transfer* – Unrestricted digital information transfer is an asymmetric non-real-time connectionless or connection-oriented service, which provides data transfer in multiples of 64 kbit/s.

*Telewriting* – Telewriting is a symmetric real-time connection-oriented service, which allows both connected users to contribute to the same document. Alterations on one side must be displayed on the other side in real-time. Requirements depend on the type of document discussed (text, image).

*Location and navigation services* – Location and navigation applications are symmetric non-real-time connection-oriented services. They require bit-rates of 16 kbit/s and 32–64 kbit/s, respectively.

### 2.2.4 Service Characteristics
These services will have to consider some general parameters and associated quality of service (QoS) values, as defined for all UMTS services and summarised in Table 2.1. Furthermore, most of S-UMTS services will also have to follow standardised requirements for their basic components (e.g. ITU-T Rec. G.721 for voice; ITU-T Rec. G.722, MPEG-1 or MPEG-2 for audio; ITU-T Rec. H.261, H.263 and MPEG-1 to MPEG-4 for video).

## 2.3 User Requirements Definition

### 2.3.1 Introduction
In general, customers have very different demands regarding types of services, their availability and quality. The available set of services to a particular user will already be predefined to a great extent with the selection of the type of the terminal, and with the user's geographic location and mobility. Thus, the availability of different sets of services is likely to be constrained by the following user and system requirements:

- the user profile;
- the type of terminal used;
- the environment in which the user is establishing the connection.

### 2.3.2 Definition of User Profiles

In general it is possible to distinguish between two major categories of user; business users and private users. Business users are characterised by the demand for a wide variety of applications and services with more stringent demands on quality and availability. However, business user demands could be further differentiated according to their profession and mobility requirement. Private users' demands, on the other hand, can be characterised by a more limited set of services and applications, mostly related to social interaction, entertainment and information services. Again these users could be sub-classified according to their interests, habits, social status, etc.

### 2.3.3 Terminal Categories

Third generation mobile systems are likely to support a range of bit-rates, the rate being dependent upon the operating radio environment (see Section 2.4.2). In UMTS, or the ITU equivalent IMT-2000, the satellite component is foreseen to provide mobile multimedia services to individual users at rates of up to 144 kbit/s.

Five terminal types are envisaged for S-UMTS: personal (hand-held, palm-top), vehicular (mounted), transportable (lap-top, brief-case), fixed and paging receivers. Each terminal classification will provide a distinct range of S-UMTS services, with capabilities for handover and roaming between networks. The maximum S-UMTS bit-rate may only be achievable using fixed or stationary, transportable terminals. Space/terrestrial dual-mode facilities will be required, as will single-mode, satellite-only handsets. Vehicular mounted terminals will not be so limited by the availability of transmitting power and antenna gain, when compared to personal terminals. Furthermore, some vehicular mounted terminals will allow throughputs up to 512 kbit/s or even 2 Mbit/s, but these type of terminals will mainly be used as gateways to local mobile networks (LMN) on means of mass transportation (e.g. aeroplane, train, coach, ship).

Transportable terminals will essentially be aimed at the international business traveller, and will be similar to the existing Inmarsat mini-M terminal. Fixed VSAT-type antennas will be used to provide communications to areas without access to the fixed network infrastructure. Paging terminals will be very low gain, receive only devices capable of receiving and displaying alphanumeric messages.

## 2.4 System Requirements Definition

### 2.4.1 Introduction

Following the definition of different terminal types and their implementation forms from the service and user requirement point of view, it is necessary to highlight, that throughputs of terminals depend on several factors, such as type of satellite orbit, frequency band, terminal mobility, etc.

The capacity limitations together with specific impairments in the S-UMTS environment make it unlikely that satellite personal communication services (S-PCS) beyond 2 Mbit/s will be delivered within the foreseeable future. Furthermore, in several radio operating environments (e.g. indoor environment, urban environment, etc.) services will have to be further restricted to lower data-rates due to severe fading effects caused by multipath propagation, shadowing and blocking.

In the following, various conditions are addressed, some of them having a significant influence on the behaviour, capacity and complexity of the satellite personal communication system. These conditions should be considered very carefully in the design phase of a communication system.

### 2.4.2 Radio Operating Environment

Satellite personal communications aim to complement terrestrial systems mainly regarding coverage area, while they are expected to offer the same or similar types of services. Some expected operating environments, however, introduce severe radio propagation impairments not well suited for satellite communications, affecting both the choice of services and terminals. As for the effect of different impairments on the satellite communication system, the following radio operating environments should be considered separately:

- *Indoor environment*: in this environment, the effect of propagation impairments are very significant, therefore only very low bit-rate services such as email and paging will be possible, regardless of type of orbit, frequency band, etc. As a consequence, this environment is only suitable for hand-held type terminals and bit-rates in the range of a few kbit/s.

- *Urban environment*: this environment is characterised by considerable fading effects due to multipath propagation, shadowing and blocking. Direct line of sight visibility is not obtainable in a high proportion of connections, which increases the probability of interruption regardless of minimum elevation angle. Typical terminals in this environment will include hand-held and vehicular-mounted with throughputs below 144 kbit/s. In any event, most of the services should be available in this environment, but their availability will be lower than in a rural environment.

- *Rural environment*: in this environment fading due to shadowing is still the major problem, but considerably less than in an urban environment. Consequently, the probability of service interruption is lower, which leads to an increase in service availability. In this environment, it is much easier to achieve direct line of sight of the satellite. This environment will host the complete spectrum of terminal types, with bit-rates limited to 144 kbit/s in most cases.
- *Fixed-mounted environment*: in the case of fixed-mounted terminals, higher transmission powers can be employed and direct line of sight should be guaranteed for most of the time. Therefore, availability of services should be very high, while bit error rate (BER) should be lower than in previous cases. Only fixed (at least during the connection) types of terminal can be used in this operating environment and throughput up to 2048 kbit/s should be available.
- *Open environment*: two specific mobile satellite environments, which do not fit in any of the previously mentioned groups, can be described as open radio operating environments. These are aeronautical and maritime environments. In both cases the restrictions regarding the size of antenna and power consumption are much less significant than in other mobile environments. Moreover, the maritime environment is characterised by a relatively low mobility, while the most attractive characteristic of aeronautical environment is the absence of atmospheric and rain attenuation, once the aircraft reaches its cruising altitude above the clouds. Both environments are especially well suited for the use of group mobile terminals with throughput up to 2 Mbit/s.

### 2.4.3 Candidate Satellite Orbits

*Overview* – In recent years the level of sophistication of satellites has advanced significantly. Indeed, space technologies, such as multi-satellite non-geostationary systems, multi-spot-beam coverage and inter-satellite links, are right now making their path into everyday commercial use. The satellite configuration in an integrated satellite/terrestrial environment has considerable scope for variation. Essentially, four types of satellite orbit can provide the space component in an integrated satellite/terrestrial network, namely: geostationary orbit (GEO); highly elliptical orbit (HEO); medium earth orbit (MEO); and, low earth orbit (LEO). When designing a satellite network, the advantages/disadvantages of each type of orbit need to be carefully considered, the choice of one orbit over another being a function of a number of complex parameters not just technical, but also economic and political. A significant impact on the choice of most appropriate orbit is the target system coverage area. Here it is worth noting that the work within COST 252 Action has focused on future generation satellite systems providing global coverage.

Figure 2.3 shows distances of particular types of orbit from the centre of the earth and their trajectories, while some of the most obvious comparisons between candidate orbits are shown in Table 2.2. Geostationary is the simplest type of orbit to implement and has been providing mobile satel-

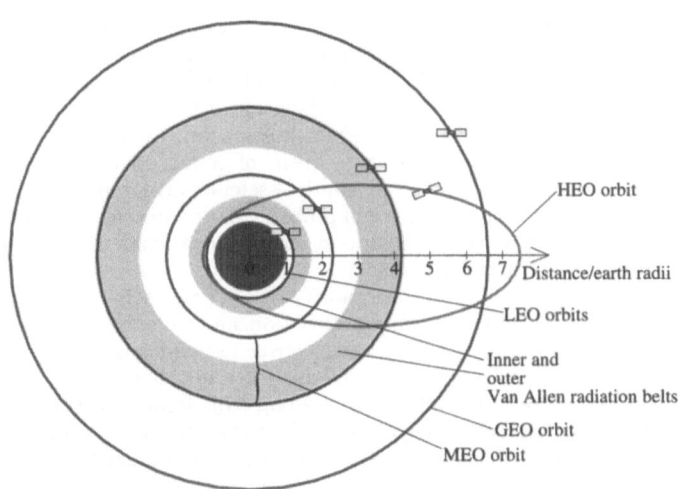

**Figure 2.3.** Satellite orbits.

**Table 2.2.** Comparison of different types of satellite orbit

|  | GEO | HEO | LEO | MEO |
|---|---|---|---|---|
| Type of orbit | Circular | Elliptical; inclined | Circular; polar or inclined | Circular; polar or inclined |
| Orbit period | 86 164.1 s | Typically 8–24 hours | In the range of 100 min | 5–12 hours |
| Orbit altitude | 35 786 km | Up to 42 000 km | 700–2000 km | 8000–20 000 km |
| Technology | established since the 1970s | Established since the 1970s | New, operational since late 1998 | New, operational 00 |
| Achievable coverage | regional / global excluding poles | Regional | Global | Global |
| Max. single-hop delay | 280 ms | Up to 210 ms | In the range of 20 ms | 80–120 ms |
| Doppler effects | Very small | Significant | Very high | High |
| Propagation path losses | Very high | Very high | Low | Moderate |
| Handovers | Not needed between satellites, Rarely between spot-beams | Not frequent | ≈ 10 min between satellites, ≈ 2 min between spot-beams | ≈ 90 min between satellites, ≈ 15 min between spot-beams |

lite communications for over twenty years in one form or another. However, this type of orbit introduces a significant delay in transmissions, which can be noticeable in speech, and generally offers poor visibility to a mobile terminal in suburban and urban areas.

Significantly, shorter propagation delays and lower path losses in LEO and MEO orbits seem to be advantageous over GEO and HEO orbits, when global coverage is to be supported. Furthermore, with LEO and MEO orbits, higher minimum elevation angles are achievable than with GEO orbit, which results in lower probability of shadowing and higher link availability. Although high minimum elevation angles in HEO orbits can be guaranteed much easier than in LEO and MEO orbits, propagation delays and path losses are even worse than in GEO orbit in addition to large Doppler shifts when the handover between satellites is performed. Moreover, terminal requirements for satellite systems utilising LEO and MEO are less stringent in terms of power consumption, which will result in lower terminal cost.

However one should take into account that not all services are delay sensitive; especially this can be confirmed for broadcast type of services, which, apart from trunk telephony, have been the main driving force in satellite communications. These type of future broadband services can be still provided over geostationary satellites, thus relaxing the capacity requirements on non-geostationary satellite systems, which are more optimal for provision of interactive delay-sensitive type of services. Hence, the future satellite personal communication network is expected to consist of satellites in different orbits (hybrid satellite constellation), possibly interconnected via inter-satellite links (ISLs).

The biggest disadvantage of LEO and MEO orbits is the large number of satellites required for global coverage of the earth, together with the increased network complexity and Doppler effects. Due to the relatively low orbital altitude (i.e. 700–2000 km for LEO and 8000–20 000 km for MEO) each satellite will only illuminate a small region on the surface of the earth. In order to provide a continuous global communication service, it is necessary to place a number of satellites in orbit, equally spaced around the earth in a number of orbital planes. These satellites can be placed in either an inclined or polar orbit, or a combination of the two. The polar orbit is a special case of the inclined orbit where the inclination is 90° to the equator. Total global coverage is only possible by deploying satellites in either polar orbits or orbits of greater than 70° inclination.

Due to the requirement for multiple satellite orbits, at least one satellite should always be in view of a terminal (although this will probably not be the case in built-up urban areas). Thus it should be possible to optimise the satellite-to-terminal link when multiple satellites are in view through the use of link diversity techniques.

However, the high orbital velocity of LEO and MEO satellites means that transmissions will be subject to a significant Doppler variation, and to maintain continuous real-time transmission some means of implementing handover between satellites is required.

In order to select the most suitable satellite constellation, a number of constellation characteristics have to be taken into account along with some other orbital parameters. The minimum elevation angle has been recognised as the most limiting parameter. It impacts the service availability time as well as delay variations. However, it also influences the satellite constellation; the higher the minimum elevation angle, the more satellites are needed to cover the same area. In the design of a

**Table 2.3.** Orbital parameters of the selected constellations

| | Constellation | | | | Satellite altitude h (km) | $\varepsilon_{min}$ (°) | Orbit period T (s) | Satellite vs (km/s) | Max. range velocity $\psi_{max}$ (°) |
|---|---|---|---|---|---|---|---|---|---|
| | N | P | Q | m | | | | | |
| Inclined MEO | 27 | 9 | 3 | {5/3} | 10 354 | 40 | 21 541 | 4.881 | 33.02 |
| polar MEO | 28 | 4 | 7 | N/A | 10 354 | 40 | 21 541 | 4.881 | 33.02 |
| polar LEO | 96 | 8 | 12 | N/A | 1 666 | 30 | 7 180 | 7.039 | 16.63 |

satellite personal communication system for the provision of real-time and non-real-time broadband services, the minimum elevation angle, at which a user sees the satellite, must be high and independent of the user's geographic location. In order to guarantee high link availability and acceptable delay variations with a reasonable number of satellites, the target value for minimum elevation angle is envisaged to be 40° for MEO constellations and 30° for LEO constellations.

*Selected satellite constellations* – Taking into account different orbital parameters and constellation characteristics, three satellite constellations have been selected for more in-depth study of various aspects having impact on system requirements definition. In order to be able to evaluate both the impact of orbital altitude and of inclination angle, two distinctive MEO constellations using polar and inclined satellite orbits and one polar LEO constellation have been selected. Orbital parameters of the three selected constellations are summarised in Table 2.3. A particular constellation is denoted by total number of satellites, N, number of orbit planes, P, number of satellites in each orbit plane, Q, and harmonic factor, m, which denotes initial distribution of satellites over the sphere.

Polar views at the three selected constellations are shown in Fig. 2.4 for: (a) the selected inclined MEO constellation, (b) the selected polar MEO constellation, and (c) the selected polar LEO constellation. Similarly, Fig. 2.6 shows coverage areas of satellite footprints on the earth for the three selected constellations.

## 2.4.4 Frequency Spectrum

Traditionally, mobile satellite systems have been operating in the L and S bands for uplink and downlink, respectively, while feeder links have been established in the C band. These frequency bands have been adequate for conventional narrowband services, but in order to accommodate broadband services a transition from traditional bands to higher frequencies in the Ka (20/30 GHz) and EHF (40/50 GHz) bands will be necessary.

These frequency bands, however, need to be allocated on a global level at World Radio Conference (WRC) for each particular satellite system. S-UMTS as a member of IMT-2000 family system, for example, is still envisaged to operate in the frequency band at 2 GHz, as depicted in Fig. 2.5.

## 2.4.5 Satellite Visibility Aspects

### 2.4.5.1 Overview

As mentioned in Section 2.4.4, satellite personal communication networks (S-PCN) offering broadband services to mobile users will have to move from traditional L and S frequency bands to higher frequencies in order to accommodate broadband channels. At these frequencies, propagation

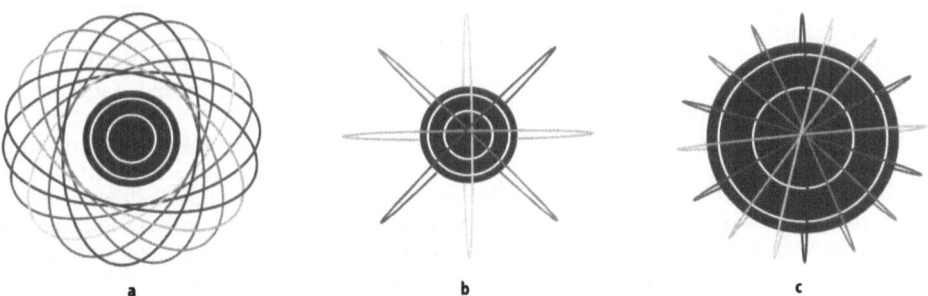

a        b        c

**Figure 2.4.** Polar view at the selected satellite constellations: **a** inclined MEO, **b** polar MEO, **c** polar LEO.

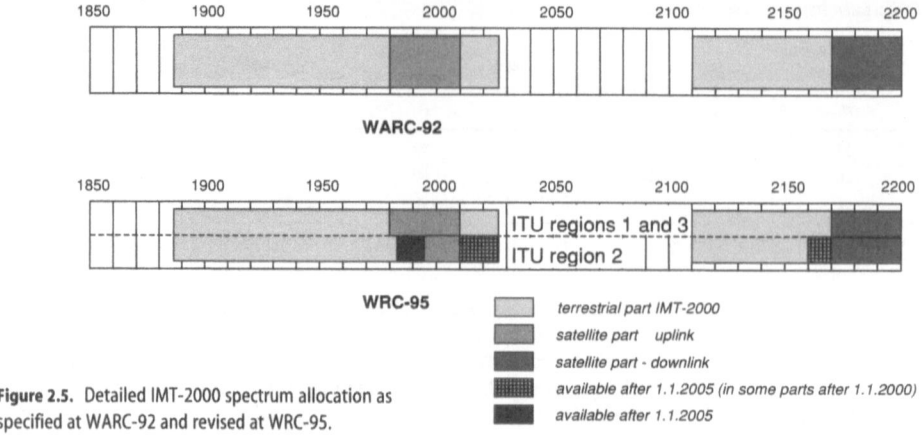

**Figure 2.5.** Detailed IMT-2000 spectrum allocation as specified at WARC-92 and revised at WRC-95.

impairments due to rain, cloud and so on, can have a significant impact on the availability of the satellite link. Different methods based on resource redundancy should be used to guarantee acceptable link availability in such a system. Such methods include:

- site and satellite diversity;
- power control;
- adaptive modulation techniques, data rate and coding;
- fade spreading.

Achievable improvement and implementation complexity of some of these methods depend on the selected satellite constellation. The efficiency of closed-loop adaptive power control, for example, greatly depends on the round trip delay. As to the site and satellite diversity, their efficiency is determined by satellite visibility parameters.

In the following, satellite visibility parameters are discussed: minimum elevation angles ($\varepsilon_{min}$) for multiple visible satellites; probability of multiple satellite visibility; probability of satellite visibility at different elevation angles; and distribution of azimuth angles for simultaneously visible satellites. As an example, simulation results are compared for the selected satellite constellations, as described previously.

### 2.4.5.2 Multiple Satellite Visibility for Link Diversity

Besides the minimum elevation angle of the highest visible satellite, the minimum elevation angle of the second/third highest visible satellite is also of a great interest, if the possibility to implement satellite diversity is to be evaluated. The best results for $\varepsilon_{min}$ of the highest and the second (and the third) highest satellites in a particular constellation are generally not obtained at the same inclination angle $\beta$. Thus, a trade-off has to be performed between the minimum elevation angle of the highest visible satellite and the probability of multiple satellite visibility.

Global minimum elevation angles for the highest three/two visible satellites and a probability of multiple satellite visibility at the selected minimum elevation angle is shown as a function of inclination angle for the inclined MEO constellation in Fig. 2.7, for the polar MEO constellation in Fig. 2.8, and for the polar LEO constellation in Fig. 2.9.

In order to guarantee $\varepsilon_{min} = 40°$ for the highest visible satellite in the inclined MEO constellation and to obtain the highest possible $\varepsilon_{min}$ for the second and the third highest visible satellite (26° and 21.5°), an optimal inclination angle for the selected constellation equals 59° (see Fig. 2.7a). Results for multiple satellite visibility show that the probability of visibility for the second and the third highest satellite above $\varepsilon_{min} = 30°$ equals 99.5 per cent and 89.5 per cent, respectively (see Fig. 2.7b).

As shown in Fig. 2.8, the distribution of satellite coverage areas in polar constellations is highly non-uniform in comparison to inclined constellation. Consequently, the minimum elevation angle for the second highest satellite in polar MEO constellation achieves only 17.5° and is independent of variation of inclination angle (see Figure 2.8a). As a result of optimisation for single/multiple satellite visibility as a function of minimum elevation angle, the inclination angle has been set to 88°. Probability of visibility of the second highest satellite above $\varepsilon_{min} = 30°$ is only 93.6 per cent (see Figure 2.8b).

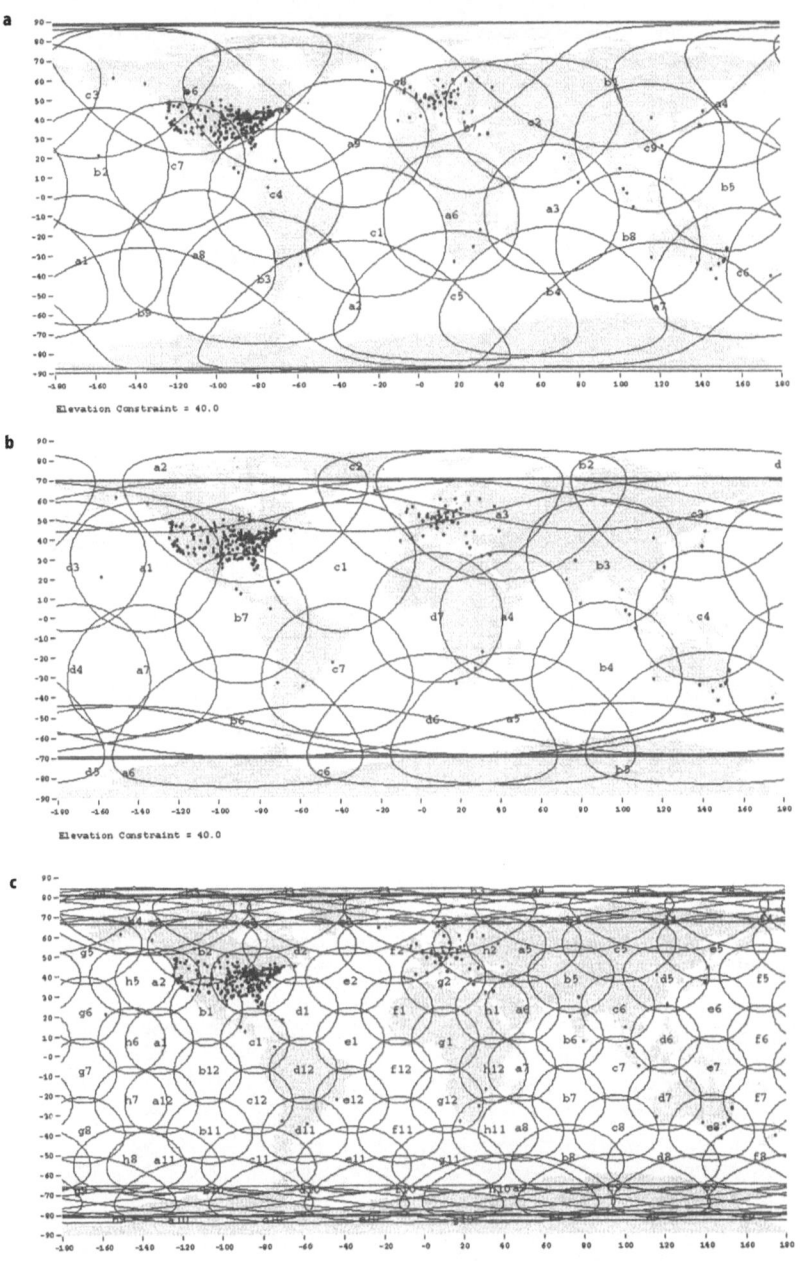

**Figure 2.6.** Coverage obtained by the selected satellite constellations: **a** inclined MEO, **b** polar MEO, **c** polar LEO.

In order to guarantee $\varepsilon_{min} = 30°$ for the highest visible satellite and to obtain the highest possible $\varepsilon_{min}$ for the second highest visible satellite (12°), an optimal inclination angle for the selected LEO constellation equals 91°. Results for multiple satellite visibility at global $\varepsilon_{min} = 20°$ show that the probability of visibility for the second highest satellite reaches only 92.7 per cent (see Figure 2.9).

### 2.4.5.3 Probability of Satellite Visibility at Different Elevation Angles

Results presented above are considering the global minimum elevation angle, which is the worst case taking into account all latitudes between 0° and 90° (north and south). Improved results can be obtained for satellite visibility, if the main area of interest is limited to the latitudes between 35° and

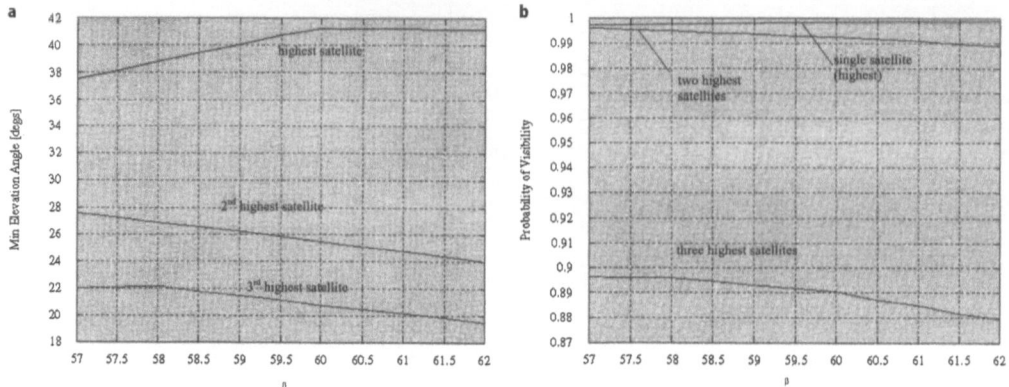

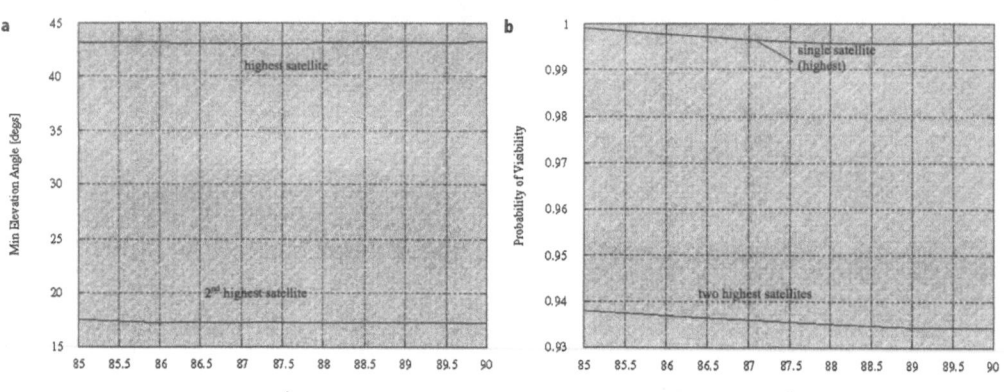

**Figure 2.8.** Selected polar MEO constellation: **a** Global $\varepsilon_{min}$ for the highest two visible satellites, **b** probability of multiple satellite visibility at $\varepsilon_{min} = 30°$.

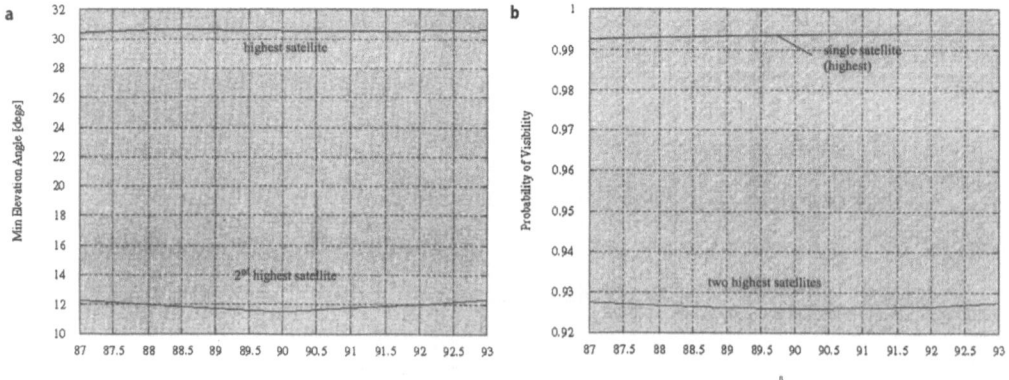

**Figure 2.9.** Selected polar LEO constellation: **a** Global $\varepsilon_{min}$ for the highest two visible satellites, **b** probability of multiple satellite visibility at $\varepsilon_{min} = 20°$.

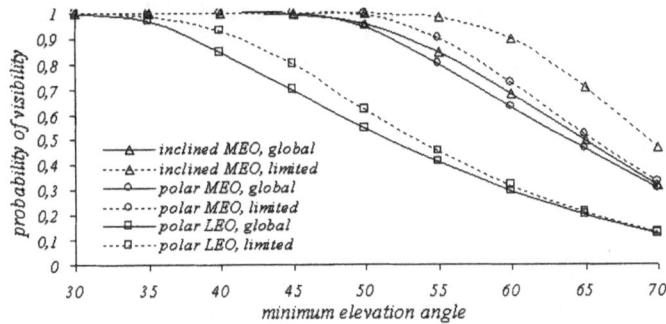

**Figure 2.10.** Probability of a single satellite visibility versus $\varepsilon_{min}$ (global and limited between 35° and 65°).

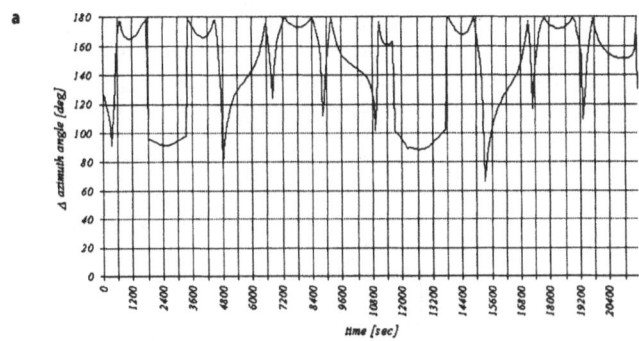

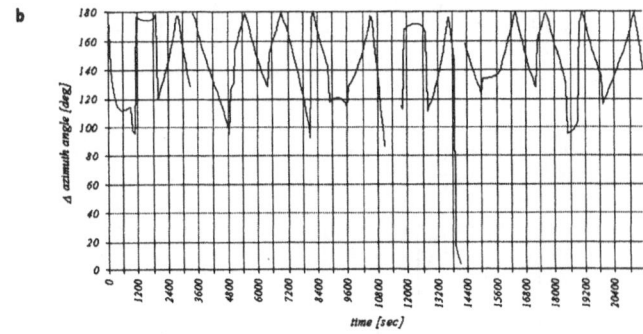

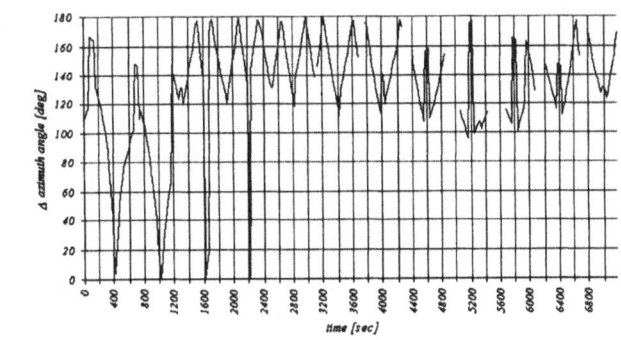

**Figure 2.11.** Difference between azimuth angles for the highest two visible satellites from London: **a** inclined MEO constellation, **b** polar MEO constellation, **c** polar LEO constellation.

65° (north and south), where most potential users are located, while the global $\varepsilon_{min}$ for the highest visible satellite remains the same. The probability of a single (the highest) satellite visibility as a function of a minimum elevation angle is given in Fig. 2.10 for unlimited and limited area of interest.

### 2.4.5.4 Distribution of Azimuth Angles for Simultaneously Visible Satellites

The distribution of azimuth angles for simultaneously visible satellites is another important satellite visibility parameter, especially for the evaluation of feasibility of satellite diversity implementation. This parameter, however, depends on the user location and dynamically changes as satellites move in their orbits. Hence, computer simulation has to be used to analyse dynamic changes of azimuth and elevation angles for the selected satellite constellations for a certain chosen location on the earth. As an example, the simulation results for difference between azimuth angles for the highest two visible satellites in the selected satellite constellations are given in Fig. 2.11 for a user located in London.

Results in Fig. 2.11a for the inclined MEO constellation show that there are at least two satellites visible above $\varepsilon_{min} = 40°$ (actually the highest above $\varepsilon_{min} = 60°$), and the difference in azimuth angle is for most of the time between 80° and 180°. As to the polar MEO constellation (Fig. 2.11b), only one satellite is visible above $\varepsilon_{min} = 40°$ during certain periods of time (broken line), when satellite diversity might not be feasible. At those moments the highest satellite is normally visible above $\varepsilon_{min} = 70°$. When two satellites are visible above $\varepsilon_{min} = 40°$, the difference in azimuth angle is mostly between 90° and 180°. Results for the polar LEO constellation, shown in Fig. 2.11c, also show that

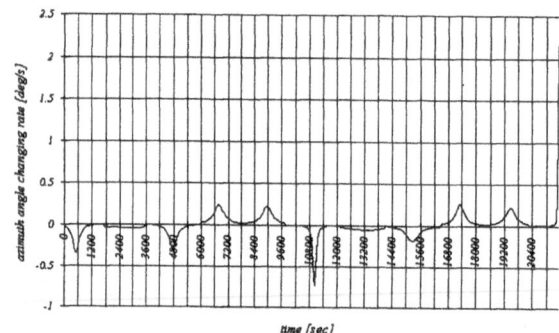

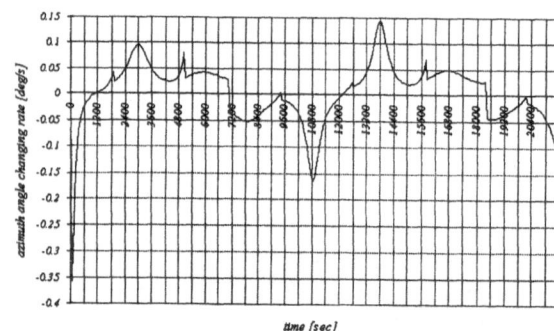

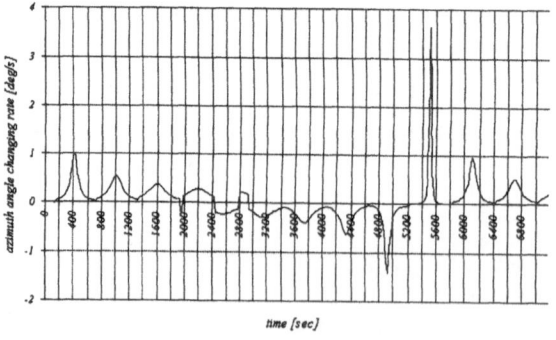

**Figure 2.12.** Azimuth angle changing rate for the highest visible satellite from London: **a** inclined MEO constellation, **b** polar MEO constellation, **c** polar LEO constellation.

**Table 2.4.** Characteristics of satellite service area for the selected satellite constellations

| | Max. distance within a satellite coverage area | Max. duration of satellite pass |
|---|---|---|
| | $D_{max}$ (km) | $T_{max}$ (s) |
| MEO | 7351.8 | 3951.4 |
| LEO | 3703.2 | 663.4 |

only one satellite is visible above $\varepsilon_{min} = 30°$ during certain periods of time. At these instants of time, however, the minimum elevation angle for the user located in London is in general above 55°. The difference in azimuth angle is in the range 100°–180° when two visible satellites are in co-rotating orbits, and in the range 0°–180°, when two visible satellites are in counter-rotating orbits.

Besides azimuth angle, at which a certain satellite is visible, an important parameter is also the rate of change of this angle, since some terminal types may be developed which use mechanic or electronic steering antennas. Hence, Fig. 2.12 gives the change rate of azimuth angle for the highest visible satellite from London.

The change rate for the selected constellations is most of the time very low. Actually, it only increases when a handover to a new highest visible satellite is performed or when the satellite passes just above the location of the user (i.e. London). The change rate is somewhat higher for the polar LEO constellation, which is due to higher velocity of satellites, but even for this constellation the azimuth changing rate is most of the time below 1°/s.

## 2.4.6 Satellite Service Area

### 2.4.6.1 Overview

The analysis of the satellite visibility parameters for the non-geostationary satellite system from a global point of view is here complemented by the analysis of a single satellite coverage area. The satellite coverage area is hereafter considered a service area, and can be further divided into satellite cells, in order to use the channel resources in a more efficient manner and to achieve the required high satellite antenna gain.

In the following, the extent of a single satellite service area and the duration of the satellite pass are defined, which, in turn, determines the time between inter-satellite handovers. Similar parameters related to inter-spot-beam handovers are also defined.

### 2.4.6.2 Single Beam Satellite Coverage Area

The single beam coverage area for a given $\varepsilon_{min}$ can easily be calculated using space geometry equations. The maximum distances within a satellite coverage area for the three selected satellite constellations are given in Table 2.4. The maximum distance also determines the maximum duration of the satellite pass, or in other words, the maximum time duration between two consecutive inter-satellite handovers.

The maximum distance within a single satellite coverage area, $D_{max}$, and the maximum duration of the satellite pass, $T_{max}$, both depend on a centre angle of the spherical coverage area, which in turn depends on the altitude of the satellite and on the minimum elevation angle. Consequently, the same results are obtained for these two parameters for both selected MEO constellations, thus from here on simulation and calculation results are represented for MEO and LEO, standing for both selected MEO constellations and for the selected polar LEO constellation, respectively.

The actual duration of a certain satellite visibility depends on the user location inside the satellite coverage area. The bigger the distance between a user and an imaginary trace of the subsatellite point on the earth, the shorter the duration of the satellite pass. The maximum distance between the user and the subsatellite trace equals half of the maximum distance within a single satellite coverage area, $D_{max}/2$. Only a user located on the subsatellite trace experiences the maximum duration of the satellite pass, $T_{max}$.

The duration of the satellite pass as a function of the distance between the user and the subsatellite track, $d$, is presented in Fig. 2.13 for the selected non-geostationary constellations. The time in which a user is served by a certain satellite decreases while the distance between the user and the subsatellite trace is increasing. Similarly service duration decreases if $\varepsilon_{min}$ is increasing.

### 2.4.6.3 Multibeam Satellite Coverage

Results shown in Fig. 2.13 apply for the service area of a single satellite utilising only one beam. Such an implementation leads to a wide coverage, but in order to achieve the required high satellite antenna

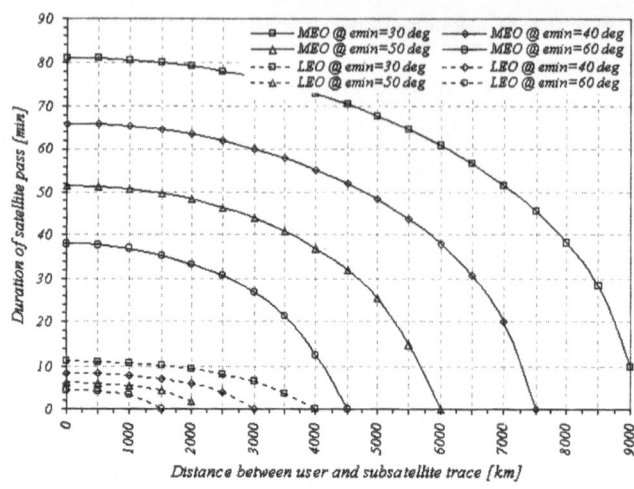

**Figure 2.13.** Single beam satellite service duration.

gain, due to the low EIRP and G/T values of mobile terminals, very large antennas should be deployed on spacecraft. Consequently, in order to obtain a relatively small antenna with the required high gain, multibeam satellites should be used. Each spot-beam footprint on the surface of the earth generates a distinctive area, hereafter referred to as a (satellite) cell. A multibeam satellite allows the coverage of the satellite service area, constrained only by a minimum elevation angle, with a number of beams. The antenna gain and the system performance increase as the beamwidth decreases; a large number of high gain beams, each of them illuminating only part of the entire satellite coverage area, allow efficient reuse of frequency spectrum, high channel density, and low transmitter power. However, the satellite payload complexity, as well as overall system complexity, increases with an introduction of multibeam coverage. Therefore, an optimal number of beams, and their beamwidth, have to be defined for each selected satellite constellation.

In order to simplify the analysis, only regular cell patterns have been considered, based on a hexagonal structure with the typical values for the number of cells equal to 1, 7, 19, 37, 61, etc. Some other widely used values for $N_{cell}$, such as 48, can also be represented by a hexagonal cell layout with an incomplete outer crown. In reality, however, cells mostly have circular or elliptical shape and the adjacent cells slightly overlap. Due to overlapping, the frequencies used in adjacent cells have to be different. In order to decrease the interference to an acceptable level, the same frequency band can be re-used several times within the same satellite coverage area, as long as cells using the same frequencies are sufficiently separated.

Multibeam satellite coverage can be obtained by different types of antenna, which also determine the shape and distribution of beams. In non-geostationary satellite systems, spot-beam footprints on the earth have in general circular or elliptical shape. In the first case, spot-beams are assumed to have the same size of footprints on the surface of the earth (equal cells), while in the second case, spot-beams are assumed to be the same size at the satellite (equal spot-beams).

According to the implementation of the satellite antenna, beams can be fixed with the spacecraft's axis, scanned, or mechanically pointed. The last solution is not desirable for large antennas. Geostationary satellites usually use an antenna with fixed beams. Non-geostationary satellites, especially at LEO altitudes, can benefit from implementation of scanning beams, which allow a significant reduction in the number of handovers and higher available gain to be obtained. The major disadvantages are more complex satellite payload and lower traffic efficiency, if channel resources are permanently associated with a corresponding cell. Scanning beams generate earth-fixed cells, while fixed beams generate satellite-fixed cells.

If the maximum steering angle of the satellite antenna with scanning beams is large enough to be capable of pointing spot-beams to the corresponding cell for the entire duration of the satellite pass, no inter-spot-beam handovers are required. In general, however, inter-spot-beam handover has to be performed when the satellite antenna reaches the maximum steering angle. At this instance of time, all cells within the satellite serving area change serving spot-beams. Since all active users in all cells of a particular satellite are to be handed over at the same time, there should be no blocked handovers, provided that all spot-beams have the same capacity. Due to constant satellite motion, this handover process becomes periodic and predictive, and leads to an efficient satellite capacity management, without extensive signalling overhead.

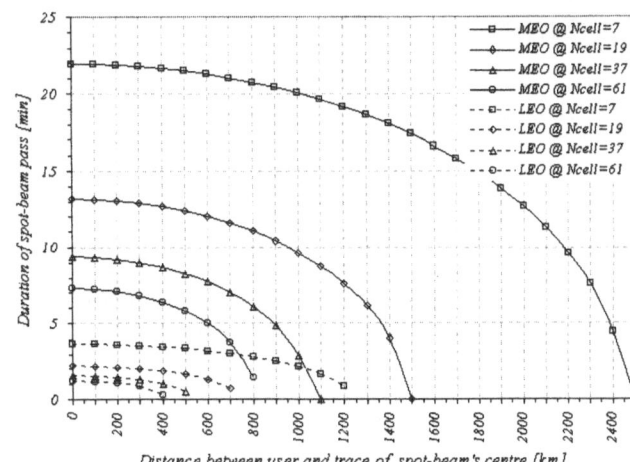

**Figure 2.14.** Spot-beam service duration.

*Distance between user and trace of spot-beam's centre [km]*

Although steering of spot-beams complicates the satellite payload, it considerably relaxes resource and handover management, which otherwise leads to significant signalling overhead, especially in lower satellite orbits. Moreover, since satellite cells are small and constantly fixed at certain geographical locations, the type of service allowed in a particular cell can be predetermined, a sort of traffic optimisation can be introduced, and it is possible to shape service areas according to national boundaries.

Fixed spot-beams are fixed with respect to the spacecraft's axes; hence they sweep the surface of the earth as the satellite travels around it. Here the channel resources are allocated to the satellite spot-beams, and not to the satellite cells. Since different channel resources have to be used in adjacent satellite cells, which are moving relative to the user with a high velocity, inter-spot-beam handovers have to be performed frequently, as well as inter-satellite handovers, when the user moves out of the satellite service coverage area. Moreover, handovers in the system with fixed spot-beams have to be performed on a user by user basis, according to the results of channel quality measurements. This leads to a substantial signalling traffic and to a higher probability of failed handovers, unless an allocation scheme with channel reservation and prioritisation of handover requests is used.

The maximum distance within a satellite cell and the time duration of the spot-beam pass can be again calculated using space geometry equations. However, now the maximum distance within a cell determines the maximum spot-beam service area, and the time duration of a spot-beam pass determines the service duration within a particular spot-beam. Figure 2.14 shows duration of spot-beam pass for the selected MEO and LEO constellations at their nominal minimum elevation angles for different number of spot-beams per satellite.

A spot-beam is characterised by its angular width $\theta$. Table 2.5 lists the beam angular width for the regular cellular patterns consisting of 7, 19, 37, 61, 91, 127, 168, 216 and 271 cells at the selected MEO and LEO constellation altitudes. Furthermore, corresponding values for the maximum duration of the satellite and the spot-beam passes (maximum duration between inter-satellite and inter-spot-beam handover) are given; for the latter, fixed spot-beams are considered. As for the scanning spot-beams, the time duration between two successive inter-spot-beam handovers depends on the maximum steering angle of the antenna and can be substantially improved, especially at LEO altitudes.

**Table 2.5.** MEO satellite antenna characteristics for different number of spot-beams

| $N_{cell}$ | | 7 | 19 | 37 | 61 | 91 | 127 | 169 | 217 | 271 |
|---|---|---|---|---|---|---|---|---|---|---|
| MEO | $\theta$ (°) | 13.26 | 8.07 | 5.78 | 4.50 | 3.69 | 3.12 | 2.70 | 2.39 | 2.14 |
| | $T_{max}$ for sat. (s) | 3951.4 | 3951.4 | 3951.4 | 3951.4 | 3951.4 | 3951.4 | 3951.4 | 3951.4 | 3951.4 |
| | $T_{max}$ for beam (s) | 2634.2 | 1580.5 | 1128.9 | 878.09 | 718.44 | 607.91 | 526.85 | 464.87 | 415.94 |
| LEO | $\theta$ (°) | | 39.94 | 24.89 | 17.98 | 14.04 | 11.52 | 9.76 | 8.46 | 7.47 | 6.69 |
| | $T_{max}$ for sat. (s) | | 663.4 | 663.4 | 663.4 | 663.4 | 663.4 | 663.4 | 663.4 | 663.4 | 663.4 |
| | $T_{max}$ for beam (s) | | 442.3 | 265.3 | 189.5 | 147.4 | 120.6 | 102.0 | 88.4 | 78.0 | 69.8 |

## 2.5 Scenario Definition

After considering a number of different approaches, it was decided that the terrestrial backbone network infrastructure would provide the main consideration in the definition of the scenarios for in-depth study. Essentially there are two options: an ATM B-ISDN backbone network infrastructure; or an IP based backbone network infrastructure.

Furthermore, the first option can be further differentiated by considering the required network interface between the mobile network (i.e. that being considered by COST 252 Action) and the backbone. Specifically, whether some form of adaptation is required (as in S-UMTS, where ATM cells are not transmitted via satellite) or whether ATM-like transmissions (with cells appropriately modified for the satellite environment) via satellite are available.

Hence, the following three scenarios were considered appropriate for further in-depth study by working groups WG2 and WG3:

*S-UMTS scenario*:   Non-ATM transmissions. Current guidelines suggest that rates of up to 144 kbit/s are possible for S-UMTS. This scenario necessitates some form of adaptation to allow S-UMTS network to interface with the B-ISDN network.

*S-ATM scenario*: Transmission of ATM-like cells via satellite, assuming a B-ISDN backbone network.

*S-IP scenario*: Internet protocol transmissions allowing direct connection to an IP backbone network.

## References
(1)   Fawley RJ, Mohorcic M, Sheriff RE (1997) Satellite-PCN service requirements. COST 252 TD(97)01.
(2)   Mohorcic M, Sheriff RE (1997) Non-geostationary satellite constellations for provision of mobile broadband services, COST 252 TD(97)06.
(3)   Wiesendanger S, Scenarios for Satellite-PCN evolution, COST 252 TD(98)03.
(4)   Wiesendanger S, Short summary of WRC 1997, COST 252 TD(98)04.
(5)   Sheriff RE, Current Activities in COST 252/WG 1: strategic scenarios and feasibility studies, COST 252 TD(98)11.
(6)   Mohorcic, M, STSM report: modelling satellite constellations using OPNET, COST 252 TD(98)20.
(7)   Mohorcic M, Sheriff RE, Non-geostationary satellite constellations: visibility aspects, COST 252 TD(98)21.

# CHAPTER 3

# Network Aspects of Dynamic Satellite Multimedia Systems

**A. Sammut (Editor), I. Mertzanis, C. Meenan, G. Sfikas**  University of Surrey
**E. Del Re, R. Fantacci, G. Giambene**  Università degli Studi di Firenze
**V. Santos M. Dinis, J. Neves**  Instituto de Telecomunicações
**M. Werner, J. Bostic, C. Delucchi**  Deutsches Zentrum für Luft- und Raumfahrt (DLR)
**M. Mohorcic, A. Svigelj, G. Kandus**  Institut Jozef Stefan

## 3.1 Introduction

The spread of wideband services to the office and home through fixed and terrestrial mobile networks will create a market for the extension of such services for people on the move and for those who work outside the office environment. Various wideband terrestrial cellular systems will serve users at both local and global level, although the scope for deployment of such systems in the early phase (2000+) will be limited to densely populated and affluent areas, resulting in a potential requirement for wireless access technologies which can connect users to the network in other areas.

Satellite systems can provide access to the network in areas that terrestrial and even proposed stratospheric systems cannot. Global standards compatible with the International Telecommunications Union (ITU) specifications will overlay and complement the multitude of radio access standards and will exist in the IMT-2000 (International Mobile Telecommunications after the year 2000) and UMTS (Universal Mobile Telecommunications system) scenarios. The IMT-2000/UMTS concept aims to provide seamless and ubiquitous advanced mobile wideband multimedia services to users throughout the world at rates up to 2 Mb/s on the terrestrial segment but limited to 64/144 kb/s on the satellite element. The satellites will be second-generation systems with integration at the service level. At the same time, current work in ACTS (Advanced Communications Technologies and Services) projects (SECOMS,[1] ASSET,[2] WISDOM[3]) exists under the mobile broadband systems (MBS)

[1] Satellite EHF Communications for Mobile Multimedia Services (Second Call for ACTS).
[2] Advanced Satellite Switching End-to-end Trials (Third Call for ACTS).
[3] Wideband Satellite Demonstration of Multimedia (Third Call for ACTS).

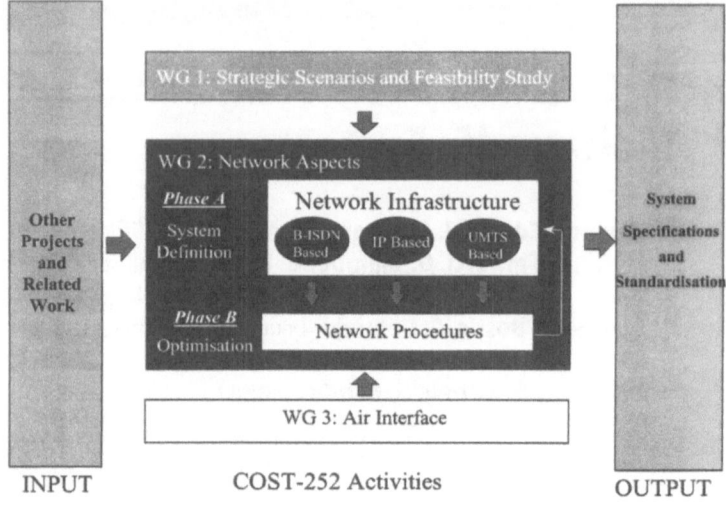

**Figure 3.1.** COST 252 WG2 working plan and interaction with the other work groups.

banner and from the Ka-band scenarios (SPACEWAY, GALAXY, EUROSKYWAY, SKYBRIDGE and TELEDESIC) which are predominantly aimed at fixed terminals and are looking to possible portable implementation, maybe up to 2 Mb/s.

Research undertaken in COST 252 has been aimed at mobile and portable systems in the access network and does not conflict with core network aspects. Thus, as far as systems are concerned, COST 252 has dealt with the satellite parts for third generation mobile UMTS/IMT-2000 as well as the portable aspects of the MBS systems. As far as time scales are concerned, UMTS had planned to standardise in 1998 and be operational by 2002 whereas portable terrestrial systems are not expected to come on line prior to 2002 and maybe not until 2005. The second-generation low earth orbit (LEO) and regional geostationary orbit (GEO) satellites will clearly fix the form for this new market in mobile/portable multimedia in the 2002–2010 time scale. Within the European Telecommunications Standards Institute (ETSI) work has already been done for UMTS through RACE and ACTS, although this has been mainly concentrated on the terrestrial segment.

### 3.1.1 Working Plan For WG2 – Network Aspects

The Working Group 2 (WG2) study has been undertaken in two main phases (Fig. 3.1). During phase A inputs from WG1-Strategic Scenarios and Feasibility study were used as the baseline for the system definition. Three main scenarios were foreseen for the satellite network infrastructure. The first one was based on the proposed wireless extensions of broadband-integrated services digital network (B-ISDN) (such as MBS or W-ATM[4] forum's specifications), the second was based on the modifications of the existing Internet protocol (IP) in order to support user/terminal mobility and Quality of Service (QoS) guarantees, and the third was based on a post UMTS integration scenario.

Phase A was defined as a period in which the integration scenarios which were identified at the preliminary stage would be studied to determine the important issues which required research effort. Phase B involved optimisation of the network procedures. Depending on the selected scenario for the network infrastructure the most important working areas would be addressed and examined in depth, by the involved partners. Outputs from Phase B would be used as feedback to the system definition stage (Phase A) in an interactive way. The outputs from WG2 are expected to contribute to the final system specification and to the standardisation of the network procedures.

### 3.1.2 WG2 Working Areas and Tasks

Prior to the beginning of Phase A, a set of preliminary areas for study were established which were then ascribed to WG2 participants according to the existing or proposed research effort within each institution. Table 3.1. illustrates the work areas and associated participants, the acronyms used for each participant are expanded in Table 1.3.

---

[4] Wireless Asynchronous Transfer Mode.

**Table 3.1.** Preliminary study areas and concerned participants

| WG2 working areas | Participant |
|---|---|
| *Network architecture (WG2100)* | |
| Satellite and B-ISDN/ IP /UMTS integration/interworking | SRU, EPF, IJS, BRA |
| On-board processing /ATM switching-affects on the satellite architecture | UFI |
| Backbone network /fixed earth station inter-connectivity | COST 253 and other related projects |
| *Protocols (WG2200)* | |
| Mobility management | SRU |
| Call control protocols and signalling , intelligent network (IN) protocols | SRU |
| Multiple Access schemes and ATM/IP/protocol compatibility | TOR, UFI, DLR, SRU, IJS, EPF |
| *Network algorithms and control (WG2300)* | |
| Broadband/multimedia services impact on networking procedures | DLR, EPF |
| ATM based/compatible routing concepts in networks with inter-satellite link, efficient traffic adaptive routing | DLR |
| Network dimensioning | DLR |
| Source traffic modelling and traffic flow identification | DLR, EPF |
| Wireless call admission control and QoS provisioning | SRU, EPF |
| Channel allocation schemes, handoff | UFI, IJS, SRU, UAV |

## 3.1.3 System Definition

### 3.1.3.1 Satellite Constellation

Satellites are foreseen to be complementary to the future fixed and terrestrial mobile networks such as the MBS and UMTS, and will provide global coverage to remote users. In order to provide high system QoS and line of sight communications, the satellites must be available at high elevation angles. The implication is that a large number of satellites is required in the constellation.

Both LEO and Medium Earth Orbit (MEO) systems have been considered by COST 252. LEO satellites provide low delay communications and low path attenuation but the satellite numbers are high even for low system minimum elevation angles. The high satellite velocity with respect to the ground, means that inter-satellite handovers are frequent, and the use of multiple spotbeam arrays means that inter-spotbeam handover will be a large signalling task for these constellations. MEO satellite constellations with satellite numbers ranging from 10 to 30 can provide coverage to users at minimum elevation angles of up to 30° (1), so in terms of elevation angle performance it can be seen that MEO outperforms LEO even with less satellites, although at the cost of delay and propagation attenuation.

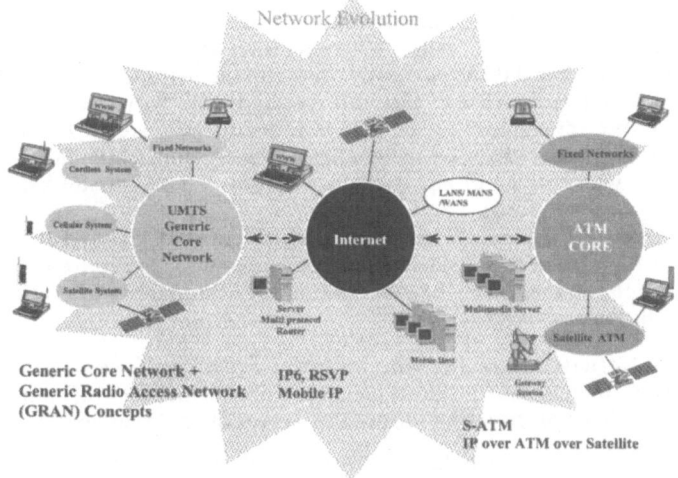

**Figure 3.2.** Core and satellite network integration scenarios.

MEO satellite systems require a small ground segment whereas a LEO system requires a large number of fixed earth station sites unless ISLs are used.

### 3.1.3.2 System Integration

Both MBS and UMTS mobile terrestrial networks rely on a very fast and reliable backbone network, which could be seen as the common platform to interwork/integrate with any other network. For the integrated satellite component, the working group effort has been focused on the S-ATM and S-UMTS scenarios in conjunction with WG1 (Strategic scenarios and feasibility). The most desirable scenario is that of high integration with the dominant terrestrial core network.

### 3.1.3.3 Satellite UMTS

The S-UMTS standards are being established through ACTS for services up to 144 kb/s. Research carried out by WG2 into the integration of UMTS with multimedia satellite systems providing services up to 2 Mb/s has been concentrated on the adaptation of the medium access control (MAC) and radio resource management (RRM) protocols which will facilitate the implementation of higher rate services.

A mobility management scheme suitable for S-PCN is described in (2) for both LEO and MEO networks. According to (3) the unified concept of UMTS can be viewed under two main drivers: the generic radio access network (GRAN) and the UMTS generic core network. Current activities within the COST 252 WG2 include the investigation of suitable MAC schemes for a wide range of multimedia applications. In (4, 5) a feasibility study of the applicability of the packet reservation multiple access (PRMA) protocol for LEO satellite systems is performed for both voice and data traffic. The objective of this study is to propose the best candidate MAC protocol that is based on the same (or similar) principles in both terrestrial and satellite systems. It is shown that when selecting optimum values for both permissions probabilities and frame duration, PRMA is advantageous in respect to time division multiple access (TDMA) in managing integrated voice and data traffic.

### 3.1.3.4 Satellite ATM

Proposals which have been presented in literature and under study in the standardisation bodies identify the structure of the future wireless ATM networks and incorporate the concept of intelligent network (IN) to solve the problems of mobility (ITU-T Recommendations). In our investigations we have examined the networking requirements of the satellite components of the future broadband mobile telecommunication networks and we have identified a possible satellite network protocol architecture that best satisfies an integrated solution with B-ISDN (6). This approach is not restricted by any other terrestrial mobile system compatibility/interoperability constraints in the way that the Satellite UMTS (7) system approach would be. The aim here has been to bring ATM functionality closer to the user terminal without violating its connection-oriented nature and the guaranteed quality of service (QoS) characteristics. The satellite system will reuse the B-ISDN protocols in its network with some adaptations, which will enable a low level of interworking, together with minimal reconfiguration of existing ATM user terminal software. The work performed within COST 252 has focused on inter satellite link (ISL) routing and performance evaluation of MAC techniques (8) for transmission of ATM to and between LEO satellites.

A general new routing concept, called "Discrete-Time Dynamic Virtual Topology Routing" (DT-DVTR) has been developed (9) to efficiently cope with the periodic topology changes of the ISL sub-network in LEO systems. A view of how the concept of the "virtual connection tree" and the "neighbouring mobile access region" can be used in future mobile satellite networks was presented in (10). Since the satellite movement can be predicted, a new approach to balance the system's call blocking probability with the call dropping probability due to unsuccessful handoffs for different type of services has been proposed.

An investigation of a hybrid-modular MAC protocol for LEO satellites (11) is presented. Both real time and non-real time traffic is considered covering different ATM service classes. This modular approach utilises both random access and reservation based multiple access techniques so that all traffic classes can be supported efficiently.

### 3.1.3.5 Satellite Network Based on the Internet Protocol (S-IP)

The adoption of IP routing for the satellite segment provides some true challenges in an end-to-end true packet network. Some of the advantages of this approach are listed below:

- IP routing can be very attractive in LEO meshed networks which adapt badly to circuit switching.
- Support of widely accepted standards that include multicasting and simplified interworking with many ground IP based networks

- Adoption of the new Internet standards such as IP6, RSVP (ReSerVation setup Protocol) and Mobile IP
- S-IP has no external protocol performance issues to worry about (such as IP over ATM)

However, research in this area is at a very early stage and most of the commercial systems have adopted a different approach. For example, Celestri and SkyBridge incorporate ATM variants for the satellite switch and Teledesic proposes fast packet switching using proprietary connection-less adaptive routing protocols.

## 3.2 S-UMTS Approach

### 3.2.1 Resource Management Schemes for LEO Mobile Satellite Systems

#### 3.2.1.1 Introduction

The integration of Mobile Satellite Systems (MSSs) with terrestrial cellular networks will pave the way for the global roaming envisaged in future third generation mobile communication systems (12). Several satellite orbital constellations have been proposed for MSSs. An interesting solution is given by Low Earth Orbit (LEO) satellites, since they permit relaxation of the constraints on the link budget, allow the use of low-power hand-held mobile terminals and provide a high level of earth coverage with smaller cells (spotbeams).

LEO satellites are not stationary with respect to a fixed point on the earth: the satellite ground-track speed, $Vtrk$, is far greater than the earth rotation speed and the user speed. Two different coverage approaches are possible for LEO-MSSs (13): (i) cells (= spot-beam footprints on the earth) are fixed on the earth and satellite antenna spot-beams are steered to point to the same area during all the time the satellite is above the horizon; (ii) cells move on the earth. This research activity has dealt with the second solution that requires specific procedures in order to manage the cell change for an active call.

#### 3.2.1.2 Description of the Compared Resource Management Schemes

Only voice traffic has been considered in MSSs: each time a call arrives at a cell $x$, it must be served by an available channel in $x$; call attempts that do not immediately find free resources are blocked and lost (Blocked Calls Cleared – BCC).

The spectrum assigned to LEO-MSSs is reduced with respect to the expected market diffusion of these services. Therefore, optimised radio resource management strategies have to be investigated for LEO-MSSs. We have assumed that the channel allocation techniques have to fulfil the following constraint (14, 15, 16, 17): two different cells on the earth may reuse the same channel provided that they are at a suitable distance, called *reuse distance*, $D$, that allows tolerable levels for the co-channel interference. We have compared two channel allocation schemes, that is Fixed Channel Allocation (FCA) and Dynamic Channel Allocation (DCA):

- With the FCA technique, a set of channels is permanently assigned to each cell, according to the reuse distance $D$. A call can only be served by an available channel (if any) belonging to the set of the cell. If an arriving call does not find any free nominal channel in its cell, the call is blocked and lost.
- Whereas, a DCA strategy allows that any system channel can be temporarily assigned to any cell, provided that the constraint on the reuse distance $D$ is fulfilled. Different DCA techniques can be defined on the basis of the strategy used to select the channel to be allocated on a cell when a new call occurs in it. The DCA technique considered in this study has been introduced in (15, 16).

Moreover, we have considered that when an active Mobile Subscriber (MS) goes out from a cell and enters an adjacent one, a new channel must be automatically assigned to it in order to have a seamless conversation. This procedure is called *handover*; it involves the re-routing of a call between two adjacent beams that may belong to either the same satellite or two adjacent satellites of the MSS. If no channel is available in the destination cell, the handover is unsuccessful and the call is dropped.

The selection of a suitable policy for managing handover requests is a central issue in defining resource management strategies. From the user standpoint, the interruption of a conversation is less desirable than the blocking of a newly arriving call. In LEO-MSSs, inter-beam handover requests are extremely frequent during call lifetime (one could expect that a call experiences an inter-beam handover every 1 min or even less) and at each beam change the call may be dropped due to an unsuccessful handover. Hence, LEO-MSSs require specific techniques that prioritise the service of handover requests with respect to the service of new call attempts in order to reduce as much as possible the call dropping probability (17, 18, 19, 18, 19, 20).

In this study we have envisaged a suitable handover scheme for LEO-MSSs based on the queuing of handover requests (16, 18, 19 and 20). When an MS with a call in progress leaves cell $x$ and enters an adjacent cell $y$, there is an area (*overlap area*) where this MS can receive a signal with an acceptable power level from both cells. The time the MS spends to cross the overlap area, $t_{wmax}$, can be used to queue the related handover request, if no channel is free in cell $y$. An inter-beam handover strategy based on the Queuing of Handover requests (QH) is essential to guarantee a suitable quality of service to mobile users in LEO-MSSs, where there are frequent inter-beam handovers during call lifetime.

Different queuing schemes can be applied depending on the way handover requests are ordered in the waiting list of a cell. The most common queuing discipline is the First Input First Output (FIFO) scheme, where handover requests are queued according to their arrival instants. We have investigated also another scheme, called Last Useful Instant (LUI) (19): this discipline relies on the fact that, when a handover request is queued, the controller on the satellite exactly estimates the related $t_{wmax}$. A new request is stored in a queue position before (after) all handover requests having a greater (lower) residual value of $t_{wmax}$. In such a way, the system tries to serve first the most urgent handover request. We can note that this ideal LUI scheme represents the best scheduling strategy for handover requests. A practical implementation of this LUI scheme in LEO-MSSs may be based on the use of a suitable positioning system which estimates the MS position at the beginning of the call and tracks the MS position during call lifetime (19).

### 3.2.1.3 Simulation Scenario and Results

In order to carry out a comparison among different resource management techniques (e.g. FCA, DCA, FCA-QH, DCA-QH) we have modelled the user mobility as follows: the relative satellite-MS motion has been approximated by only the satellite ground-track speed (i.e. vector $V_{trk}$), due to the high value of $V_{trk}$ with respect to the other motion component speeds. Hence, the relative motion has a fixed orientation with respect to the cellular layout irradiated on the earth by satellites. Let the track of the relative motion be disposed as shown in Fig. 3.3: all users have parallel and straight trajectories with respect to the cellular network illuminated by satellites on the earth.

On the basis of both these assumptions and a specific cell shape (i.e. hexagonal or circular cell shape on the earth) (19, 20), we have characterised the distribution of the MS sojourn time in a cell, the distribution of the channel holding time in a cell and the handover process offered to a cell. Moreover, we have also determined the average number of handover requests per call, $n_{h0}$, (in the absence of call blocking) and we have found that it is related as follows to a suitable parameter $\alpha$ that measures the user mobility ($T_m$ denotes the mean call duration and $R$ denotes the cell radius) (19):

$$n_{h0} = \frac{4}{3\alpha}, \quad \text{where } \alpha = \frac{\sqrt{3}\,R}{V_{trk}\,T_m} \tag{3.1}$$

Figure 3.4 presents the behaviour of the handover probabilities as a function of parameter $\alpha$ (19).

It is worth noting that the mobility model proposed in this study is valid for whatsoever type of MSs, since $V_{trk}$ is far greater than the speed of every kind of MS, it does not matter if it is pedestrian, vehicular or flying in an aeroplane.

On the basis of this mobility model, we have developed a simulation tool to evaluate the performance of different resource management schemes in LEO-MSS. In the case of FCA schemes we have also developed a performance analysis that has been validated by simulation results (19, 20). The following parameter values have been assumed for the numerical evaluations:

- the reuse distance is $D = \sqrt{21}\,R$ (where $R$ is the cell side);
- the average call duration is $T_m = 3$ min;
- the simulated cellular network is parallelogram shaped and folded onto itself with 7 cells per side;
- a number of 70 channels is available to the system; then, 10 channels/cell are available with FCA;
- the IRIDIUM mobility case is considered (i.e. $\alpha$ . 0.27).

Let $P_{b1}$ denote the blocking probability for new call attempts and $P_{b2}$ denote the handover failure probability. From Fig. 3.5 and Fig. 3.6, we note that DCA schemes outperform the corresponding FCA ones and that the allocation schemes that envisage the queuing of handover requests (i.e. FCA-QH LUI, FCA-QH FIFO, DCA-QH LUI, DCA-QH FIFO) significantly outperforms the corresponding

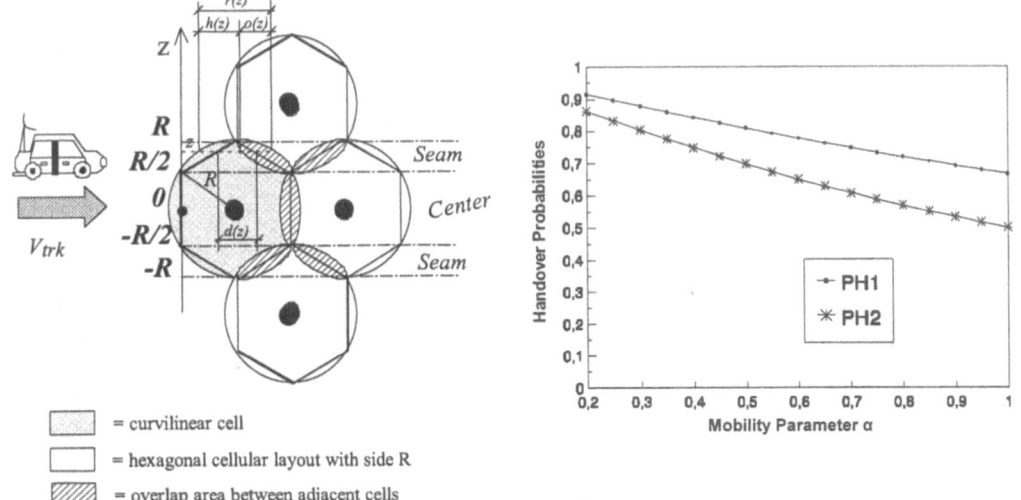

= curvilinear cell

= hexagonal cellular layout with side R

= overlap area between adjacent cells

**Figure 3.3.** The geometry of the cellular network with a hexagonal layout.

**Figure 3.4.** Handover probability for a call from its source cell, $P_{H1}$, and handover probability for a call from a transit cell, $P_{H2}$.

schemes without queuing (i.e. FCA with No Priority Scheme for handovers, FCA NPS, and DCA with No Priority Scheme for handovers, DCA NPS). Let us refer to the behaviours of $P_{b2}$ shown in Fig. 3.6. With FCA-QH, the advantages of the LUI discipline with respect to FIFO are practically negligible. Whereas, we have obtained that the LUI scheme permits to reduce $P_{b2}$ as regards the FIFO policy in the DCA-QH case. The reason for these different behaviours has to be searched in the way the handover queuing is managed by FCA-QH and DCA-QH. With FCA-QH, each cell has its queue. Whereas, DCA-QH requires that the system manages a virtual global queue formed by the handover requests waiting for service in all the cells. This global queue contains a greater number of handover requests than each single queue of FCA-QH. Hence, a specific ordering discipline has a greater impact on DCA-QH rather than on FCA-QH. This is an interesting advantage of the DCA approach that makes it particularly suitable for LEO-MSSs.

### 3.2.1.4 Conclusions

In this research activity we have investigated different resource management strategies in LEO-MSSs. A mobility model has been defined that removes the approximations made in other works. Since the service of handover requests must be prioritised with regard to new call attempts, two different handover queuing schemes have been considered: FIFO and LUI. We have observed that the DCA technique outperforms the corresponding FCA, therefore, it does not matter which handover prioritisation scheme is adopted. Finally, we have shown that DCA-QH with the LUI queuing discipline achieves a good performance and, hence, it is a very attractive scheme for LEO-MSSs.

## 3.2.2 Mobility Management in Satellite-UMTS

### 3.2.2.1 Introduction

When a mobile communications network based on the GSM (21) network topology carries a mobile-terminated call, the call is routed to the Mobile Switching Centre (MSC) that is associated with the user's last registered Location Area (LA). As an LA is associated with one MSC only, this routing strategy does not involve any further inter-MSC re-routing during call set-up. Any further inter MSC handovers during the call are due only to the mobility of the user.

In satellite personal communication networks (S-PCN) it is possible that each Fixed Earth Station (FES) in the ground network will have MSC functionality. On receiving a user-terminated call, it is preferable to route the call to an FES that has a high probability of maintaining the call throughout its duration. Inter-FES handover during a call is highly undesirable as the handover is likely to involve international re-routing of the call to the new FES. The re-routing of the call will have an associated delay and possibly different call charge rates, as the new FES may be owned by a different network operator. Unlike terrestrial networks, an inter-MSC handover can be initiated not only by a mobile

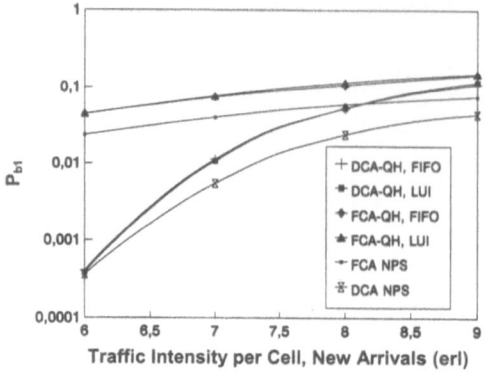

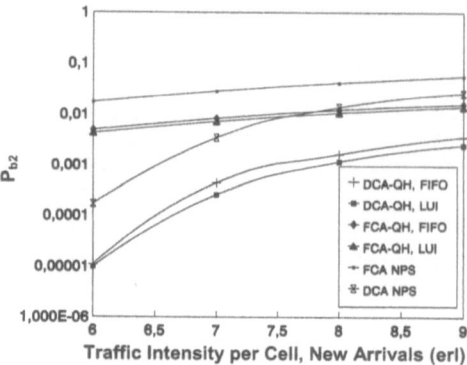

**Figure 3.5.** $P_{b1}$ performance for FCA NPS and DCA NPS, FCA-QH and DCA-QH both FIFO and LUI (IRIDIUM case).

**Figure 3.6.** $P_{b2}$ performance for FCA NPS and DCA NPS, FCA-QH and DCA-QH both FIFO and LUI (IRIDIUM case).

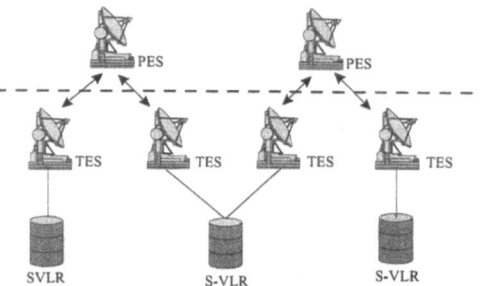

**Figure 3.7.** S-PCN network architecture.

**Table 3.2.** Constellation parameters

|  | ICO-like | Globalstar-like |
|---|---|---|
| N° sats. | 10 | 48 |
| N° orbit planes | 2 | 8 |
| Inclination | 45° | 52° |
| Altitude (km) | 10 349 | 1414 |
| FES min elev. | 5° | 10° |
| UT min elev. | 10° | 10° |

User Terminal (UT) but also by a stationary terminal. This is caused by the dynamic connectivity property of non-GEO satellites. As satellites move in and out of view, a mobile UT should attempt to use the satellite that is providing the most optimal channel, even if this satellite does not have direct connectivity with the FES that is currently handling the call. To restrict the UT's choice of satellites to those only offering connectivity with the current FES could compromise the Quality of Service (QoS), which is directly related to the signal quality between the user terminal and satellite.

### 3.2.2.2 Network Architecture

We assume that the ground segment of the satellite network consists of two types of FES. The first type of FES is the Primary Earth Station (PES). PESs are responsible for the co-ordination and control of satellite resources. We assume that there are enough PESs distributed around the globe to ensure that each satellite is always in view of at least one PES. The second type of FES is the Traffic Earth Station (TES). The TES is responsible for carrying service traffic to terrestrial networks such as GSM, PSTN and ISDN. The results in this paper refer to the TES type of earth station.

Table 3.2. below lists some constellation parameters of the two systems that have been considered in this study.

### 3.2.2.3 Instantaneous Coverage Area

In *bent pipe* S-PCN systems, a UT can communicate with an FES only through a satellite. Connectivity between a mobile user terminal and an FES is provided when the FES is within the area described by a satellite's FES minimum elevation angle and when the UT is within the coverage area described by the same satellite's UT minimum elevation angle. As satellites in a non-GEO orbit are continually moving relative to the surface of the earth, the area around the FES where connectivity is currently provided is continuously changing with time. Figures 3.8 and 3.9 show two instantaneous connectivity areas at time $t = 0$ and $t = 6$ hours for an FES at 20° latitude in the ICO-like system. The white section is the coverage area, whereas the black section is the area where connectivity between UT and FES is not possible.

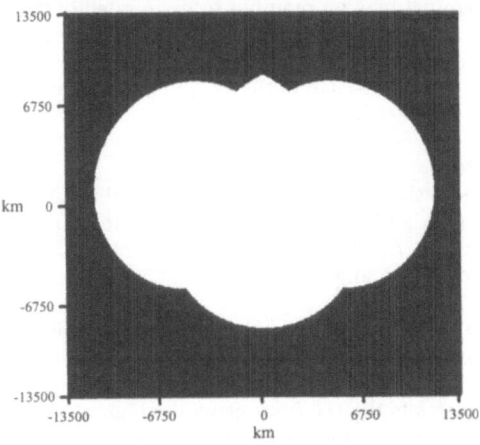

**Figure 3.8.** ICO-like FES instantaneous coverage area at $t = 0$ hours.

**Figure 3.9.** ICO-like FES instantaneous coverage area at $t = 6$ hours.

The above diagrams demonstrate clearly how the effective service area around the FES, where a terminal can connect with the FES via a satellite, is continuously changing in time. This type of dynamic service area contradicts the terrestrial definition where an MSC has a static service area.

### 3.2.2.4 Guaranteed Coverage Area

In order to reuse the terrestrial approach to call routing in an S-PCN system, the concept of the Guaranteed Coverage Area (GCA) was introduced in (22). The GCA is defined as the area around the earth station where a mobile terminal can always connect to the earth station via at least one satellite that is above the minimum elevation angles of both the UT and FES. Although the instantaneous coverage area of the FES shown above will be larger than for the GCA, outside the GCA region connection with the FES cannot be guaranteed. Therefore, while a terminal is roaming in the GCA of an FES, calls to the mobile are to routed to that FES. A network of FESs has been presented in (22) for the Globalstar-like and ICO-like systems respectively. The FESs are arranged to have minimal overlap between GCAs, while ensuring that the earth's land mass is completely covered by the GCA's.

The GCAs for an FES at 40° latitude in the ICO-like and Globalstar-like systems are shown below in Figures 3.10 and 3.11.

The size and shape of the GCAs varies with latitude due to the constellations' elevation angle distribution, which also varies with latitude. Guaranteed connection with the FES through any satellite above the minimum elevation angle of the terminal requires that the FES minimum elevation angle is less than the UT minimum elevation angle. This is the case with the ICO-like system where the difference in minimum elevation angle between the FES and UT results in a circular area with radius 550 km around the FES, wherein a terminal can connect to the FES through any visible satellite.

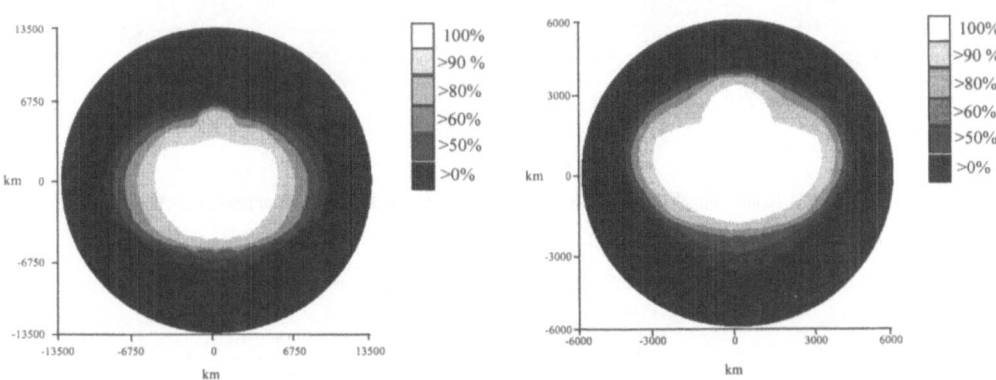

**Figure 3.10.** ICO-like GCA.

**Figure 3.11.** Globalstar GCA.

Therefore, while roaming in the GCA the mobile terminal may be required to use a satellite that is not optimal (i.e. highest elevation angle) or to use a satellite which is currently not available due to shadowing, which becomes more prevalent at lower elevation angles (23, 24). These two cases could potentially lead to a forced inter-FES handover, while inside the GCA of the FES, an alternative result is that the call could be dropped.

### 3.2.2.4.1 Optimum Satellite Connection Area

As described above, the GCA requires that the terminal should connect through a satellite which is currently providing a connection to the expected FES, and not necessarily a satellite which the terminal would be most appropriate in terms of elevation. As shadowing becomes more prevalent at lower elevation angles, the satellite with the highest elevation angle to the terminal is the one which is most likely to offer good channel conditions.

Figures 3.12 and 3.13 below show the areas around an FES at 40° latitude, in both the ICO-like and Globalstar-like systems respectively, where a UT can connect to the FES through the satellite with the highest elevation with respect to the UT. When Figs 3.10 and 3.11 are compared with Figs 3.12 and 3.13 respectively, the extent to which the GCA restricts the UT from using the highest satellite becomes clear. When Figs 3.11 and 3.13 are compared, it is clear that optimum satellite coverage area extends to a higher latitude than the GCA, due to the fact that a satellite is not always available above 70° latitude in the Globalstar-like system, therefore the GCA does not include this region in its definition, as connection with the FES cannot be guaranteed. However the optimum satellite coverage area is concerned only with the highest satellite losing connectivity with the FES. Although above 70° latitude a terminal may not always have a satellite visible, when one or more satellites are visible, connectivity with the earth station is guaranteed through the highest satellite, in the region shown in Fig. 3.13.

### 3.2.2.4.2 Optimum Satellite In-call FES Handover

In this section the probability that the highest satellite with respect to the terminal will lose connectivity with the FES during a call and therefore induce an FES handover is evaluated. Calls within the 100 per cent connectivity area will always be completed without loss of FES connectivity. To evaluate loss of connectivity with an FES during a call in the region where connectivity is less than 100 per cent, we must consider the duration of the connection periods in relation to the call length. Consider a connectivity period $i$, where the user terminal is connected to an FES via the highest satellite for a period $Ct$ sec. If we assume that the call distribution length is negative exponential with a mean of $m$ sec, and that the starting point of each call is uniformly distributed within the connection period, then the probability of a call completing within the connection period is given by Eq. 3.2;

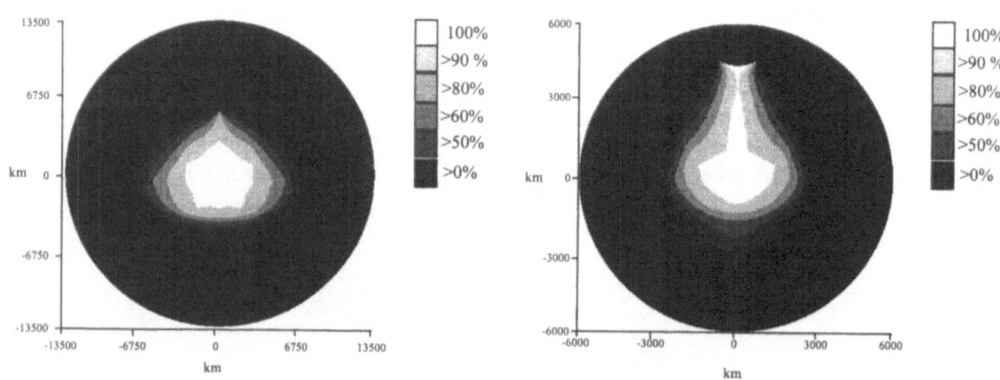

| Figure 3.12. ICO-like highest satellite connectivity. | Figure 3.13. Globalstar highest satellite connectivity. |

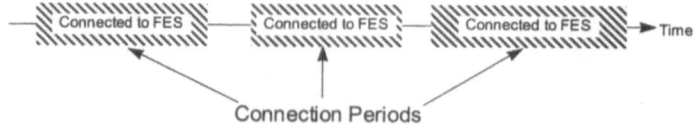

**Figure 3.14.** Sequential connectivity periods between UT and FES.

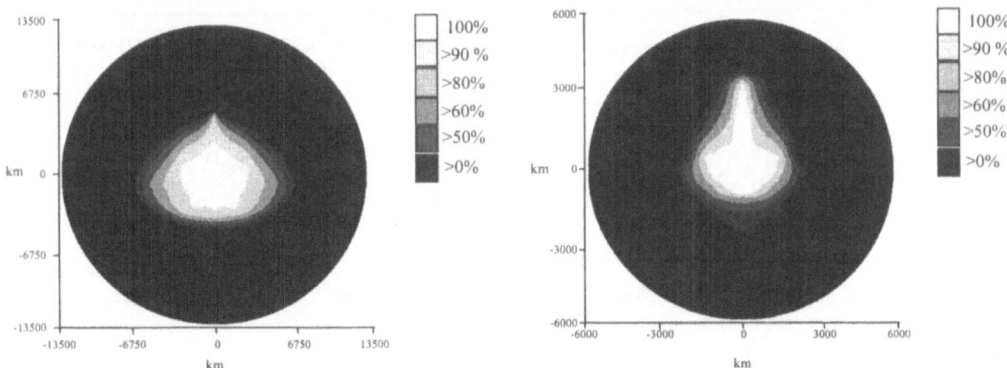

**Figure 3.15.** ICO-like call completion rate without loss of connectivity with FES @ 40° latitude.

**Figure 3.16.** Globalstar-like call completion rate without loss of connectivity with FES @ 40° latitude.

$$P_{\text{complete}}\{i\} = \frac{1}{mCt\{i\}} \int_{\tau=0}^{Ct\{i\}} \int_{t=0}^{\tau} e^{-t/m} \, dt \, d\tau \tag{3.2}$$

The overall probability of a call successfully completing without loss of connectivity with the FES, requires that the call begins and is completed within the connectivity period.

Therefore the probability of a call completing successfully within a series of $N$ different connection periods, as in Fig. 3.14, is given by Eq. 3.3;

$$P_{\text{noFES handover}} = \sum_{i=1}^{N} P_{\text{complete}}\{i\} \, P_{\text{connect}}\{i\} \tag{3.3}$$

Where $P_{\text{connect}}\{i\}$ is the probability of the connection period $i$ occurring. Shown below are the successful call ($m = 90$ s) completion rates for FES at 40° latitude in the Globalstar and ICO-like systems.

When Fig. 3.12 is compared with Fig. 3.15 is it clear that there is little difference between the connection probability and successful call completion probability, indicating that the connectivity periods are significantly longer than the call duration. However, this is not the case when Fig. 3.13 and Fig. 3.16 are compared. The successful call completion rate area is noticeably smaller than the connection probability area, indicating that the connection periods are nearer to being of the same order as the call duration periods. This is due to the fact that the average duration of an ICO-like satellite pass is much longer than that of a Globalstar-like satellite. Also, comparing Figs 3.13 and 3.16, the successful call completion rate above 70° latitude is not 100 per cent although the connection probability is. This is due to the fact the call cannot be completed because the terminal loses connectivity with the satellite, as the Globalstar like system does not provide continuous satellite

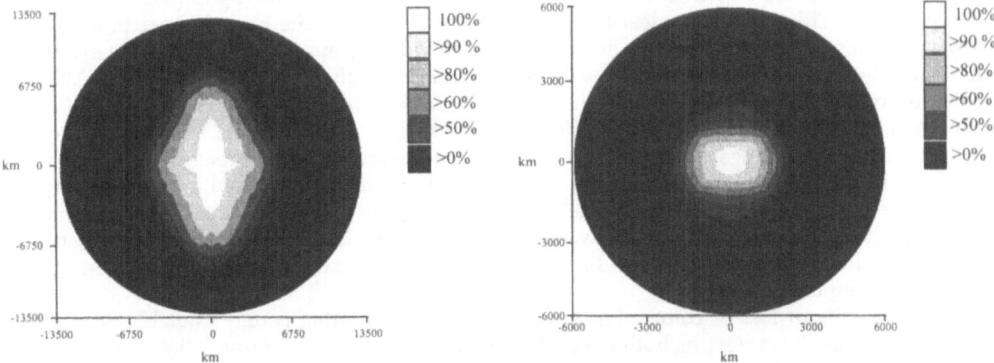

**Figure 3.17.** ICO-like Call completion rate without loss of connectivity with FES @ 0° latitude.

**Figure 3.18.** Globalstar-like Call completion rate without loss of connectivity with FES @ 0° latitude.

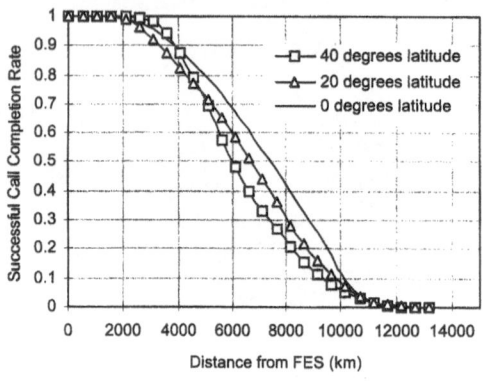

**Figure 3.19.** ICO FES handover probability vs. distance.

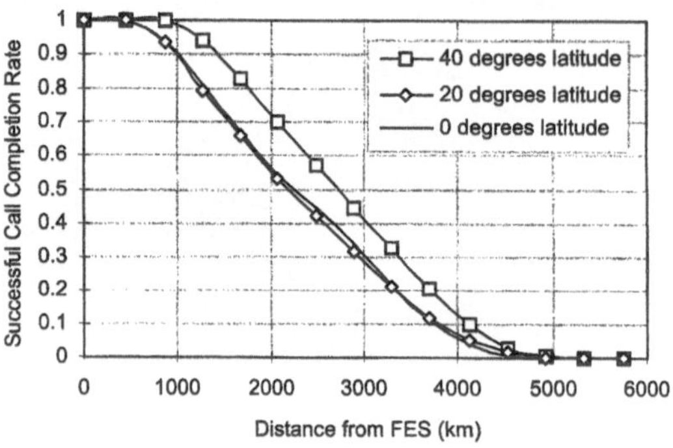

**Figure 3.20.** Globalstar FES handover probability vs. distance.

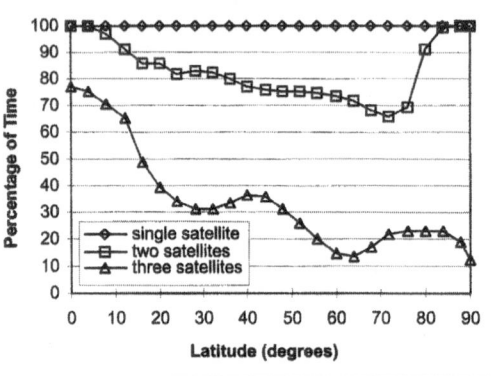

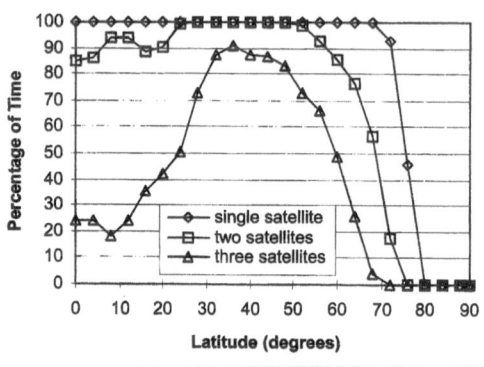

**Figure 3.21.** ICO satellite diversity.

**Figure 3.22.** Globalstar satellite diversity.

coverage above 70° latitude. Figures 3.17 and 3.18 show the optimum satellite call completion probability maps for FESs at 0° latitude in the two systems. It is noticeable that the Globalstar system has a much smaller area at 0° latitude, this is due to the lower mean elevation angle of the highest satellite around this latitude. Although the same is true of the ICO mean elevation around these latitudes, the 100 per cent area at 0° appears to be as large as the 100 per cent area at 40° latitude. This is because the mean elevation angle of the highest satellite between 0° and 40° latitude is higher than between 40° and 80° latitude, which results in larger 100 per cent connectivity area. This is explained further in the next section.

Figures 3.19 and 3.20 show the call completion rates without FES handover against distance (averaged over azimuth) from FES's at 0°, 20° and 40° latitude.

So far in the discussion we have only considered the satellite with the highest elevation angle to the mobile UT. This satellite is clearly the most likely to provide the best channel to a terminal. However, both satellite constellations described here provide significant levels of satellite diversity in an attempt to increase the quality and availability of service to the user. Figures 3.21 and 3.22 show the visibility statistics for one, two and three satellites in the two systems.

It can be seen that neither system provides 100% visibility to secondary and tertiary satellites over the complete latitude range, so any call on either of these systems has a probability of being terminated due to the satellite moving below the minimum elevation angle allowed for connectivity with the UT. Therefore, when evaluating the call holding probability area around an FES two factors can now cause the mobile to lose connectivity. However, we are only interested in the probability that the terminal will lose connectivity due to a break in the link between satellite and the FES through which the UT has been communicating.

The probability of loss of connection to an FES by a satellite that is only available for a limited time, can be found by evaluating both the probability of call completion through the satellite, without considering the FES, and the probability of call completion though the satellite to a given FES, where connectivity with the FES can be broken by either the satellite losing connectivity with the FES or the mobile terminal.

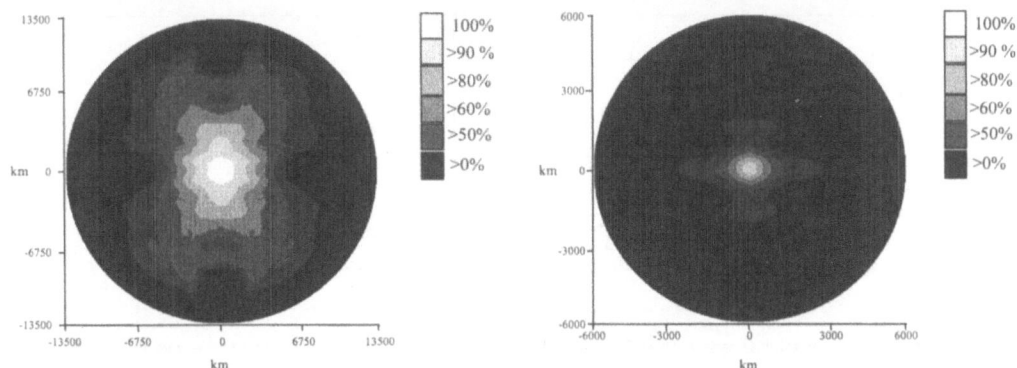

**Figure 3.23.** Probability of no FES handover from second highest ICO satellite (FES @ 0 latitude).

**Figure 3.24.** Probability of no FES handover from second highest Globalstar satellite (FES @ 0 latitude).

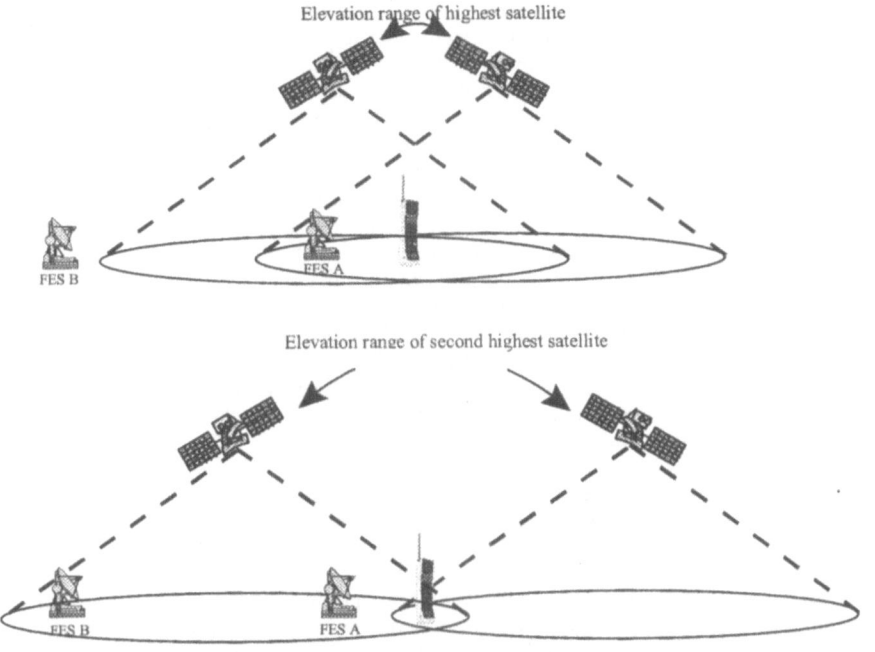

**Figure 3.25.** Highest and second highest satellite connectivity.

The results shown in Figs 3.23 and 3.24 demonstrate that the second highest satellite provides connectivity over a larger area than the highest satellite, as expected.

In general, satellites at lower elevation angles will provide lower percentage connectivity to an FES over a wider area than satellites at higher elevation angles. However, satellites at higher elevation angles can provide a higher percentage connectivity over a larger area than lower elevation angle satellites. This characteristic is illustrated in Fig. 3.25, through the larger footprint overlap between high elevation satellites. Consider FES A, it is clear from Fig. 3.25 that the highest satellite will provide connectivity between FES A and the UT for a higher percentage of the time than the second highest satellite. However, when FES B is considered, the highest satellite cannot provide connectivity between the earth station and the terminal due to its limited elevation angle range above the UT. However, where FES B is concerned, connectivity can potentially be provided through the second highest, due to the fact that the satellite has a much lower elevation angle to the mobile terminal, therefore potentially providing connectivity with earth stations further away from the mobile.

In the ICO case the area covered is significantly larger than in the Globalstar case due to the larger

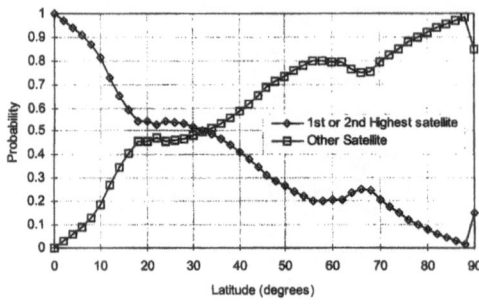

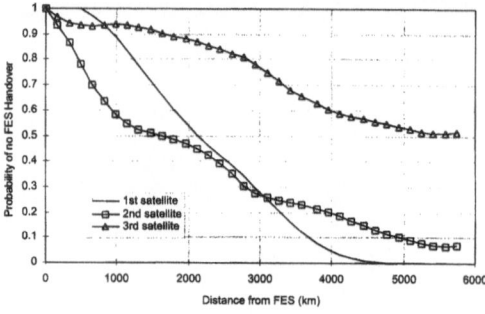

**Figure 3.26.** Second highest satellite relative to an FES at 0° latitude.

**Figure 3.27** Probability of no FES handover vs. distance from FES.

footprint of the MEO satellites. When Fig. 3.23 is examined, there appears to be an anomaly in the results approximately 6500 km (± 58° latitude) directly above and below the FES. The results indicate that the connectivity through the second highest satellite does not necessarily decrease as a mobile terminal moves away from the earth station. Consider Fig. 3.26, each line represents the probability that the second highest satellite at a specified latitude is either the first or second highest satellite at 0° latitude, or is the third or lower satellite at 0° latitude in the ICO system. From the diversity graph for ICO (Fig. 3.21) it can be deduced that at zero degrees latitude an FES always has visibility with two satellites, visibility with more (lower) satellites however is not guaranteed. Around 58° latitude, the probability that the second highest satellite seen by a UT is the first or second highest seen by the FES at 0° latitude goes through a minimum, while the third to tenth highest satellite probability increases, as visibility to the third or lower satellites is not guaranteed at 0° latitude, this explains why the connectivity results shown in Fig. 3.23 fall and rise again around ± 58° latitude.

The reason for this occurrence is difficult to explain as it depends on from which plane in the constellation the second highest satellite originates and its position relative to the FES.

The graph shown above in Fig. 3.27 relates to an FES at 0° latitude in the Globalstar system. The highest satellite offers the greatest probability of FES connectivity at relatively close distances to the FES. As the UT moves away from the FES the probability that connectivity to the FES is provided by the lower satellites dominates. Although lower satellites appear to offer a high degree of connectivity with the FES it should be noted that the potential connectivity between them and a mobile terminal is considerably lower than that of the highest satellite (Fig. 3.22).

### 3.2.2.5 Mobility Management

Due to the dynamic motion of satellites in a non-GEO constellation, adoption of the GSM approach to mobility management results in a large increase in associated signalling. To overcome this problem, an alternative scheme was proposed in (25). In this scheme, mobile terminals make a location update after moving a predetermined distance from their last point of contact with the network. The user's movement is monitored by a satellite based positioning system. The area which is described by the locus of the radius of the maximum distance before location update, from the point of last update, is known as the users location area (LA).

Upon receipt of a user terminated call, the network must page the UT through all spotbeams that provide coverage over the LA. This is necessary as the terminal could be anywhere inside the LA and therefore could be monitoring the paging channel on any spotbeam providing instantaneous coverage to that area. The size of each user's location area can be varied to reach an optimum trade-off between location updating frequency and paging signalling load, depending on the user's known mobility characteristics. This results in a significant reduction in mobility management signalling when compared with GSM type method, it does however assume that the mobile terminal will have the ability to measure its position while in idle mode. It should also be noted that once the mobile terminal makes a location update the new location area is a circle of a given radius LAR centred at the point of the new location update. This results in the location area being positioned relative to the user's location and not to any given network entity as is the case with GSM. As a result the location area cannot be guaranteed to be contained wholly within the coverage area of one FES. Therefore the S-PCN system must evaluate which FES's are suitable to have the terminal registered with at location updating time. As shown earlier, we can evaluate the probability of a user terminated call being successfully completed without FES handover at a given distance and azimuth away from the FES. Given that the network has no knowledge of when the next call will arrive for the terminal, and there-

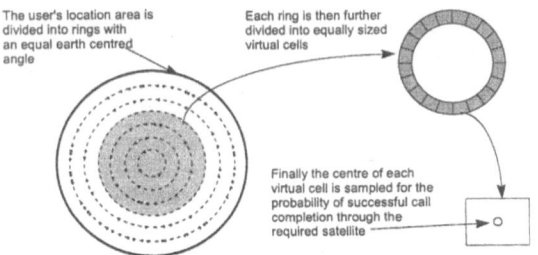

Figure 3.28. Virtual cells.

**Table 3.3.** User mobility parameters

| Number of users | 20,000 |
|---|---|
| Direction of travel | Constant between 0 and $2\pi$ |
| Speed | 100–150 km/h |
| Location area radius | 200 km |
| Call arrival rate/user | 1 call/hr |
| Simulation time | 24 hours |
| FES position | 40° lat., 0° long. |

The user's location area is divided into rings with an equal earth centred angle

Each ring is then further divided into equally sized virtual cells

Finally the centre of each virtual cell is sampled for the probability of successful call completion through the required satellite

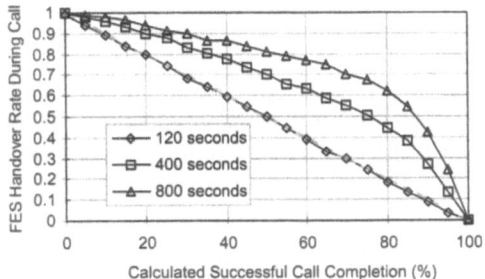

Figure 3.29. Globalstar-like FES handover rate vs. $P_{FES}$.

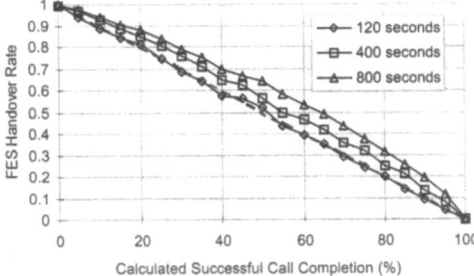

Figure 3.30. ICO-like FES handover rate vs. $P_{FES}$.

fore, where the terminal will be in the LA when the call arrives, the probability of successful call completion without FES handover must be evaluated over the entire LA, which is given by:

$$P_{\text{FES}} = \int_{\theta=0}^{2\pi} \int_{r=0}^{\text{LAR}} P_{\text{NoFES handover}}(r, \theta)\, P_{\text{user}}(r, \theta)\, r\, dr\, d\theta \tag{3.4}$$

$P_{\text{FES}}$ represents the probability that a UT making a location update will successfully complete a user terminated call without losing connectivity with the given FES, at some point in time after the location update. The function $P_{\text{user}}(r, \theta)$ relates to the user's position probability distribution within the location area.

The above function describing $P_{\text{FES}}$ can be evaluated by using the Virtual Paging Cell (VPC) method introduced in (26) and the data shown in Figs 3.15 and 3.16 (for an FES at 40° latitude). In this method, at location updating time the location area is div. ed into VPC's, the probability of the user being located in one of the cells some time later (e.g. when a mobile terminated call arrives) is evaluated (See Fig. 3.28).

Once the probability of the user being in a cell when a user terminated call arrives has been calculated, the probability of completing a call without FES handover can be found for the centre of the cell from the data in Fig. 3.15 and 3.16. For this study it has been assumed that each user will travel in a constant direction from the centre of the location area.

The $P_{\text{FES}}$ function was evaluated by simulation and compared with FES handover rates for various mean call duration in the Globalstar-like and ICO-like systems. The parameters used to described each UT's mobility and call arrival rate are listed in Table 3.3.

The results for mean call durations of 90,400 and 800 s are shown below in Figs. 3.29 and 3.30.

For a call duration of 120 s, both sets of results demonstrate a close match to the expected result, as the data used to calculate $P_{\text{FES}}$ was based on a mean call duration of 120 s. The effect of increasing the call duration results in a higher than expected FES handover rate. The effect is much more dramatic in the Globalstar-like system, this is again due to the relatively shorter duration of satellite passes in this system.

### 3.2.2.5.1 FES Location Updating Area

Given that we have presented the probability of a call being successfully completed without FES handover at location updating time, we can evaluated the size of the areas around the FES where a mobile terminal can make a location update with a given location area radius with a probability $P_{\text{FES}}$ greater than a specific threshold value. The effective service area of the FES is larger than the FES

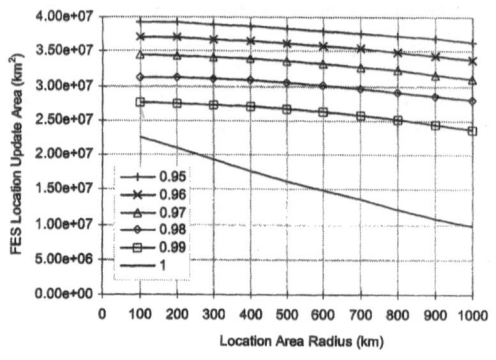

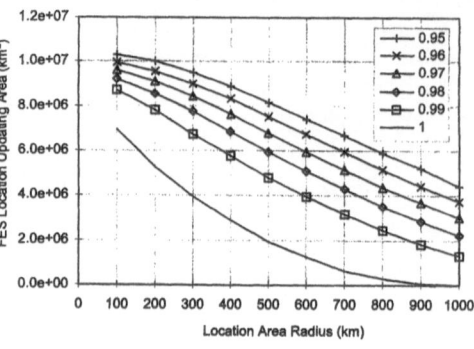

**Figure 3.31.** ICO-like FES location updating area vs. location area radius.

**Figure 3.32.** Globalstar-like FES location updating area vs. location area radius.

location updating area as the terminal may move to a point in the location area and make a call successfully, where a location update may result in a probability less than the threshold probability for that FES. The two areas are intrinsically linked as a larger location update area will result in a larger service area.

### 3.2.2.5.2 Primary Satellite
In this section we examine the area around an FES where a UT can make a location update with a probability of call completion via the highest satellite being above a threshold range.

The results shown in Figs 3.31 and 3.32 relate to an FES at 40° latitude. They demonstrate the impact that the size of the location area has on the area around the FES where a location update can be made with a value of $P_{FES}$ above a given threshold. The large increase in the FES location update area when the probability threshold is decreased from 1 to 0.99 is due to the fact that with a threshold less than 1.0 the location area is no longer completely bound by the 100% connectivity area in Fig. 3.16. This effectively means that in determining the user's optimum location area radius, the user's position relative to the FES should also be considered, as well as the call arrival rate and mobility of the user.

### 3.2.2.5.3 Secondary and Tertiary Satellites
The results in the previous section only consider the highest satellite to the mobile terminal. At location updating time it is also possible to determine the probability of inter-FES handover from the second, third, etc satellites. To reduce the number of dimensions in the results in this section, we examine the variations in FES location updating area using only one location area radius.

Figure 3.33 shows the variation in the FES (0° latitude) location update area size with respect to probability of future call completion through the first and second highest satellite for a location area with 300 km radius. It can be seen from the graph that for a high second satellite probability threshold the first satellite probability threshold has little impact on the FES service area. However, for lower second satellite thresholds (e.g. 0.9), the highest satellite probability threshold begins to have a larger influence on the location update area. The reason for this trend is as explained earlier regarding lower satellites offering more connectivity than higher satellites with increasing distance from the FES.

Figure 3.34 shows the location update area for an FES at 0° latitude in the Globalstar like system, again using a location area radii of 300 km. In this case the location update area is virtually completely determined by the probability threshold of the second highest satellite. It is only when the second highest satellite probability threshold drops below approximately 0.75 does the highest satellite probability threshold being to have an impact on the size of the FES location update area.

The domination of the second satellite threshold is due to the relatively low availability of the second satellite around the 0° latitude region in the Globalstar system. In latitude regions, where the second highest satellite visibility is more prevalent (e.g. 40° latitude for the Globalstar system), the results are similar in characteristics to those for the ICO system at 0° latitude.

### 3.2.2.6 Conclusions
In the study presented in this section, the GCA approach to call routing in a dynamic satellite personal communications network was investigated. The GCA method was shown to prevent the mobile terminal from using the highest satellite for a significant proportion of the time, at certain distances from the FES. Consequently we presented the Optimum Satellite Connectivity area as the area around

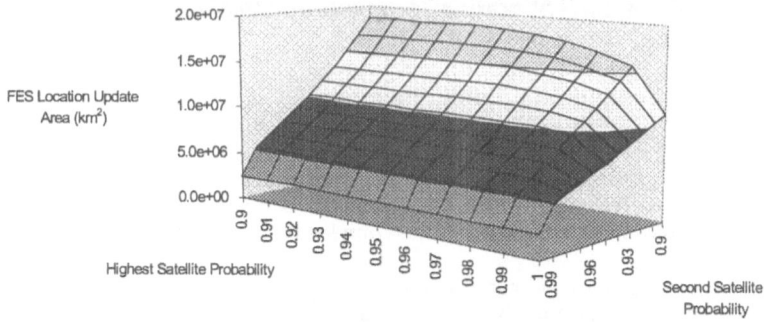

**Figure 3.33.** First and second highest satellites vs. ICO FES service area.

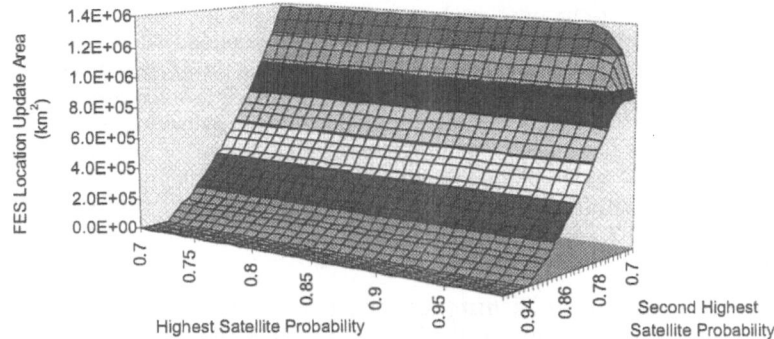

**Figure 3.34** First and second highest satellites vs. Globalstar FES location update area.

the earth station, where the satellite with the highest elevation to the mobile terminal will provide connectivity to the FES. We then presented inter-FES handover probabilities for FESs at various latitudes for a typical mobile call length distribution. The connectivity provided by the second highest satellite was analysed and at lower probabilities was shown to provide connectivity with an FES over a larger area than the highest satellite.

A mobility management scheme suitable for S-PCN based user positioning was revisited. A novel method of calculating, at location updating time, the probability of a future user terminated call requiring FES handover, for a given location area radius was illustrated. Results demonstrated the effect of longer mean call duration on the accuracy of the algorithm. Finally, we showed how the size of the area around the FES, where a terminal can make a location update with a probability of no future in-call FES handover being required above a certain threshold value varies with location area radius. The results show that second highest satellite can have a large impact on the FES service area, whereby the QoS specified by the service providers in terms of inter-FES handover probability or call dropping due to loss of FES connectivity determines the size of such an area.

The results can be interpreted in different ways depending on the network architecture and the ownership of earth stations. Consider firstly the case where each operator owns an earth station. It would be envisaged in this scenario that inter-FES handover would not be supported unless there are specific agreements between operators. Therefore, operators of FESs would wish to have users remain registered at their FES, only while the probability of any future call being completed without handover to another FES is high. The second scenario considered, is where all the FESs are owned by one operator or where in-call FES handover agreements exist. In this case the results could be used to reduce in-call FES handovers by registering the mobile terminal at the optimal FES with an appropriate location area size at location updating time.

### 3.2.3 A New Dynamic Channel Allocation Technique and Optimisation Methods for Mobile Satellite Systems

#### 3.2.3.1 Introduction

The capacity increase of a mobile satellite communication system assumes an important role taking into account the scarcity of available spectrum for this type of networks. An increase in capacity could be obtained by improving the use of the available spectrum namely by the selection of an appropriate multiple access technique. Another possibility is to use new DCA techniques that improve the use of the radio resources adding however complexity in the allocation and de-allocation processes.

MSS (Mobile Satellite System) networks implemented over NGSO (Non-GeoStationary Orbits) present some particularities that must be studied in order to assess its implications in the resource management process. In (27) is demonstrated that DCA techniques are suitable to be used as resource management methods in this type of systems due to its inherent capability to respond to time variant unbalanced traffic load patterns.

A large variety of DCA techniques can be found in the literature, some of those however can not be used because they have been conceptualised for mobile terrestrial cellular systems. The MP (Maximum Packing) channel allocation technique presents the lowest blocking probability in comparison to all the studied radio resource management algorithms. Its complexity precludes however its implementation in real and reliable mobile satellite networks.

Thus, the main goal of this paper is to propose a new DCA technique named Simplified Maximum Packing, analyse and evaluate its performance by simulation under uniform and non-uniform traffic load patterns. The core of the performed work was the implementation of this algorithm and the development of new analytical models to evaluate the MP performance, space state and system GoS (Grade of Service).

#### 3.2.3.2 Channel Allocation Techniques for NGSO Systems

As previously mentioned a possible method for increasing MSS capacity is to employ DCA techniques to manage the radio resources. However, NGSO systems present some particularities and challenges to a possible introduction of DCA techniques, with respect to other mobile cellular systems that must be highlighted:

- Almost all the mobile network hardware equipment such as: antennas, filters, amplifiers and switching elements are centralised in the satellites.
- The satellite hardware cannot be physically changed during the system lifetime. The maximum capacity is incorporated on the satellites since the construction. Only reconfiguration is available to the MSS network operator.
- It is necessary to implement some kind of frequency planning in multibeam satellite antenna to avoid co-channel interference (28).
- The continuous satellite orbital motion and the heterogeneous spatial traffic distribution lead to time-variant unbalanced traffic load patterns.
- The satellites high mobility and their corresponding short time visibility lead to a high number of inter-beam and inter-satellite handovers.

From the above sentences we conclude that each satellite manages all the radio resources assigned to its spotbeams in order to avoid interference and to respond to the uneven traffic patterns. This inherently centralised network architecture allows the implementation of more efficient and advanced DCA algorithms aiming to the desired GoS. However, only those that present a small level of overheads in terms of signalling and processing power will be presented and analysed in the following sections.

##### 3.2.3.2.1 Channel Allocation Techniques Overview

Several channel allocation policies can be used in a mobile satellite cellular network to assign a channel to a call. In the simplest case, the network uses the FCA (Fixed Channel Allocation) technique, in which a fixed predefined number of channels are permanently assigned to each spotbeam. Therefore, only a limited small number of calls can be simultaneously accepted in each spotbeam. The calls that find all the resources in use will be blocked and cleared. The FCA technique can be seen as a group of individual cells, where the blocking probability follows the Erlang B formula.

On the other hand, in the dynamic channel allocation techniques the channels are grouped together in a central pool, so all the channels are potentially available in any cell as long as the reuse distances are satisfied. The new calls are allocated to channels that can guarantee no interference to the existent calls. During the call duration these channels are removed from the pool and returned when the call ends. One of the most important aspects of DCA is its adaptability to the traffic variations leading to more effective use of the spectrum.

There are several ways to assign channels to calls using the DCA concept, namely by choosing randomly one non-interfering channel, by choosing the first channel in a predefined ordered list, or by choosing the channel that better matches the ongoing calls interference cell patterns, etc. The results obtained from the original DCA algorithm with random assignment show that this technique offers bad performance in overload situations. This undesirable behaviour occurs because cells that use the same channel are, on average separated by a distance larger than the minimum co-channel reuse distance. To overcome this drawback several concepts have been proposed, namely: channel ordering, channel reassignment, the use of cost functions, the combination of dynamic and fixed techniques and dynamic parameters adjustment. The SMP channel allocation technique presented in this paper will use some of these concepts to achieve a lower blocking probability.

### 3.2.3.3 Simplified Maximum Packing

The MP technique (29) is an idealised DCA algorithm that will only block a call if there are no possible reassignments in order to free a channel for the new call. This algorithm tries to discover a possible solution to serve each call at each moment. The large amount of information required by this technique, in order to search for all possible reallocations make this algorithm impractical for implementation.

#### 3.2.3.3.1 Simplified Maximum Packing Algorithm

The SMP technique (30) was constructed based on different cost functions associated with the allocation and de-allocation procedures, similar to those used in (31), and on the reassignment of ongoing calls concept. The allocation algorithm presented in Fig. 3.35 assigns, if possible the available nominal channel, which minimises the cost associated with the allocation procedure. The allocation algorithm is based on the fact that this channel belongs or not to the nominal set of channels used in that cell and in the amount of interference caused in the neighbour cells.

Initially, at the call set-up this technique operates as any normal DCA technique. It accepts a call only if there is at least one channel that is not used in the cells interference region. First the allocation algorithm will search for available nominal channels in that particular cell. If so, the system will determine among them the channel that presents the lowest allocation cost. This corresponds to a channel whose interference cells are as much packed as possible with the interference cells of the ongoing calls supported by the same channel. In other words, there is a tendency to dynamically use the FCA model as much as possible. If there is not any available nominal channel in that cell the algorithm will search for non-nominal channels. Once again, the SMP technique allocation algorithm will select a non-nominal channel that minimises the same earlier mentioned cost function.

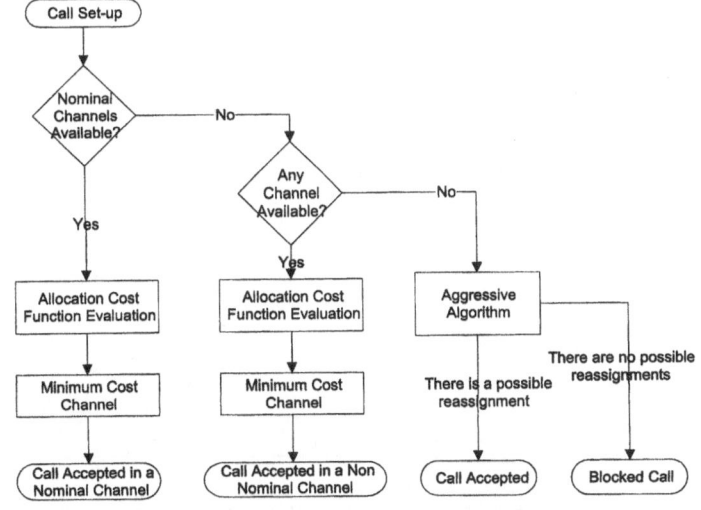

**Figure 3.35.** SMP technique allocation algorithm.

However, if the SMP technique allocation algorithm fails in obtaining a channel, that call is not necessarily blocked. Instead, the algorithm will determine the set of channels that are blocked in that cell due to the existence of a single call in the cell interference region. The system will try to reallocate the calls in progress carried in each one of these channels, in order to free a channel in the original cell. If this procedure is not successful, the reassignment is abandoned and the new call is blocked. This part of the algorithm was based on the PPADCA (Persistent Polite Aggressive DCA) technique presented in (32).

To increase the SMP performance, when a call ends in a cell, due to a call termination or handover, a channel is released in compliance with a de-allocation criterion. The de-allocation algorithm presented in Fig. 3.36 must chose a channel in order to reduce the mismatching with the optimal FCA channel distribution. Now calls in non-nominal channels and in high cost function channels have priority over nominal and low cost function channels. Thus, if the same cost function is used the most costly situation on the allocation phase corresponds to the least costly in the de-allocation phase.

### 3.2.3.4 Optimisation Methods

As mentioned in the introduction the aim of any mobile cellular system is to provide the users with the best quality of service appealing to an efficient use of the available radio resources. The main difficulty on the resource allocation process arises when a non-uniform traffic distribution is offered to the system cells. In these conditions the resources must be wisely distributed among the cells in order to minimise the system blocking probability. It seems logic that the optimal allocation must respond to the demand for the usage for those radio resources. The complexity of this problem some times referred as a NP optimisation problem motivates the study and analysis of several optimisation techniques. From those many implementations that have been proposed in the literature over the last years, two methods have been chosen as candidates to evaluate the optimal radio resource distribution over the network cells: the SA (Simulated Annealing) (33, 34, 35) and the GA (Genetic Algorithms) (36, 37, 38). The aim of this analysis is to perform a comparative study between them in terms of convergence speed, computational effort and the obtained results.

#### 3.2.3.4.1 Simulated Annealing

The SA algorithm is based on the simulation of physical phenomena that occurs during the solidification of some type of fluids. The main advantage of this algorithm is the possibility to find a new optimal point after a local minimum of the function has been found, accepting states for which the function value are worst than the previously optimum guess.

The SA is an optimisation technique that has been extensively used in the resolution of problems related with the radio resource allocation in mobile cellular networks. The interest in this particular optimisation technique arises from some of its attractive proprieties, namely:

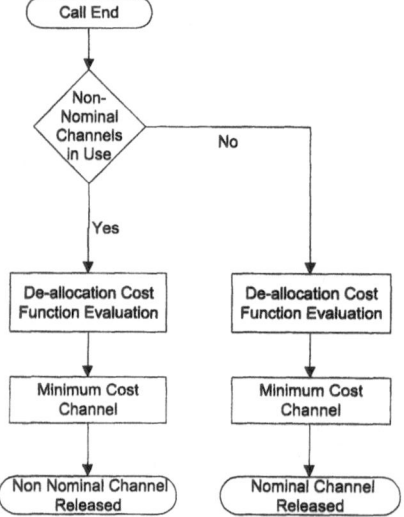

**Figure 3.36.** SMP technique de-allocation algorithm.

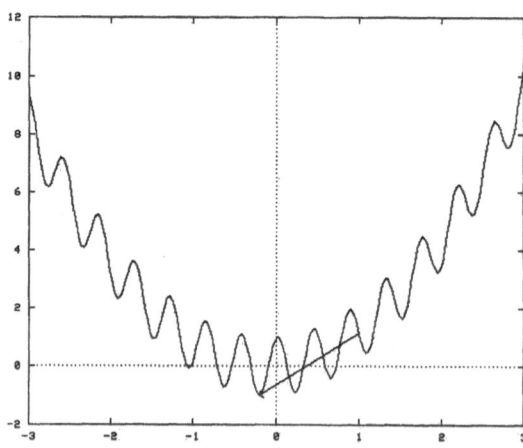

**Figure 3.37.** Cost function minimum evaluation using the Simulated Annealing algorithm.

- process cost functions with multiple degrees of non-linearity's and discontinuities;
- process cost functions with multiple boundary conditions and others limitations on the domain values;
- it is conceptually easy to implement and needs only a small quantity of code when compared with other non-linear optimisation algorithms;
- statistically an optimal solution is guaranteed.

The main advantage of this algorithm is the possibility of continuing to search for the function optimal point, after its local minimum has been found as shown in Fig. 3.37.

The originally proposed SA algorithm may be described as follows:

Sets the initial temperature and the reduction temperature factor
Generates randomly the initial point and its corresponding function value
While the stops criteria have not been reached
{
      While the statistical balance have not been fulfilled
      {
            Generates a new solution
            Evaluates the function value
            If the new function value is an optimum
                  Update optimum
            Else
                  If the Boltzman criteria was fulfilled
                  Accept temporary value
      }
      Update temperature
}

The SA it goes from one solution to another if the new guess value of the solution is closest to the optimum point. However, if the new guess is far away from the previously reached optimum solution there is still a possibility for this value to be accepted if the following relation, known as the Boltzman criteria, is fulfilled.

$$e^{-\Delta E/T} \geqslant p \qquad (3.5)$$

with,

$T$  –  parameter that controls the algorithm's evolution, by analogy with the physical system it is usually designated by temperature.

$\Delta E$ –  represents the difference between the cost of the new solution and the cost of the last one.

$p$  –  random value uniformly distributed over 0 and 1.

The sequence of the generated solutions at a given temperature could be mathematically modelled by and homogenous Markov chain. It is known that the SA algorithm converges asymptotically to the cost function global optimum if we perform an infinite number of monotonically decreases of the temperature. However is almost impossible to obtain the theoretical equilibrium for a given temperature. In general a limited number of trials are performed, those approximations are usually designated as quasi equilibrium. Another important factor that heavily conditions the performance of the SA algorithm is that in general the temperature decreases monotonically.

$$T_i = \alpha \cdot T_{i-1} \qquad (3.6)$$

The constant $\alpha$ can be fixed or variable, it depends on of the type of algorithm used in the cooling process.

### 3.2.3.4.2 Genetic Algorithms

Genetic algorithms are search and optimisation techniques based on the natural evolution of the species. These algorithms implement a stochastic search method that randomly samples the state space. The new guess for the optimal value of the cost function (either minimum or maximum), is built taking into account the best solutions of the earlier trials. This algorithm implements this codifying the parameters into binary strings. After that they are combined with others sequences, in order to obtain a new guess for the solution of the problem. This method eventually converges to the solution or to a point very close to the global solution. The genetic algorithm implemented in this paper has the following structure:

1. Create randomly a set of parameters that define the initial point in the space state that will be used as the argument of the function that we want to optimise. Evaluate the function value for this particular initial point.
2. Convert the set of parameters into binary strings. The complete set of a binary string is called a "Chromosome" and a particular bit is known as a "Gene".
3. Repeat steps (1) and (2) as many times as necessary, recording each population element in an array. The initial set of individuals is named the "Parents" population.
4. Select two good "Parents" chromosomes according to its cost function value. This is performed using the natural selection process that tries to ensure that parents with good genes will mate and combine almost the time and parents with bad genes do not.
5. Combine the genetic material of the chosen individuals, taking some genes of each "Parent". After the crossover a new binary string or "Chromosome" is obtained.
6. Expose the "Children's" to a low mutation rate, switching or flipping some bits of the binary sequence.
7. Repeat the step (4) through (6) until there are as many "Children" as "Parents".
8. Replace the existing "Parents" population by the new "Children" population.
9. Repeat steps (4) through (9) until every member of the population looks the same. This means that the algorithm has reached a stationary point.

### 3.2.3.4.3 Analytical Description of the Radio Resources Optimisation Process

Besides the SA and GA optimisation methods description presented above, other topics must be envisaged, in order to be able to apply these methods that is the cost function definition. The logical choice is to choose a cost function that directly evaluates the system GoS. As first approximation we can assume a system that uses the FCA technique.

Equation 3.7 presented in (39) is the modified Erlang B formula that describes the behaviour of a system that employs the FCA technique for some particular conditions, a non-uniform traffic load distribution and an uneven resource allocation pattern. This cost function will be extensively used along this paper to evaluate the optimisation methods performance.

$$P_b = \frac{1}{L} \cdot \sum_{k=1}^{n} \frac{A_k^{C_k+1}}{C_k!} \cdot \left[ \sum_{i=0}^{C_k} \frac{A_k^i}{i!} \right]^{-1} \quad \text{with} \quad L = \sum_{k=1}^{n} A_k \qquad (3.7)$$

Where the variable $A_k$ represents the traffic values in cell $k$ and $C_k$ the number of available resources in that particular cell. One can assume that these traffic values $A_k$ have been obtained from statistics previously collected or from some traffic simulation developed for this propose. With respect to the number of channels per cell evaluation a new matrix formulation has been developed.

The first step is to numerate sequentially and univocally all the cells in the system. After that is necessary to evaluate the number of different possible reuse configurations based on the cluster size and on the values of the corresponding reuse parameters. For reuse factors whose reuse parameters

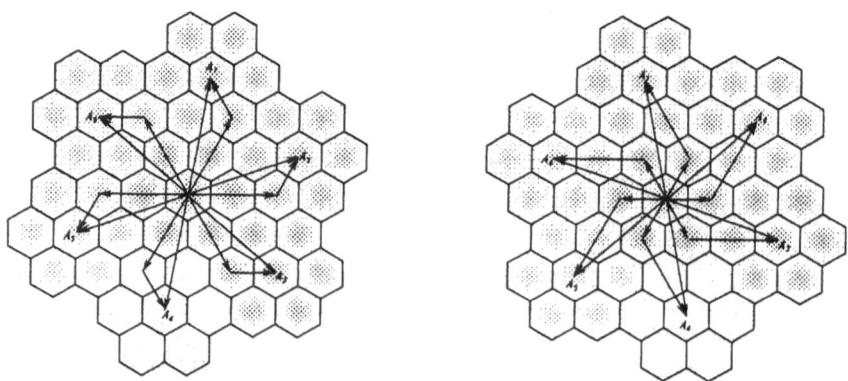

**Figure 3.38.** Two possible patterns for $K = 7$ for different reuse parameters a) $(i, j) = (2,1)$; b) $(i, j) = (2,1)$.

$i$ and $j$ satisfy one of the following equities $i = j$ or $i = 0$ or $j = 0$, we have only $K$ possible reuse configuration patterns, otherwise the number of possible reuse configuration doubles as can be seen in Fig. 3.38.

The next step consists in the construction of a rectangular matrix $R$ represented in Eq. 3.8 of dimensions $M \times N$, where $M$ represents the total number of possible reuse configuration and $N$ the number of cells in the system. The matrix is constructed in the following way: each row of the matrix describes the spatial localisation of the cells that belong to a particular reuse configuration. The matrix elements can only take two different values 1 and 0, corresponding respectively to the existence or not of that reuse pattern in that particular cell.

$$R = \begin{bmatrix} 1 & 0 & 0 & 0 & 0 & 0 & \cdots & 0 & 0 & 1 & 0 & 0 & 0 & 0 \\ 0 & 1 & 0 & 0 & 0 & 1 & \cdots & 0 & 0 & 0 & 1 & 0 & 0 & 0 \\ 0 & 0 & 1 & 0 & 0 & 0 & \cdots & 0 & 0 & 0 & 0 & 1 & 0 & 0 \\ 0 & 0 & 0 & 1 & 0 & 0 & \cdots & 0 & 0 & 0 & 0 & 0 & 1 & 0 \\ & & & \vdots & & & & & & & \vdots & & & \\ 1 & 0 & 0 & 0 & 1 & 0 & \cdots & 0 & 0 & 1 & 0 & 0 & 0 & 0 \\ 0 & 1 & 0 & 0 & 0 & 1 & \cdots & 0 & 0 & 0 & 1 & 0 & 0 & 0 \end{bmatrix} \tag{3.8}$$

Finally, a vector $Q$ of dimensions $1 \times N$ whose elements are non-negative integers representing the number of channels assigned to each one of the possible reuse configuration was defined. The sum of the vector $Q$ elements must be equal to the number of available channels in the system.

$$Q = [q_1 \quad q_2 \, q_3 \quad \cdots \quad q_{n-2} \, q_{n-1} \quad q_n] \quad \text{with } q_i \in N_0 \tag{3.9}$$

The number of channels per cell $C_k$ is obtained by the matrix product of $Q$ by $R$.

$$C = Q \times R \tag{3.10}$$

### 3.2.3.5 Simulation Assumptions and Parameters

In the performed analysis, the scenario described by the following assumptions and parameters was assumed:

- the call arrival process follows a Poisson distribution;
- call duration is exponentially distributed with a mean time of 3 min;
- blocked calls are lost and cleared;
- the system consists of 49 cells with a cluster size of 7;
- fading and co-channel interference effects are not taken into account;
- the system has a total of 70 available channels;
- the simulation results are obtained after statistical equilibrium is achieved.

Figure 3.39 shows a traffic pattern that is typically used for non-uniform traffic load. The numbers contained inside each cell represent the Poisson arrival rates (number of calls per hour). They range from 20 to 200 calls/h corresponding to a mean offered traffic value of 4.06 Erl, and a standard

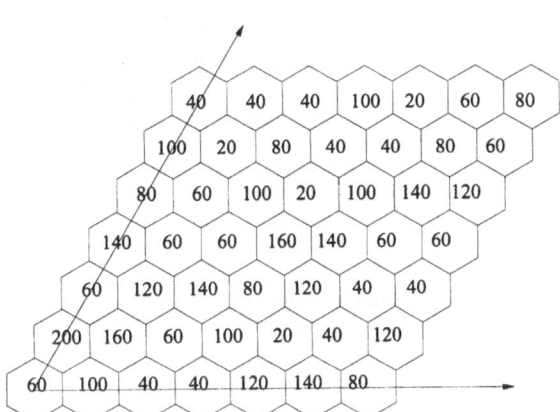

**Figure 3.39.** Number of calls per hour in a non-uniform traffic pattern.

**Table 3.4.** Optimisation methods performance

| Traffic increase (%) – mean value (Erl) | FCA | FCA with SA | FCA with GA | Optimal solution (*vector* **Q**) |
|---|---|---|---|---|
| 0 % – 199.0 Erl | 0.044455 | 0.019424 | 0.020940 | ( 2 5 0 1 6 7 4 5 5 7 8 7 6 7 ) |
| 20 % – 238.8 Erl | 0.077378 | 0.040640 | 0.041156 | ( 3 6 0 2 8 9 4 4 4 6 7 6 5 6 ) |
| 40 % – 278.6 Erl | 0.114571 | 0.068911 | 0.070452 | ( 3 6 0 2 8 9 4 4 4 6 7 6 5 6 ) |
| 60 % – 318.4 Erl | 0.153194 | 0.102454 | 0.103565 | ( 3 6 0 2 8 10 4 4 4 6 7 5 5 6 ) |
| 80 % – 358.2 Erl | 0.191463 | 0.138767 | 0.139997 | ( 3 5 0 2 9 10 4 4 4 6 7 5 5 6 ) |
| 100 % – 398.0 Erl | 0.228388 | 0.176637 | 0.176923 | ( 3 5 0 2 9 11 4 4 3 6 7 5 5 6 ) |

deviation of 2.16 Erl, assuming a mean call duration of 3 min. The simulation results obtained from the optimisation methods SA and GA over non-uniform traffic patterns are.

### 3.2.3.5.1 Optimisation Methods Results

Several simulation were performed in order to evaluate the performance of the proposed optimisation methods: SA and GA over non-uniform traffic patterns. All those simulations follows the conditions and assumptions presented in the last section. The obtained results for the SA, GA and the classic FCA technique with uniform number of resources are presented in Table 3.4 as function of the traffic load increase with respect to the initial traffic pattern presented in Fig. 3.39.

From the results obtained, we conclude that when the channels distribution per cell obtained by the optimisation process is used, a superior performance is obtained in comparison to the uniform radio resources utilisation. It is also observed that the optimisation SA algorithm always converges to the same solution, which is assumed as the real optimal solution. On the other hand, the solution obtained with the GA algorithm is different from simulation to simulation. These results differ some permutations apart from the optimal solution. The solution obtained from the SA algorithm always presents the best result from all other algorithm implementations. The only drawback of the SA algorithm is the large computational effort necessary to obtain good results.

A reduction of 22.66 and 56.31 per cent in the system blocking probability was observed for overload conditions and for the initial traffic value respectively when the number of radio resources by cell obtained by the SA optimisation method was used instead of the usual uniform radio resources distribution.

From the performed simulation we conclude that the convergence velocity of the optimisation methods depends on a number of parameters specific to each method. In the SA algorithm it depends on the initial temperature, the temperature-cooling factor and the number of the cost functions evaluation performed at the same temperature. On the other hand, the GA depends on the population size, the number of generations considered and the chosen selection method.

## 3.2.3.6 MP Performance Evaluation

### 3.2.3.6.1 The Dimension of the MP Performance Evaluation Problem

In mobile cellular networks that adopt the FCA technique as a resource management method the existing radio resources are divided among the cells of the different clusters in a predefined and permanent way. Having in mind that in this channel allocation technique the cells are independent from each other, we conclude that the performance evaluation problem as only one dimension. In this case the blocking probability of each cell is given by Erlang B formula where it is only necessary to known the number of available channels and the offered traffic load values per cell.

DCA techniques employ a different philosophy in the radio resources utilisation. In addition to the knowledge of the channel usage in a particular cell, the system must also be aware to the resource utilisation in the neighbouring cells. The acceptance of a new call depends on the global state of the system i.e. the number of calls accepted in each cell and the information with respect to the channels used in supporting ongoing calls.

The MP algorithm however precludes the necessity of that information because a call is admitted only if there are possible reassignments of the existing calls, in general in order to free a channel in that cell. From the last sentence we verify that the dimension of the MP performance evaluation problem in a mobile cellular system is equal to the number of cells.

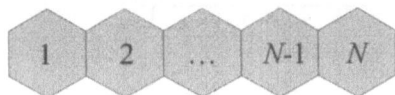

**Figure 3.40.** System cells identification.

### 3.2.3.6.2 MP Mathematical Model Definition

To analyse the MP technique performance in a cellular system, we chose the terminology and procedures similar from those presented in (40). The vector $\mathbf{x} = (x_1, x_2, ..., x_N)$ identifies the state of the cellular system at a given time, where the vector's $x_i$ components represent the total number of active calls at cell $i$. The index $i$ whose values are in a range of 1 to $N$ identifies the cells of the system.

The global space state for the MP technique Z is limited by the values that the $x_i$ components can take, which themselves depend on the frequency reuse constraints, the number of available channels, the number of cells and its layout. These limitations on the system state space can be expressed as:

$$\sum_{j \in C_i} x_j \leq M \quad \forall \ C_i \tag{3.11}$$

The space state is built by all the states that represent occurrences in the maximum number of calls inside any cluster containing that particular cell and is less or equal to the total number of channels of the system. In Eq. 3.11 $Ci$ represents an arbitrary cluster that contains the cell and $M$ represents the number of available channels on the system.

### 3.2.3.6.3 MP Space State Evaluation

This algorithm allows the evaluation of all possible states Z of the system using only the number of cells, the maximum number of available channels, and the interference pattern constraints between the several cells of the system. The states of the space Z will then be used in conjunction with the cells offered traffic values to evaluate the probability of occurrence of each state. The blocking probabilities per cell will be determined by adding all the states probabilities that correspond to a situation for which it is impossible to accept a new call in that cell. This procedure is used to measure the performance of a particular system by evaluating the system GoS.

Due to the impossibility of previously knowing the total number of states for the system we have chosen a dynamic data structure to record the states values. A linked list to group the various states has been chosen for this effect. Each linked list element has two distinct fields, one contains the state value and the other contains a pointer for the next element of the list, if such element exits.

The algorithm always begins with a $N$ dimension null state $(0, 0, ..., 0)$ which corresponds to a situation where the system does not support any call. The remaining system states will be determined from that initial state. The algorithm chooses the first element of the list to be the studied element and generates new states from that state corresponding to the arrival of individual calls to each cell of the system, mathematically this is expressed incrementing one unit to each of the vector's co-ordinates $\{(1, 0, ..., 0); (0, 1, ..., 0); ...; (0, 0, ..., 1)\}$. The next step is the validation of the generated states. This procedure consist in verifying if the ongoing calls, represented by the vector co-ordinates, could be supported by the existing system radio resources without violating the interference constraints. This is performed appealing to Eq. 3.11. Taking into account the specificity's of the proposed algorithm it is necessary to verify if the proposed new state already exists on the list as a result of a same number of calls arriving in a different order. If the new state is a valid one, i.e. it does not violate Eq. 3.11, and if it is not already in the list, it will be inserted at the end of the list.

After all generated states have been analysed for a particular studied state a new one is chosen. The new state is compared with all the existing states in the linked list that have the same number of ongoing calls until the end of the list is reached. The algorithm stops when the end of the linked list is reached. This corresponds to a situation in which the last studied element does not produce valid states.

Consider as an example to illustrate the algorithm a system with three cells, four channels and a reuse factor $N = 2$. If a call arrives to one of the extremity cells and there is an available channel to serve that call, this channel is selected being locked in that particular cell and in the central cell in order to avoid interference. On the other hand, if a channel is assigned to the central cell to serve a call, this channel must be locked in all the cells. Fig. 3.41 presents all the states that belong to the space Z obtained from the algorithm earlier presented. For this particular system 55 states have been identified.

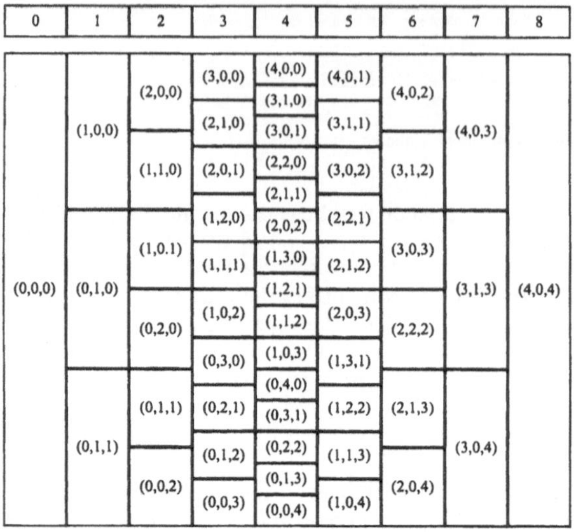

| 0 | 1 | 2 | 3 | 4 | 5 | 6 | 7 | 8 |
|---|---|---|---|---|---|---|---|---|
| | | | (3,0,0) | (4,0,0) | (4,0,1) | | | |
| | | (2,0,0) | | (3,1,0) | | (4,0,2) | | |
| | (1,0,0) | | (2,1,0) | (3,0,1) | (3,1,1) | | (4,0,3) | |
| | | (1,1,0) | (2,0,1) | (2,2,0) | (3,0,2) | (3,1,2) | | |
| | | | | (2,1,1) | | | | |
| | | | (1,2,0) | (2,0,2) | (2,2,1) | | | |
| | | (1,0.1) | | | | (3,0,3) | | |
| | | | (1,1,1) | (1,3,0) | (2,1,2) | | | |
| (0,0,0) | (0,1,0) | | | (1,2,1) | | | (3,1,3) | (4,0,4) |
| | | | (1,0,2) | (1,1,2) | (2,0,3) | | | |
| | | (0,2,0) | | | | (2,2,2) | | |
| | | | (0,3,0) | (1,0,3) | (1,3,1) | | | |
| | | | | (0,4,0) | | | | |
| | (0,1,1) | | (0,2,1) | (0,3,1) | (1,2,2) | (2,1,3) | | |
| | | | (0,1,2) | (0,2,2) | (1,1,3) | | (3,0,4) | |
| | | (0,0,2) | | (0,1,3) | | | | |
| | | | (0,0,3) | (0,0,4) | (1,0,4) | (2,0,4) | | |

**Figure 3.41.** Space of states Z for a 3 cell, 4 channel system with $N = 2$.

### 3.2.3.6.4 Cell and System Blocking Probability Evaluation

After all the states have been identified the next step is to evaluate the occurrence probability of each state. The occurrence probability of the state $x$, $p(x)$ in statistical balance is given by the following expression:

$$p(x) = \begin{cases} \dfrac{A_0^{x_0}}{x_0!} \cdot \dfrac{A_1^{x_1}}{x_1!} \cdot \ldots \cdot \dfrac{A_n^{x_n}}{x_n!} \cdot p(0, 0, \ldots, 0) & \text{for } x \in Z \\[2mm] 0 & \text{for } x \notin Z \end{cases} \tag{3.12}$$

This set of equations is not however sufficient to evaluate the occurrence probability of the states. We must also introduce the statistical normalisation equation for this effect.

$$\sum_{x \in Z} p(x_0, x_1, \ldots, x_n) = 1 \tag{3.13}$$

Figure 3.42 presents the occurrence probability values for each of the system states, obtained analytically using Eqs. 3.12 and 3.13 and via simulation, to a particular offered traffic value. From the obtained results we conclude that the simulations have confirmed the proposed model.

In order to evaluate the systems and cells blocking probability we must identify the states that

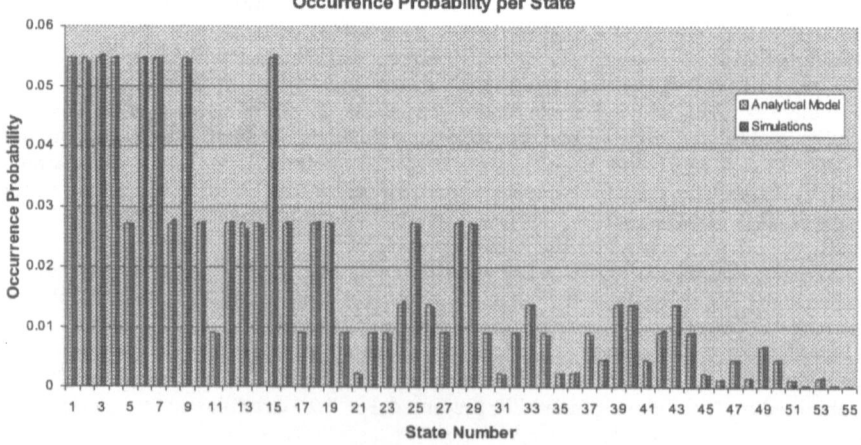

**Figure 3.42.** Space state Z occurrence probability values.

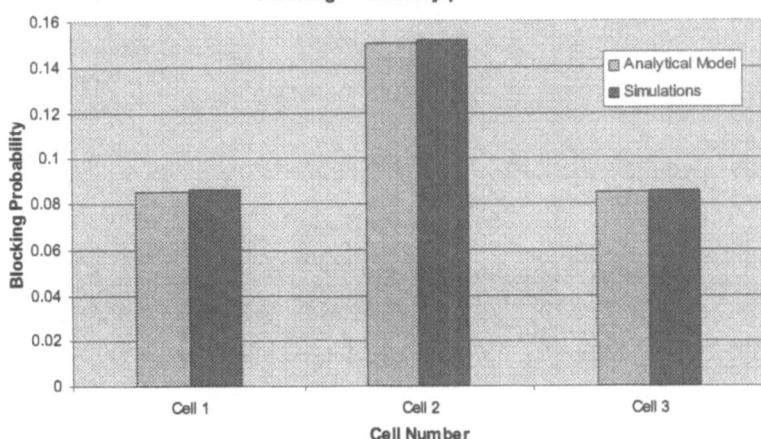

**Figure 3.43.** Space state Z occurrence probability values.

correspond to a blocking situation. A call will be blocked if it finds the system in a state for which the maximum number of ongoing calls in at least one cluster that contains this particular cell is equal to the number of available channels. This occurs whenever the system is in a state $x \in B_i$ defined for the following expression:

$$x \in B_i : \sum_{j \in C_i} x_j = M \quad \forall \, C_i \tag{3.14}$$

It can verify that the cells in the system extremity have a small number of blocking states than the central cell. This arises from the fact that the boundary cells do not have a complete set of interference cells. In the example shown in Fig. 3.41, the subspaces $B_1$ and $B_3$, that correspond to a blocking situation in the extremity cells, present 15 states less than the central cell blocking subspace state $B_2$ (25 states). The state co-ordinates that correspond to a blocking situation have been underlined.

The blocking probability of a particular cell $i$ is given by

$$P_{b_i} = \sum_{x \subset B_i} p(x) \tag{3.15}$$

Figure 3.43 shows the blocking probabilities per cell obtained appealing to the described analytical model and by simulations using the SMP technique. The obtained values correspond scenario described by the assumptions earlier presented system with three cells, four channels, reuse factor equal to 2 and a offered traffic value per cell equal to 1 Erl.

Finally, the system blocking probability could be obtained using the following equation

$$P_b = \sum_{i=1}^{n} A_i \cdot P_{b_i} \bigg/ \sum_{i=1}^{n} A_i \tag{3.16}$$

The system GoS value was evaluated analytically using the blocking probabilities and traffic values per cell as explained in Eq. 3.16, the obtained results was $P_b = 10.69$ per cent and experimentally by means of the conducted simulations we obtain the following value $P_b = 10.81$ per cent.

### 3.2.3.7 SMP Technique Results

This section presents an evaluation of the SMP technique performed via simulation under two different traffic scenarios: uniform and non-uniform load distributions. The performance evaluation criteria used to estimate the GoS is the system blocking probability, i.e. the ratio of the number of blocked calls to the total number of call attempts.

Initially, several items of the SMP technique are analysed namely: the nominal and non-nominal channel utilisation on the allocation algorithm and the assessment of the success rate of the aggressive algorithm. In this first study only uniform offered traffic patterns have been considered.

Figure 3.44 shows that under low traffic conditions the cells nominal channels carry approximately 75 per cent of the total traffic. However, as the traffic load increases we observe an increase on the traffic value supported by the non-nominal cells channels. For a particular traffic load value around 9.5 Erl the percentage of traffic carried by the nominal and non-nominal channels is approximately

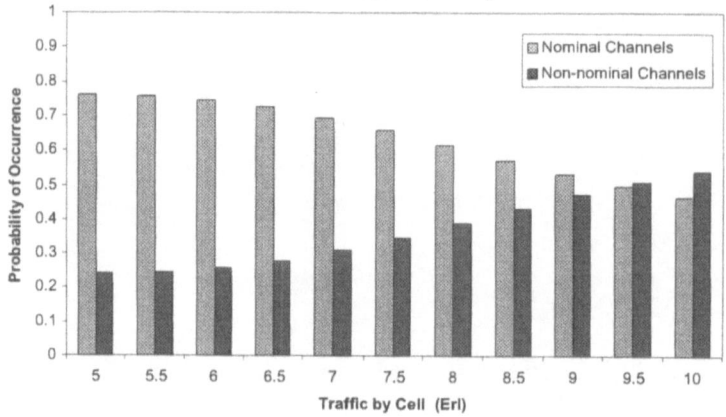

**Figure 3.44.** Nominal and non-nominal channels utilisation on the SMP technique.

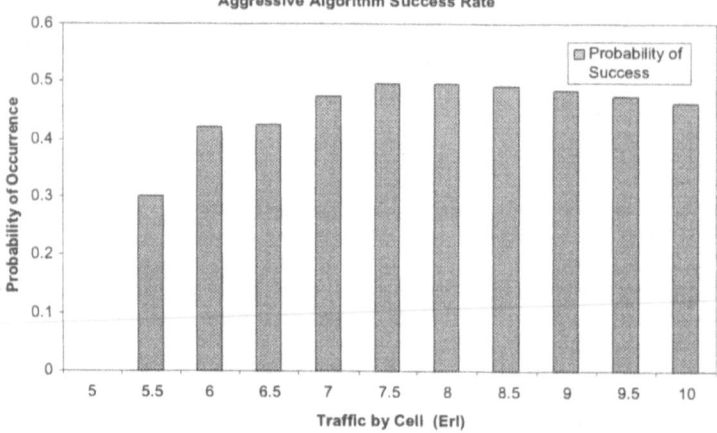

**Figure 3.45.** Aggressive algorithm success rate.

the same. This situation is similar to the observed on the borrowing channel techniques in overload conditions. In these situations a channel borrowing induces multiple borrowings reducing the cells nominal channels utilisation.

From the analysis of the aggressive algorithm used on the SMP technique we can observe, a reduction on the system blocking probability resultant from the acceptance of some calls that normally would be blocked due to the lack of available resources.

From the analysis of Fig. 3.45, we can verify that approximately 50 per cent of the calls blocked by the conventional allocation algorithm proposed in (31) are now accepted appealing to an additional reassignment of the ongoing calls performed by the aggressive algorithm. This value is approximately constant independently of the offered traffic value per cell. This fact explains the performance increase of the SMP with respect to others techniques presented on the literature.

Next, a comparative study of the proposed SMP algorithm, the classical FCA technique and the DCA cost function under uniform traffic load patterns is presented. The blocking probability results obtained via simulation for each one allocation techniques are depicted in Fig. 3.46. From the results we conclude that the SMP technique presents always the lowest blocking probability. This improvement results from the nominal channel concept and the utilisation of the aggressive algorithm.

One of the most important characteristics of the DCA technique is its intrinsic capacity to dynamically assign the resources to the cells aiming to fit the usual non-uniform traffic patterns found in real mobile systems. Besides the DCA technique's natural capacity to match the demand of resources, we have included an additional degree of intelligence in the system appealing to the inclusion of the optimisation methods results. The technique that uses both the nominal channels patterns from the SA optimisation method, and the concepts previously presented for the SMP technique, is named SMP Enhanced.

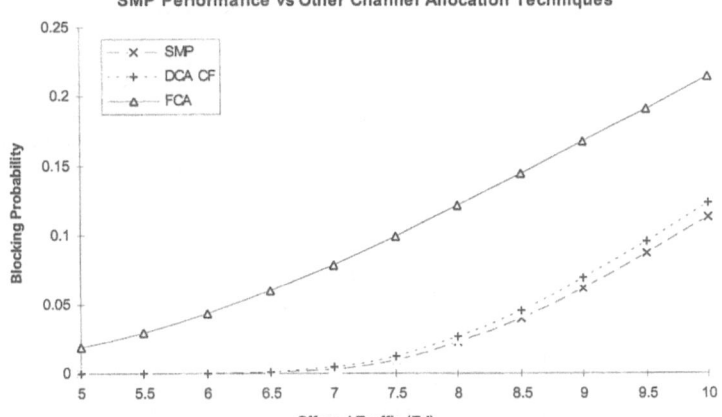

**Figure 3.46.** Blocking probability results for some channel allocation techniques.

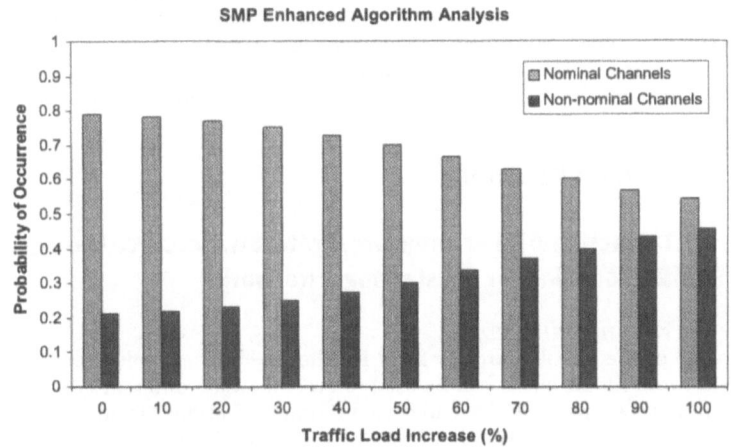

**Figure 3.47.** Blocking probability results for some channel allocation techniques.

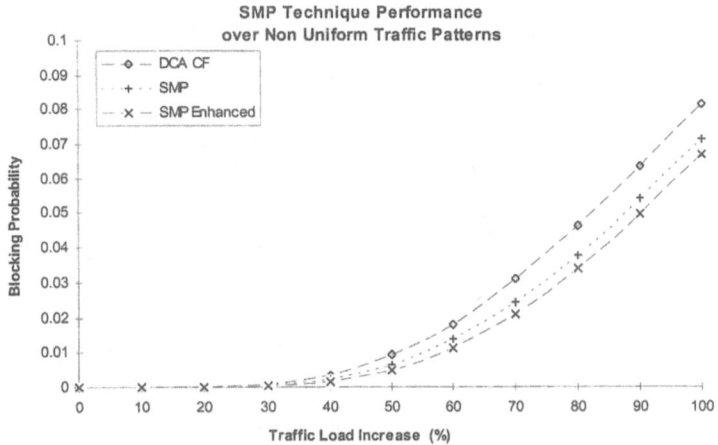

**Figure 3.48.** DCA techniques performance under non-uniform traffic loads patterns.

The results presented in Fig. 3.47 correspond to the nominal and non-nominal channel utilisation of the SMP Enhanced technique for the non-uniform traffic pattern presented earlier. The tendency observed on the graphic is the same as the observed previously for uniform traffic patterns: high nominal channel percentage of utilisation for low traffic values and an increase of the non-nominal channel usage as the traffic increases.

Figure 3.48 depicts the DCA cost function, the simple and enhanced SMP technique performance as a function of the load increase over the initial non-uniform load values presented in Fig. 3.39. From the obtained results we can observe a large reduction of the blocking characteristics when the simple SMP technique is used with respect to the DCA cost function. However an additional improvement could be obtained if we use the resources pattern resultant from the optimisation methods as in the SMP Enhanced technique.

### 3.2.3.8 Conclusions
FCA is the classic method for assigning radio resources to the cells but cannot respond efficiently to the spatial and temporal traffic variations presented in NGSO systems. The DCA techniques are appropriated to be used in these particular situations. From the simulation results we could verify that the SMP technique presents the best results for all the techniques addressed. The use of cost functions and the possibility of channel rearrangement at the beginning and end of a call privileges the use of nominal channels and reduces the average reuse distance. It was observed that the SMP Enhanced technique achieve an efficient use of system radio resources under non-uniform traffic patterns outperforming clearly the others DCA techniques. We also conclude that the newly proposed mathematical model is expected to be extremely useful to evaluate the MP technique GoS, as the classical Erlang B formula is for the FCA technique case. That algorithm was validated via simulations and constitutes the first known complete mathematical model that evaluates a DCA technique performance. The main disadvantage is the high computational effort needed to evaluate the MP technique performance even for systems with a small number of cells and channels.

## 3.3 S-ATM Approach

### 3.3.1 Satellite-ATM Architecture, Protocols, Resource Management and Traffic Modelling for Geostationary Networks

#### 3.3.1.1 Introduction
One of the possible architectures for the satellite segment is based on the satellite-ATM (S-ATM) network model and it can be used to accommodate different types of user terminals. Current work in the European ACTS (Advanced Communications Technologies and Services) (41) such as SECOMS[5], ASSET[6], WISDOM[7] and ACCORD[8] (42) investigate the possibility of a S-ATM model for the satellite segment using new generation multi-spotbeam satellites with advanced on-board processing capabilities. At the same time, large number of Ka-band (30/20 GHz) proposals such as: SPACEWAY, ASTROLINK (43), CYBERSTAR (44), TELEDESIC (45), N-STAR (46), WEST (47) and EUROSKYWAY (48) target at high data rate multimedia services mainly to fixed and possibly portable user terminals.

As suggested in (6), for the satellite network a new modified version of the ATM protocol layer (known as S-ATM) is required. This proposal has been further exploited and a highly integrated broadband Satellite-ATM network scenario is proposed (49). The implications on the system design, the protocol architecture and the resource management are discussed in this report.

#### 3.3.1.2 Broadband Satellite Network Architecture
The satellite constellation is the initial parameter that affects the design of a global network infrastructure. Most of the proposed systems, concentrate on the Geostationary Orbit (GEO) (see Table 3.5). For example: in Europe EUROSKYWAY by Alenia Aerospazio (Italy) and WEST (GEO component) by Matra Marconi Space (France/UK), in the United States ASTROLINK by Lockheed Martin, SPACEWAY by Hughes and CYBERSTAR (GEO component) by Loral Space and in Japan N-STAR by

---

[5] Satellite EHF Communications for Mobile Multimedia Services
[6] ACTS Satellite Switching End-to-end Trials
[7] Wideband Satellite Demonstration Of Multimedia
[8] ACTS Broadband Communication Joint Trials and Demonstration

**Table 3.5.** Examples of some broadband satellite Ka(Ku)-band proposals

| System | Origin | Constellation | Satellite payload | On-board switching | Access schemes |
|---|---|---|---|---|---|
| SPACEWAY | USA | GEO | Regenerative | ATM-based | MF-TDMA |
| ASTROLINK | USA | GEO | Regenerative | ATM-based | MF-TDMA |
| CYBERSTAR | USA | GEO and LEO | Regenerative | Packet switching | MFTDMA/ CDMA |
| TELEDESIC | USA | LEO (288) | Regenerative | Packet switching | MF-TDMA |
| CELESTRI* | USA | LEO (63) | Regenerative | ATM-based | MF-TDMA |
| N-STAR | Japan | GEO | Regenerative | ATM-based | TDMA |
| WEST | Europe | GEO and MEO | Regenerative | ATM-based | MF-TDMA |
| EUROSKYWAY | Europe | GEO | Regenerative | Packet switching | MF-TDMA |
| SKYBRIDGE† | Europe | LEO (64) | Transparent | No | CDMA |

*Merged with TELEDESIC.
†Ku Frequency Band.

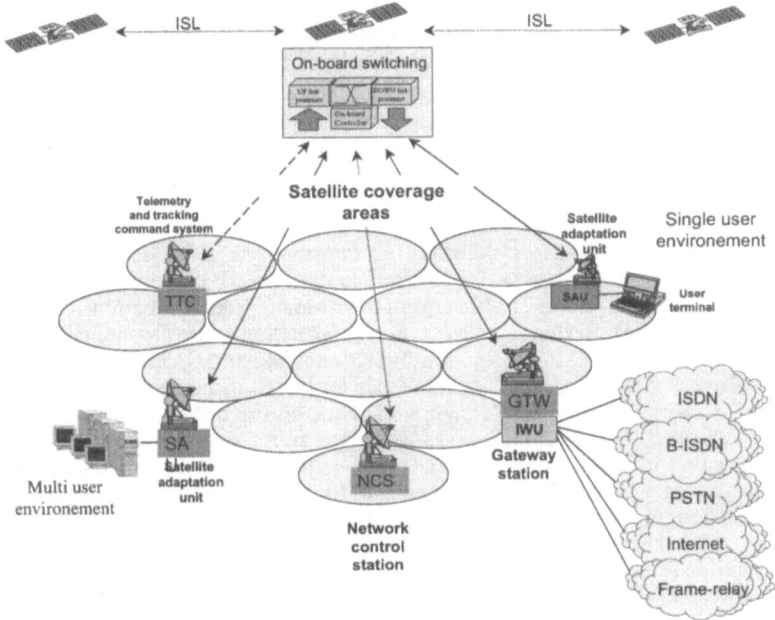

**Figure 3.49.** Satellite network architecture and connectivity with the fixed networks.

NTT. The non-GEO systems are: TELEDESIC that proposed to use 288 Low Earth Orbit satellites (LEO) and CELESTRI (LEO 63) which was recently merged with the TELEDESIC proposal. At the same time, there are proposals based on hybrid constellation approach using GEO and non-GEO components; e.g. WEST and CYBERSTAR. As far as the satellite ground network is concerned, the number and the location of the Land Earth Stations (LES) depends on the constellation parameters and on the provision of Inter-Satellite Links (ISLs) as well as the feeder link bandwidth requirements.

Another parameter that influences the selection of the satellite network architecture, is the dependency on the terrestrial network infrastructure. The satellite links are essential for inter-station signalling, when there is no terrestrial infrastructure deployed. For example, most GEO systems do not need ground station interconnection through terrestrial links, whereas non-GEO systems require only a few satellite links to the LESs, when ISLs are used; otherwise they highly depend on a fast backbone network. In GEO systems ISLs mainly handle the traffic between different regions of the earth and bypass the terrestrial links, whereas in most non-GEO constellations ISLs are essential in order to reduce the number of LESs.

In a typical broadband satellite system (see Fig. 3.49) the following network entities exist:

- *User Terminals* (*UT*): Several different protocol standards, such as: ATM User Network Interface (ATM-UNI), Frame Relay UNI (FR-UNI), Narrowband Integrated Services Digital Network (N-ISDN) Basic Rate Interface (BRI), N-ISDN Primary Rate Interfaces (PRI), Transmission Control Protocol/Internet Protocol (TCP/IP). They are connected to the Satellite Adaptation Unit (SAU) through one of the supported standard interfaces.

- *Satellite Adaptation Unit* (*SAU*): This is a non-standard, specially designed unit, responsible to provide access to the satellite network. It performs all the necessary user terminal protocol adaptations to the satellite protocol platform. The SAU also includes all the physical layer functions such as channel coding, modulation/demodulation, the Radio Frequency (RF) parts and the antenna section. A set of different types of terminals, with a variety of transmission capabilities, is usually offered by a satellite network. Starting from minimum transmission rates of 16 Kbps of even less, they can cope with maximum transmission rates of 144 Kbps or 384 Kbps for personal type user terminals, or even 2048 Kbps and higher for fixed type terminals with larger antennas. All of the supported terminals share the same access scheme and protocols stacks.

- *Payload* (*P/L*): Full on-board satellite signal re-generation is assumed in most of the future broadband satellite systems. The on-board satellite processing units perform multiplexing, demultiplexing, channel coding/decoding and ATM "like" switching, using a multi-spot beam configuration. The on-board satellite switch includes only part of the functions that a ground ATM switch performs. To minimise the payload complexity, the call set-up signalling and the CAC functions are performed on the ground.

- *Gateway stations* (*GTW*): These are the LESs that provide connectivity to the external networks. Most of the future proposed broadband satellite systems support interconnection to the fixed B-ISDN, Frame Relay, N-ISDN, PSTN and Internet, via the proper interworking units at the gateway. The level of interworking between the GTWs and the terrestrial networks depends on the type of traffic that the satellite network carries. The SAU provides an access interface very similar to the standard ATM-UNI, therefore the required interworking at the terminal and the GTW is minimised. The signalling protocols for call and connection control can be reused (based on the ITU-T Q.2931 standard) and the traffic and network management functions can share common characteristics. All ATM service classes can be supported and directly mapped into the satellite air interface through the SAU. Finally, the interconnection interfaces with all the other public terrestrial networks should be based on standards for interworking with B-ISDN. In GEO systems the placement and number of GTWs mainly depends on the traffic demand. A large number of gateways is expected in geographical areas where the traffic demand is high; these gateways are always connected with the same satellite(s). However, in non-GEO systems, the number and placement of the gateway stations depends on some additional system design characteristics, such as: constellation design, use (or not) of ISLs and the overall end-to-end delay constraints. For example, in a global MEO system with no inter-satellite links, a total number of less than ten gateways can provide full connectivity to the land masses most of the time. A LEO system will require tens to hundreds of gateways but this number can be reduced with the use of ISLs.

- *Network Control Station* (*NCS*): A central entity, used in a GEO satellite system (usually one per satellite) that provides overall control of the satellite network resources and operations. This node is responsible for allocating radio resources to the LESs/GTWs according to a long term resource planning scheme. The NCS is responsible for performing most of the resource management, call management and user/terminal mobility management functions, in addition to authentication, registration, deregistration and billing. In some systems, these operations are performed in more than one ground stations in a distributed manner.

### 3.3.1.3 *Satellite Access Interface and Protocol Stacks*

Two main scenarios for the satellite access network protocol can be envisaged. The first one uses ATM cell encapsulation and satellite specific protocols for establishing and managing user connections, whereas the second one provides a highly integrated solution with the ATM protocol stack. The ATM protocol considerations over the air interface and the on-board satellite switch architecture are different for each protocol platform. However, there are still a lot of open issues for investigation before a decision can be reached. A more general satellite packet switching approach increases the system flexibility to accommodate any protocol standards without being restricted to adopting an ATM satellite switching solution. The second approach provides a highly optimised protocol architecture, if ATM is adopted as the transport mechanism for the future broadband communication systems.

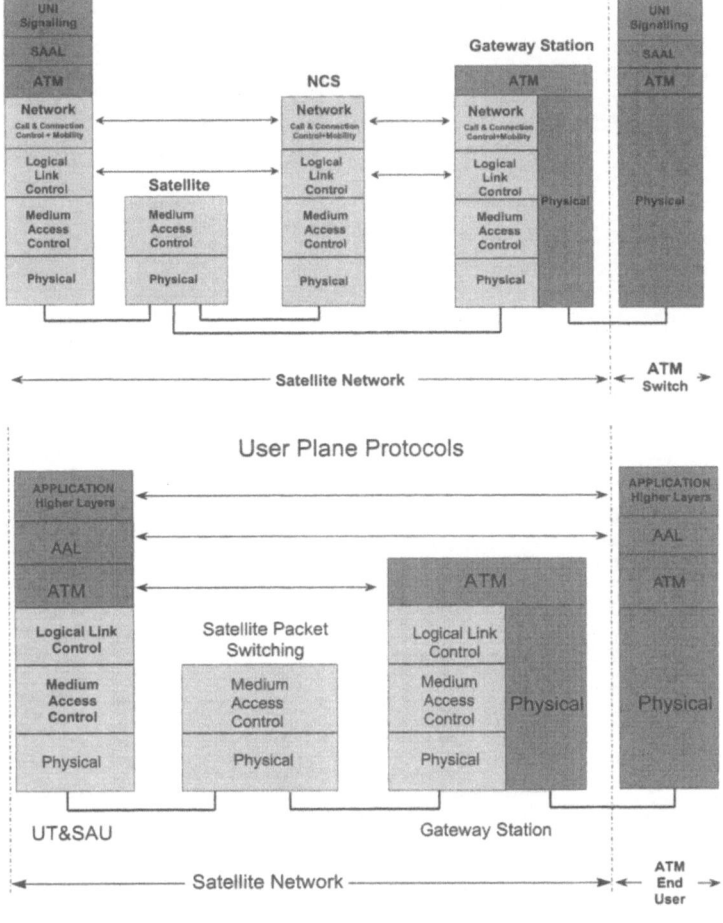

**Figure 3.50.** Protocol reference model for ATM encapsulation over a satellite specific protocol platform.

#### 3.3.1.3.1 ATM Protocol Encapsulation

Protocol encapsulation is a simple and easy to implement technique for passing arbitrary protocol information through the network entities that could not otherwise interpret specific information. In this scenario (Fig. 3.50), the satellite protocol platform is designed to transparently support different user terminal standards through a proprietary, satellite specific, interface. The satellite access protocol is terminated at the GTWs. Therefore, no modifications to the existing protocol standards are necessary. This approach appears to be very attractive in systems that need to accommodate several different types of user terminals with a variety of protocol standards. Circuit switching, packet switching or even hybrid solutions for the on-board satellite processor, can be implemented for networks that use this type of protocol platform.

An example of the return up-link frame structure of a satellite system that supports ATM protocol encapsulation is given in (50). This approach is the one that has been selected within the framework of the European ACTS project SECOMS and is currently proposed to be used in EUROSKYWAY. The minimum amount of data that a terminal can transmit is a single burst of 80 bytes for the duration of one Multi-Frequency Time Division Multiple Access (MF-TDMA) time slot. In SECOMS, the shortened Reed Solomon RS (80, 64, 16) code is selected in the return uplink direction to protect the satellite packet (53 bytes of packet payload and 11 bytes packet header) from channel errors. However, in this approach it is very difficult to offer an optimum performance to any particular protocol standard, resulting in protocol inefficiencies (related to the increased packet overheads). Therefore, a more ATM oriented approach is proposed to be used.

## Control Plane Protocols

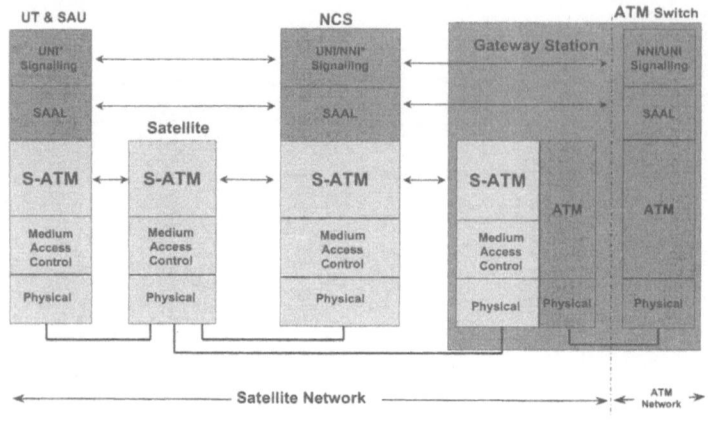

## User Plane Protocols

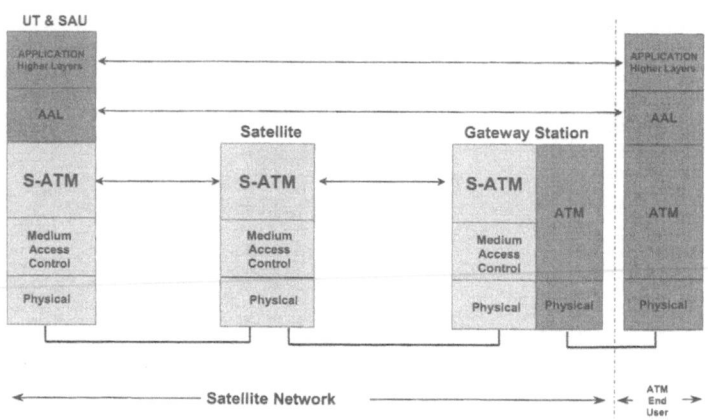

**Figure 3.51.** ATM based protocol reference model for control and user planes.

### 3.3.1.3.2 S-ATM Access Interface

The proposed protocol stacks for both the control and the user plane in a satellite ATM network are shown in Fig. 3.51. The S-ATM layer which replaces the standard ATM layer includes all the required modifications (on the cell header and its functions) that need to be addressed for the space segment.

The fields that are present in the S-ATM cell header carry essential routing and control information for the satellite segment. The dimensioning of the Satellite Virtual Path Identifier (S-VPI) and the Satellite Virtual Channel Identifier (S-VCI) fields depends on system parameters such as: the satellite capacity, the number of spot-beams, the satellite terminal transmission granularity and the switch architecture. The Medium Access Control (MAC) using MF-TDMA or Code Division Multiple Access (CDMA) (51) based on a Demand Assignment Multiple Access (DAMA) protocol and the radio physical layers reside below the S-ATM. Signalling for call control is based on the Q.2931 protocol standard and is terminated on the ground segment. In the case of mobile or portable terminals, future standards by ATM Forum or ITU-T for B-ISDN signalling and Intelligent Network (IN) concept, can be supported in a highly integrated network environment. The different network entities that are involved in the communication path in an end-to-end ATM connection are shown in Fig. 3.51. The interface of the UT to the SAU is a standard ATM-UNI and the interface of the SAU to the satellite access network is called Sat-UNI which is distributed amongst different satellite network entities.

### 3.3.1.3.3 S-ATM Layer

The common transport layer inside the satellite network boundaries is the S-ATM with similar functions to the standard ATM layer. However, traditional ATM cell headers introduce an unnecessary

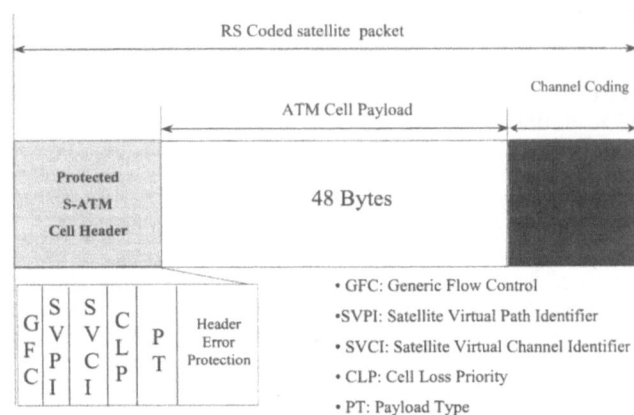

 (already placed above)

**Figure 3.52.** S-ATM packet structure.

RS Coded satellite packet

Channel Coding

ATM Cell Payload

Protected
S-ATM
Cell Header

48 Bytes

| G F C | S V P I | S V C I | C L P | P T | Header Error Protection |

• GFC: Generic Flow Control
• SVPI: Satellite Virtual Path Identifier
• SVCI: Satellite Virtual Channel Identifier
• CLP: Cell Loss Priority
• PT: Payload Type

overhead into the system, which, unlike fibre based systems, is both bandwidth and power limited and operates at much higher BER. A modified ATM cell header is proposed to be used in the satellite air interface. The S-ATM packet, equivalent to an ATM cell, is shown in Fig. 3.52. The S-ATM header combines both the functions of the MAC header and the ATM header.

The 4 bits for the GFC field that are present in the standard ATM cell header, could be removed, since their usage is currently not specified. However, the placement of the GFC bits in the S-ATM header might prove to be essential for the operation of the future satellite ATM components. One reason for this, is the compatibility to future standards; a second is that these bits could be used internally by the satellite access network to operate a dedicated resource request signalling protocol. The standard ATM Virtual Path Identifier (VPI) and Virtual Channel Identifier (VCI) fields are replaced by the Satellite VPI/VCI (SVPI/SVCI) fields which are used by the on-board satellite switch. The PT and the CLP fields, need to be present in the S-ATM cell header, because they both carry essential information related to the particular cell. Finally, extra coding is required to protect the S-ATM header bits, since from an erroneous cell no information can be extracted in the case of a Partial Packet Discard (PPD) mechanism implementation on-board the satellite switch. If the PPD algorithm is implemented on the ground, the detection of errors in the header is not sufficient. The header error correction and the erroneous cell routing to the correct destination in the ground network (with an appropriate indication) is necessary so this cell can be used by the PPD mechanism (in addition to the Cell Misinsertion Ratio, CMR reduction).

### 3.3.1.3.4 Error Recovery Mechanism for Non-Real Time Data Services

#### 3.3.1.3.4.1 Data Link Layer Protocol Considerations
A satellite specific Data Link Control (DLC) protocol could be used to convey information between the satellite network layer entities across the air interface. To support a flexible satellite air interface with low data rate granularity and to minimise any MAC layer packetisation delays, small, fixed size Protocol Data Units (PDUs) of 53 bytes need to be provided by the network layer protocol to the underlying DLC; therefore, mapping exactly one ATM cell per DLC PDU seems to be the best solution (for the return up-link direction only). In such a case, the DLC header field can be placed next to the MAC layer header on every frame which is transmitted over the air interface. The satellite MAC layer packet consists of a coded satellite packet header, 53 bytes packet payload, plus the extra channel coding bits. Since the S-ATM packet has a size compatible to the fixed ATM cell size, the use of a DLC protocol to improve the link throughput performance due to the impairments of the air interface would increase the system complexity. Difficulties exist because the DLC functions (sliding window, timers, acknowledgement transmission) increase the packet overhead, if it takes place per S- ATM cell, or because reassembly (and later segmentation) of several S-ATM cells is required when more than one S-ATM cells are used to create a larger DLC-PDU. The use of a protocol similar to (SSCOP), at the AAL, also complicates the protocol stack, since an end-to-end protocol has to be terminated at the GTW. Furthermore, additional buffer space is required at the GTW when an "out of order" cell is expected, after a "selective reject" message has been issued; this happens because in the fixed network the ATM cells must be received in the same order they have been transmitted. The size of this buffer is the cell rate in a certain connection, multiplied by the Round Trip Delay (RTD). The advantage of the multiplexing of more than one connection is not so attractive in this case, since the

ATM layer already offers multiplexing capability. Another drawback to this approach is the increase in the necessary processing power for the management of the window size and the creation of timers. Moreover, the use of such functions makes the reason for introducing ATM (its simplicity) obsolete, since for each connection an instance of the DLC protocol should be created at the GTW station. As a result, the use of an DLC layer is not so attractive for a system that is being designed for ATM support over the satellite.

### 3.3.1.3.4.2 Retransmission Mechanism Based on Partial Packet Discard (PPD)

A more ATM oriented solution can be given to this problem, if we assume that retransmissions are useful only when non real-time traffic is considered. In this case, we can assume that the receiving end-user is responsible to issue acknowledgements and retransmissions of higher layer PDUs. However, if we develop a mechanism that can detect the erroneous cells at the satellite switch, these cells and consecutive ones that belong to the same higher layer PDU can be dropped and hence reduce the traffic on the fixed network. The increase in the complexity of this approach, when compared to the DLC case, is lower since no additional processing power is required and less overhead is employed per ATM cell. Only an indication of the last ATM cell of a higher layer PDU is required in the ATM cell header and an additional state per Virtual Channel (VC). However, such a mechanism introduces an extra delay in the reception of the "retransmission" message (not so critical since non real time data are considered) and reduces the throughput on the radio link (measured in cells, without considering the overhead per cell, introduced by the DLC), because more than one ATM cells are retransmitted, every time a higher layer PDU retransmission is requested. From this point of view, the throughput on the radio link is equivalent to the "selective retransmission protocol", when a DLC protocol is used and to the "go back N" protocol, when end to end retransmissions are performed (N is the number of ATM cells per higher layer PDU) which results in a much less complex protocol scheme for both the transmitter and the receiver (52). This mechanism has been proposed in (53) as PPD.

### 3.3.1.3.4.2.1 PPD Implementation for AAL5

The implementation of PPD is straightforward for the AAL5 (an AAL designed for data communication) in fixed ATM networks. Since PPD is signalled on a per VC basis, a switch that uses it, once it drops a cell from a VC (due to buffer overflow), it continues to drop cells from the same VC until it receives the ATM layer User to User (AUU) parameter set in the ATM cell header, indicating the end of the AAL packet; this packet is not dropped. Because the AAL5 does not support the simultaneous multiplexing of packets on a single VC, the AUU can be used to indicate the boundaries of the packets. This is a disadvantage of the PPD mechanism, since the fixed network throughput is being reduced by carrying cells with no useful information. Although the same mechanism can be used in satellite ATM networks, there is a small difference; the main reason for dropping cells is the high BER of the physical link rather than the buffer size at satellite switch. To overcome this problem additional protection is required in the header to (at least) detect if an erroneous cell has a correct header. An error correcting code will increase the performance of the PPD, when the PPD function is not located in the satellite switch. In this case, the erroneous cell will be forwarded to the correct place (with a certain probability related to the code used to protect the header and the resulting CMR) and the header will provide the related information to the PPD mechanism.

### 3.3.1.3.4.2.2 PPD Implementation for AAL3/4

A different approach is used for AAL3/4, the other AAL designed for data communication. This is due to the multiplexing of different connections on the same VC (using the same SVPI/SVCI values) and to the fact that AAL3/4 does not use the AUU bit, in the ATM header. Two alternative approaches exist to the DLC solution (in both cases, the acknowledgements are sent from the end stations).

The first approach uses the AAL5 on the air interface only. When the AAL3/4 is required for a connection, it can be performed at the GTW switch. The disadvantage of this approach is that the initial switch should implement the protocol stack up to the AALs and should perform the translation of an AAL5-PDU to an AAL3/4-PDU.

The second approach requires the modification of the AAL3/4 only at the transmitting satellite terminal side. Initially, the transmitting UT must create all the ATM cells from one higher layer data packet; in this way all the ATM cells belong to the same connection. Then, the AAL3/4 at the UT is allowed to process another higher layer PDU. Before the transmission of the last ATM cell of the connection, the AUU bit must be set, similar to the AAL5 case. No modification on the AAL3/4 protocol at the receiving terminal is necessary however, every time the AUU bit is set by the transmitting satellite terminal to indicate the last ATM cell of a higher layer PDU, this bit must be reset at the GTW station (which requires extra processing). If the fixed terminal transmits information using AAL3/4, no modifications are necessary, both at the fixed and satellite terminals.

Both of these mechanisms permit the use of the PPD algorithm, for the AAL3/4 protocol case. Unlike the PPD mechanism for the AAL5, no transmission of the last ATM cell, of a higher layer PDU is necessary; the AAL3/4 at the receiver, has the ability to identify the PDUs to which the ATM cells belong.

### 3.3.1.3.4.3 Comparison Between the PPD and DLC Approaches on the Radio Link

In this section, it is assumed that the Bit Error Rate (BER) of the radio link (after the physical layer coding techniques have been applied) is known, the length of the S-ATM cell payload is p ($48*8 = 384$) bits, and the S-ATM header length is h bits. Since the BER is known, the Packet Error Rate (PER) can be calculated from Eq. 3.17, where, "length" is the higher layer PDU length. For the DLC case, this is one ATM cell (p+h bits), and for the PPD case a variable, depending on the size of the higher layer. The throughput of the radio link (in packets /packet transmission time) is given by Eq. 3.18), (assuming that the errors are random and all of them can be detected). For the DLC case, the overhead corresponds to the additional information required per ATM cell for the operation of the DLC functions, and for the PPD case, the overhead corresponds to the required redundancy in the header fields for error correction (or just error detection)

$$PER = 1 - (1 - BER)^{length} \tag{3.17}$$

$$(\text{Radio Link Throughput}) = (1 - PER) \times (\text{Overhead}) \tag{3.18}$$

For the operation of the DLC protocol an overhead of at least five octets, is required as in the extended version of the HDLC protocol, to accommodate the higher rates required in ATM. For the PPD approach, the overhead of the same CRC-16 field per PDU can be used, for error detection. However, for the header fields two options exist:

- An additional CRC (e.g. CRC-8, similar to the HEC, without the correction state) could be used for the header error detection. If the header has been correctly received, its information is used in the PPD algorithm, since they correspond to valid SVPI/SVCI values.

- A BCH code, could be used instead of the CRC-8 as a second option. Provided that the errors are not bursty, not an invalid assumption if interleaving takes place, its detecting capabilities are not equal to the CRC-8, but it offers the advantage of error correction.

Table 3.6 presents a comparison between the DLC and the different PPD approaches. Figure 3.53 displays the radio link throughput, based on Eq. 3.17 and Eq. 3.18, the relevant assumptions as well as the information in Table 3.6. For the PPD, several higher layer PDU packet lengths are presented, along with three BCH header error correcting codes. If the BER on the radio link is lower than $10^{-6}$ (after the physical layer coding), the PPD approach has better behaviour than the DLC approach.

**Table 3.6.** Comparison among the DLC and PPD solutions

| Approach | Header size (bits) | Overhead (bits) | Error correction |
|---|---|---|---|
| DLC | Any | 16 + 24 = 40 | — |
| PPD, CRC-8 | Any | 16 + 8 = 24 | — |
| PPD, BCH(15,11) | 11 | 16 + 5 = 21 | 1 |
| PPD, BCH(31,21) | 21 | 16 + 11 = 27 | 2 |
| PPD, BCH(31,26) | 26 | 16 + 6 = 22 | 1 |

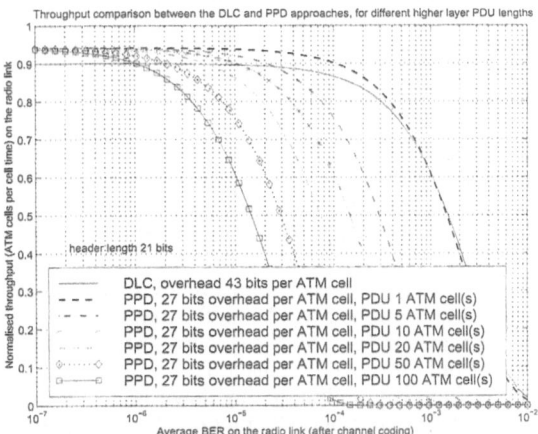

**Figure 3.53.** Throughput comparison between the DLC and PPD approaches, for different lengths (the best and the wo st cases are shown in Table 3.6).

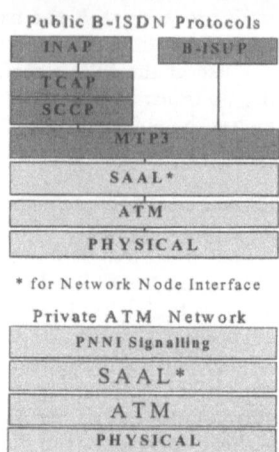

Figure 3.54. ATM based protocol stacks for NNI.

However, even if the BER becomes higher than $10^{-6}$, for a higher layer PDU length up to 20 ATM cells, the PPD mechanism still outperforms the DLC approach (for BER approximately up to $10^{-5}$).

The DLC case assumes that the Go-Back N has been used and the DLC-PDU consists on one ATM cell. Similar behaviour to the PPD mechanism is expected, when a DLC protocol, based on SSCOP, is used. In such a case, no retransmission per cell can take place and the performance will thus be very similar. However, the overhead introduced by SSCOP is on average one-half cell per higher layer PDU, when they fit in an SSCOP PDU.

### 3.3.1.4 Protocol Considerations for Satellite and B-ISDN Integration Scenario

#### 3.3.1.4.1 Network Node Interface
For the core network signalling the public B-ISDN Network Node Interface (NNI) or the ATM Forum's specifications for PNNI can be used. The Broadband-ISDN User Part (B-ISUP) (54) is being further developed to support multipoint virtual paths, virtual channels and multiconnection calls, bandwidth negotiation during setup and renegotiation of bandwidth of all active connections along with several additional capabilities. The Intelligent Network Application Part (INAP) (55) is the protocol that will be used in Intelligent Networks to provide: Location Information (retrieval and updating), Authentication, Call Routing, Handover, Charging and Operation and Maintenance. The interaction between B-ISUP and INAP is required by the call handling procedure in order to allow the IN service logic to influence the processing of IN calls. As shown in Fig. 3.54, INAP sits on top of Transaction Capabilities Application Part (TCAP) which provides functions and protocols for non circuit-related communication between nodes. TCAP uses the services of the connection-less Signalling Connection Control Part (SCCP) and Message Transfer Part 3 (MTP-3) which is needed for routing signalling messages to other network nodes.

### 3.3.1.5 Resource Management and Control Functions
One of the main areas of interest in broadband teletraffic engineering is the provision of multimedia services with QoS guarantees to a large number of users while maintaining high system resource utilisation. In systems where user applications can share common network resources some level of connection admission control is required in order to provide fairness and to guarantee that the requested QoS criteria are met. In a broadband satellite network, a two-level call acceptance control algorithm must be considered; the first at the MAC layer and the second at the ATM layer.

As shown in Fig. 3.55, the resource management and control functions are distributed between the space and the ground segment. The on-board satellite switch is responsible for providing full connectivity from any up-link to any down-link spot beam and it is controlled by the switch control unit and the Call Control (CC) and the CAC units. The physical location of the blocks that implement the CC and the CAC functions can be on the ground in order to reduce the on-board processing requirements. In such a way, all the UNI signalling overhead and the call state machine implementation can be directed to the NCS on the ground. Depending on the switch implementation and the information carried within the S-ATM cells, frequent routing table updates such on a per call basis could be

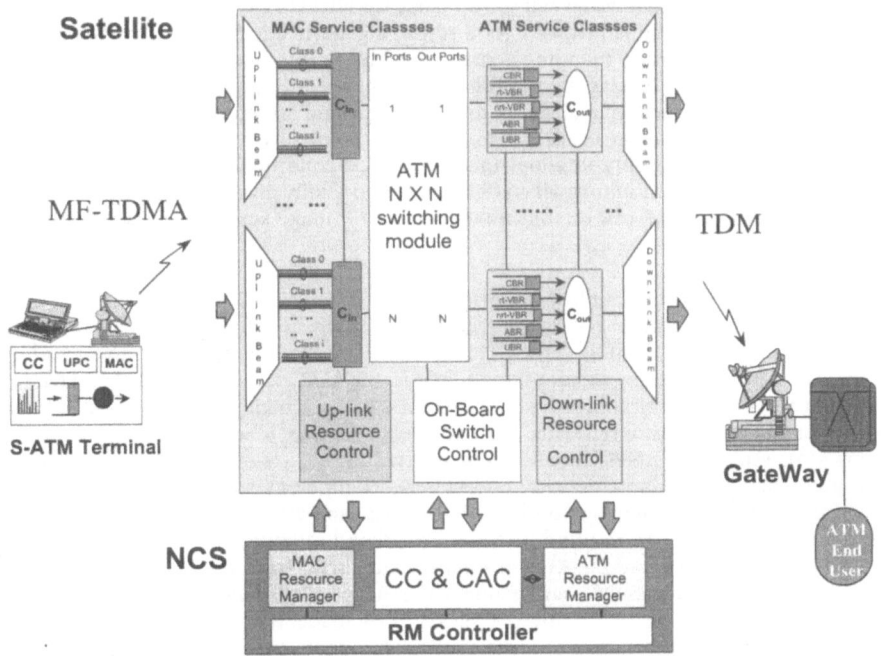

**Figure 3.55.** Resource management and control functional block diagram of a S-ATM.

avoided. As a result, the on-board switch control processing requirements are minimised. The up-link and down-link resource control units are responsible for the incoming and outgoing traffic management and control the input and the output ports of the ATM switch respectively. The overall network resource management is performed by the Resource Management (RM) controller which supervises the operation of the MAC and the ATM resource managers at the NCS. The RM controller dynamically measures the network performance criteria (GoS, QoS for all service classes) and instructs accordingly the MAC and ATM resource managers.

At the S-ATM terminal side, the control plane protocols are responsible for establishing and maintaining each connection. The bandwidth on the air interface is controlled by the MAC layer and is shared among all active users terminals. The usage parameter control (UPC) is a control function that could be placed either at an UNI or NNI (in such a case is referred as NPC) interface in order to monitor the conformance of existing traffic contracts. Although such a monitoring algorithm is quite essential in networks where traffic contract violations could happen by having misbehaving traffic sources that could transmit cells in excess of the negotiated cell rates, in a S-ATM network the incoming traffic is regulated by the allocated MAC bandwidth units. A MF-TDMA access for the up-link and TDM for the down-link is considered in our study for the satellite-ATM terminals. For gateways and other high data rate terminals that play the role of traffic concentration units, Time Division Multiplex (TDM) access is selected to be most efficient for both the up-link and the down-link. As a result, the existence of UPC or NPC at the Gateway side is more essential.

### 3.3.1.5.1 MAC in GEO Satellite Networks

As in any other wireless system, MAC regulates the incoming traffic at the satellite network access points. Several studies (56, 57, 58, 59, 60) based on Demand Assignment Multiple Access (DAMA) (61) or DAMA-variants (62, 63) using a combination of random access and reservation based schemes have been conducted. One of the most important designing targets for a MAC scheme is to maximise the resource utilisation while maintaining low delay guarantees. In addition, the protocol stability and low complexity of the control algorithm need to be considered. However, a large class of MAC protocols are not applicable for satellite communications (64). As a result, the flexibility in designing a MAC scheme in a realistic broadband satellite system is limited mainly by the long propagation delay, the on-board satellite processing capabilities and some design constraints of the air interface (i.e. maintain synchronisation, channel reconfigurability). In most of the broadband Ka-band proposals at least a minimum up-link bandwidth reservation takes place for most of the supported service classes. In addition, the initial call set-up delay for all ATM service classes is always the same, assuming that UNI signalling is involved. As a result, a generic and simple approach is needed for the air interface and the service mapping according to the source characteristics is done only at the ATM layer. All services can be classified at the MAC layer according to the number of slots that are assigned in the up-link direction in addition to the time period for which the resources are allocated (i.e. permanent, semi-permanent, on demand). By having a two level control of the service mapping, we increase the system flexibility to adopt to various load conditions, simplify the traffic management actions, provide fairness among all traffic classes and finally give different options for charging. In such a system, CAC at the call establishment phase will make sure that certain QoS criteria are met.

### 3.3.1.5.2 *The Adopted Generic-DAMA MAC Scheme*
This investigation focuses on the wireless QoS provisioning in a S-ATM network. Therefore, the combined effects at both the MAC and the ATM layers will be examined. In order to do so, it is necessary to first clearly identify any assumptions made at the MAC layer. From all the MAC proposals applicable for GEO Satellite Networks a generic-DAMA based scheme has been selected as the most appropriate one to be used as a reference (65). The reason is that, when comparing the performance of various MAC schemes for GEO satellite networks with multi-class service support, there is not a single winner. Their performance depends on the network topology, the traffic demand and type (i.e. symmetrical, asymmetrical, aggregation level, burstiness etc.) and the complexity of implementing the control algorithm. Keeping in mind that in a broadband satellite-ATM network various types of traffic need to be supported and the ATM on-board switch will route traffic coming from single user terminal up to high data rate multi-user terminals that need to share the same MAC control protocol, a tradeoff in performance versus complexity needs to be made. Recent studies that give a comparison of widely known MAC schemes seem to support this idea (64, 66). Implicit reservation is used (i.e. content in a slotted Aloha channel) for the first bandwidth request since it is the most appropriate scheme for a large number of terminals.

### 3.3.1.5.3 *Mapping ATM Service Classes into the Air Interface*
Five distinct service categories have been specified by the ATM Forum to accommodate all the different applications: CBR, rt-VBR, nrt-VBR, Unspecified Bit Rate (UBR) and ABR. We assume that all service classes except the UBR share the same pool of up-link resources. In order to guarantee fairness and a fixed GoS for the other services, the UBR resources in the air interface should be taken by another pool of resources with the use of a moving boundary. Since there are no minimum rate guarantees for the UBR service or any strict delay requirements, a hybrid approach which combines both random access and reservation based resource allocation at the air interface could be used. In this way, some level of multiplexing at the MAC layer can be achieved for a large number of bursty sources, as long as the system remains in a stable state and there is enough buffering space in the network. At the ATM layer, an effective CAC algorithm should take into account all these requirements in order to provide certain levels of QoS to all services classes that share the same down-link.

During the call set-up phase, the CBR, rt-VBR and nrt-VBR services are assigned a fixed number of up-link slots which remain constant for the duration of the call. Therefore, some multiplexing gain can be achieved only at the ATM layer by limiting the available resources at the ATM switch output queues. For the ABR service, a more flexible access scheme is assumed since it does not have delay requirements as strict as the rt-VBR. An ABR call is accepted or blocked at the MAC layer according to its Initial Cell Rate (ICR), or Minimum Cell Rate (MCR) requirements. After the call is accepted, all the slots that remain free in the up-link direction can be shared among all ABR users in order to satisfy the instantaneous requirements for increased bandwidth. If we want to maintain certain call blocking rates for all the supported service classes at the MAC layer, the available slots which are shared among all ABR services should be used in such away, as not to affect the new call blocking rate of any service type. For bidirectional services each link is setup independently from each other but in our study we did not take the forward traffic requirements into consideration.

### 3.3.1.5.4 *A Two-Stage Connection Admission Control Algorithm*
As shown in Fig. 3.55, the connection admission control unit runs the decision process during the call establishment phase based on the current information that it receives by the MAC resource manager and the ATM resource manager. The information that is available to the CAC by the MAC resource manager is the available up-link resources at a given spot-beam, the resource utilisation statistics and where applicable the collision rates on the random access channels. The ATM resource manager reports the current status of the switch queue outputs and buffer overflows statistics (i.e.

CLR). Based on this information and the predefined values of the number of the active connections of a certain type that can be supported at any output port, the CAC decides whether to accept or reject a call of a certain type. Most of the studies that investigate the problem of connection admission control in ATM networks concentrate on the equivalent bandwidth concept and the allocation of bandwidth of an outgoing link, provided that certain QoS criteria are satisfied. In (67), a review of the CAC strategies based on the Gaussian approximation, the large deviation techniques and the diffusion- based techniques is given. As it was stated clearly by the authors, despite the substantial effort on this filed, further research is required on the important issue of real time traffic management. The concept of using dynamic versus static resource allocation strategies is given in (68) while in (69, 70) the service separation and the division of the overall traffic into multi-class flows is examined. Furthermore, the analysis of a decision-based CAC using pre-calculated thresholds is given in (71) and in (72) the concept of assigning multiple effective bandwidths to a given connection corresponding to different priority levels is described. However, what is not considered in most of the existing studies is the combined dimensioning problem of the MAC up-link and ATM down-link resources.

### 3.3.1.5.5 Up-link Call Blocking Probability for Multi-Rate Services
The determination of the call blocking probabilities of ($I$) different service classes each one requesting for constant rate of ($c_i$) from a shared link of capacity ($C_{IN}$) can be performed with some widely used multi-rate models if a complete sharing admission policy is assumed. Each service class ($i$) is characterised by an arrival process which is assumed to be Poisson with mean call arrival rate ($\lambda_i$) and call holding times following a general distribution with mean call holding times ($1/\mu_i$). The analysis of multi-rate models is well presented in existing literature and product form solutions for the state probabilities can be found in (73, 74). However, the solution to this problem becomes intractable as the number of states and service classes increase. Therefore, a recursive solution that was given to this problem by (75) simplifies the calculations for a large number of service classes and system states. Both the unnormalised probability $P_{un}(m)$ of ($m$) occupied bandwidth units and the normalised probability $P(m)$ are given by:

$$
P_{un}(m) = \begin{cases} 1 & : m = 0 \\ 0 & : m < 0 \\ \dfrac{1}{m} \displaystyle\sum_{i=1}^{I} P_{un}(m - m_i) \cdot m_i \cdot \dfrac{\lambda_i}{\mu_i} & 0 < m \le P(m) = P_{un}(m) \left( \displaystyle\sum_{m=0}^{M} P_{un}(m) \right)^{-1} \end{cases}
\tag{3.19}
$$

where $M = C_{IN}/dc$, $m_i = c_i/dc$ and ($dc$) represents the greatest common divisor of ($c_i$). The blocking probability per class can be calculated from:

$$
P_{b_i} = \sum_{m=M-m_i+1}^{M} P(m)
\tag{3.20}
$$

### 3.3.1.5.6 CAC Problem Formulation
Each connection in an ATM network is associated with a set of traffic parameters in addition to a set of negotiated and non-negotiated QoS parameters (76). During the call establishment phase, one important network operation is the evaluation of the QoS impact of the new call on the accepted calls that share common resources. CAC is the network function that decides if a new call should be accepted or not. A new call can be blocked if there are not enough resources either at the MAC or at the ATM layer. So the CAC problem is to try to maximise the MAC and ATM resource utilisation while keeping certain GoS and QoS guarantees. In addition, an effective CAC algorithm should provide fairness to all services classes and connections and should be fast enough to run in real time.

From all the above statements two general conditions must be satisfied for a set of multi-class connections in a S-ATM network model. The first satisfies the admissible load in the up-link direction at MAC layer ($C_{IN}$) and the second the down-link output ATM rate ($C_{OUT}$)

$$
\sum_{j=1}^{J} \sum_{i=1}^{I} m_i \cdot N_i^{kj} \le \frac{C_{IN}}{dc} = M^k
\tag{3.21}
$$

$$
\sum_{i=1}^{I} a_i \cdot N_i^{j} \cdot u_i^{ATM} \ (N_i^{j}, \varepsilon_i) \le \frac{C_{OUT}}{dc} = A^j
\tag{3.22}
$$

Where (*i*)      is service class and (*I*) the number of service classes.

       (*k*)      is the call origination spot-beam and (*K*) the number of up-link beams.

       (*j*)      is the call destination spot-beam and (*J*) the number of down-link beams.

       ($M^k$)      represents the MAC up-link capacity (in normalised bandwidth units) of a spot-beam (*k*).

       ($A^j$)      represents the ATM down-link capacity (in normalised bandwidth units) at the destination down- link (*j*).

       ($m_i$), ($\alpha_i$)      represent the assigned resources (normalised bandwidth units) per connection of type (*i*) at the MAC (up-link) and ATM (down-link) layers respectively.

       ($N_i^{kj}$)      represents the total number of connections of type (*i*) that are originated from spot beam (*k*) and are directed towards spot-beam (*j*).

$u_i^{ATM}(N_i^j, \varepsilon_i)$ represents the *utilisation factor* of the assigned resources $\alpha_i \times N_i^j$ in the down-link ATM multiplexer for guaranteed loss probability ($\varepsilon_i$). It is expressed as a function of the total number of active connections $N_i^j = \sum_k N_i^{kj}$ per traffic class that are directed towards a given downlink-beam (*j*).

NB: For simplicity reasons and without loss of generality in the following analytical and simulation results, it is assumed that the S-ATM model consists of a single up-link and down-link beam (i.e. *J* = 1, *K* = 1). This assumption represents a uniformly distributed traffic load and call destination pattern among all beams. However, this methodology is still applicable and could be extended provided the non-uniform traffic distributions and certain call destination distribution are known.

### 3.3.1.6 Traffic Modelling and Performance Evaluation

Analytical models can become very complex when considering multi-rate services at both call and cell level and a simulation model that runs at cell level requires extremely long processing period to produce accurate results at call level. Therefore, a hybrid approach was followed in the performance evaluation that combines both analysis and simulation. At call level, both analysis and simulation were used for multi-rate, multi-class services and incorporate look-up tables to provide the ATM resource utilisation statistics for different ATM service classes. These tables can be generated by using proper analytical or simulated traffic source modelling techniques depending on the service type.

The main objective of this work that was first reported in (68) and later in (77, 78, 79), was to investigate the performance of the ABR service in a S-ATM network for which not accurate traffic models have been developed. In some studies that came to our knowledge, there is an assumption of a given traffic model for ABR service or a traffic mix of several traffic sources. On the contrary in our approach, certain traffic scenarios are assumed only for the service classes that have priority over the ABR service. Therefore, a means to quantify the aggregated available bandwidth at a given shared down-link is developed. The outcome of this investigation is to derive a new methodology for buffer dimensioning in order to guarantee certain QoS criteria for the ABR service. No special attention is given to the UBR service since it is given less priority than the ABR services. If the available system resources are more than the ABR maximum requirements, then the unused resources are left for the UBR service in a fair share algorithm.

#### 3.3.1.6.1 CBR Service

A fixed number of up-link slots are allocated for a CBR call; this number corresponds to the source Peak Cell Rate (PCR). No multiplexing gain can be performed at the MAC layer from terminals that share the same up-link resources due to the real time constraints. However, depending on the activity of the source, multiplexing gain can be achieved at the ATM layer. For the simulation it was assumed that voice traffic was supported over the CBR service. Each source was modelled as an ON–OFF process with negative exponential call holding time and the analysis presented in (80) was used to obtain the utilisation of the allocated resources as a function of the number of users, for different QoS (expressed in terms of CLR and maximum queuing delay, at the ATM switch).

#### 3.3.1.6.2 RT-VBR

Real-time VBR services are treated similarly to the CBR service at the MAC layer, since this type of service has specified delay and jitter requirements. In addition, a VBR source the PCR, the Sustainable Cell Rate (SCR) and the Maximum Burst Size (MBS) need to be specified. Therefore, for the resource allocation at the ATM layer two possibilities exist (81). The first one uses the PCR so there are always available resources and the second one takes advantage of the statistical multiplexing using the SCR or any other value between the PCR and SCR in order to increase the efficiency. For the simulation the second approach was selected and it was assumed that video traffic was supported over the rt-VBR service. The traffic can be modelled as in (82). Similar to CBR, the resource utilisation can be calculated as a function of the number of users and the expected QoS, in order to consider the multi-plexing gain at the switch.

### 3.3.1.6.3 NRT-VBR

The nrt-VBR service is intended for non-real time applications that accommodate bursty traffic and similar to the rt-VBR is characterised in terms of PCR, SCR and MBS. For the ATM cells that are transferred within the traffic contract, the application expects low CLR. The resource allocation at the MAC layer is proposed to be similar to the CBR service (circuit switch allocation) but using the SCR instead of the PCR. Therefore, any multiplexing gain can be obtained at the ATM layer depends strictly on the detail characteristics of the traffic sources. The SCR is used at the ATM layer to allocate (ATM) resources to each active connection. Since there is no strict delay or delay variations requirements, the user terminal is responsible for managing its own transmission rate, depending on the terminal buffer size.

### 3.3.1.6.4 ABR Service

During the connection set-up phase, the following traffic parameters are negotiated for the ABR service types: the Initial Cell Rate (ICR), the Minimum Cell Rate (MCR) and the Peak Cell Rate (PCR). An ABR source is allowed to transmit cells at an Allowed Cell Rate (ACR), which varies over time depending on the network congestion status. The mechanism to change the source ACR is based on the transmission of RM cells that traverse end-to-end from the source to destination (forward direction) and are looped back to the originating source in the backward direction. The contents of the RM cells can be changed by the destination or the intermediate nodes by altering the Congestion Indication (CI) bit and the Explicit Rate (ER) fields and the source ACR is adjusted according to these fields. The network congestion indication mechanism depends on the switch architecture and can be Explicit Forward Congestion Indication (EFCI)-based switch or ER-based switch. A set of time-out mechanism exists in this type of flow control to overcome the problems arising from erroneous cells. The efficient operation of a rate-based congestion control mechanism depends on the careful switch buffer dimensioning procedure in order to avoid cell loss. In general, ER based switches perform better than the EFCI but are harder to be implemented. For long end-to-end control loops such as those that are included in a GEO satellite network, it is suggested that smaller control segments should be considered (at least two segments must be created; one for the radio part and one for the fixed part of a connection. In this study, only the radio parts of all ATM connections were considered and no particular algorithm (either at the MAC or at the ATM layer) that supports bandwidth on demand for ABR services has been taken into account. Therefore, both in the performance analysis and in the simulation model the main target was to calculate the resources that remain free to be used by the ABR services at both the MAC and the ATM layers as a function of the total offered load under different traffic scenarios.

### 3.3.1.6.5 UBR Service

UBR is considered as a "best effort" service, therefore it should not be considered in the same way as any other ATM service type. The placement of CAC is to ensure that QoS guarantees of all the existing connections are satisfied before any new connection is admitted to the network. QoS provisioning to multi-class, multi-priority service classes is supported in the S-ATM network model (see Fig. 3.55) by the on-board satellite output scheduler. A movable-boundary between the UBR and all the other service resources for the air interface is envisaged. This boundary should be dynamically adjusted to the traffic mix in order to satisfy both fairness and the GoS/QoS criteria for all traffic classes. Assuming that the UBR service uses its own resources [9], the average message delay can be calculated either statistically, using a simulation model, or analytically, using the lower and upper delay bounds for the G/G/m queue (83, 84, 85). At the ATM layer, all service classes including the ABR should have priority over the UBR therefore, UBR traffic should be carried out when there is no other type of traffic present.

---

[9] A similar concept, for the internet traffic, was presented by Professor Jon A Crowckroft, Department of Computer Science, UCL, UK, in a talk he gave to CCSR, University of Surrey, with subject: "QoS provisioning in the internet". Using one bit in the IP packets, priority can be offered to these packets, when compared to the rest of the traffic. He proposed bandwidth splitting between the two "classes" of traffic in order to allow the lower priority traffic to utilise some the resources.

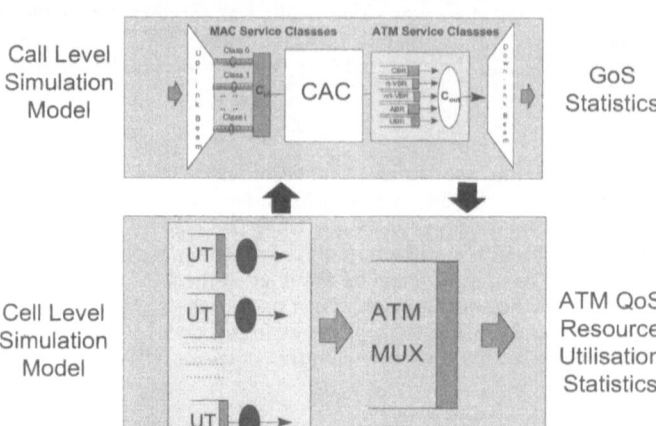

**Figure 3.56.** Simulation model block diagram.

#### 3.3.1.6.6 Simulation Modelling Approach

A simulation model was implemented in OPNET modeller in order to evaluate the performance of multi-class, multi-rate services and compare simulation with analytical results, where applicable. The block diagram of a S-ATM model that combines both the cell level statistics at the ATM layer and the call level statistics at the MAC layer is shown in Fig. 3.56. The model is split into two main tools that communicate with each other through statistical files generated by each simulator. At the call level simulation a service class is characterised by:

- Mean call arrival rate (from the total number of users per class).
- Call arrival distribution (negative exponential is assumed).
- Mean call duration.
- Call duration distribution (negative exponential distribution is assumed).
- Service rate at MAC level (MAC slots).
- ATM class: CBR, rt-VBR, nrt-VBR, ABR, UBR.
- Service rate at ATM layer (ATM class specific).

Combined results for both the MAC and the ATM resources utilisation can be produced assuming that the resource utilisation statistics for the different ATM classes are imported to the call level simulator from the cell level simulator or can be calculated analytically.

At the cell level simulator each ATM traffic source is modelled independently and runs at cell level. This model is designed to produce statistics about the MAC and the ATM buffer occupancies. The following statistics characterise each ATM traffic class:

- Number of sources.
- Traffic source packet inter-arrival statistics.
- MAC service rate.
- ATM service rate.

One of the most important areas of research in the past years is the modelling and characterisation of the ATM traffic. Several different algorithms and traffic models have been proposed and adopted so far by many researchers in their simulations. There is no such single model that can be used accurately for all traffic sources. Most of these models are usually applicable to a single type of traffic with certain characteristics. Therefore, some widely accepted traffic models were used to characterise different traffic types.

#### 3.3.1.6.7 Traffic Scenarios and Simulation Approach

Four different service classes were identified. Each service class is characterised by a number of up-link slots that are requested by the MAC layer and an ATM rate which is used for the ATM layer resource utilisation calculations. The main objective of this work is the evaluation of the resources

**Table 3.7.** Simulation service scenarios, definition of classes and parameters

| Service class | MAC Slots | ATM parameters | ATM service class | Traffic scenario | Mean call arrival rate ($\lambda_i$) |
|---|---|---|---|---|---|
| Class 0 | 1 | PCR = 1, factor = 0.4 | CBR, Voice, CLR = $10^{-3}$, 30ms max. queuing delay | I and II | Variable |
| Class 1 | 2 | MCR = 2 | ABR | I | Variable |
| | 4 | MCR = 4 | ABR | II | Variable |
| Class 2 | 4 | SCR = 4 | nrt-VBR, Data | I and II | Variable |
| Class 3 | 10 | PCR = 10, SCR = 4, MBS = 300 cells | rt-VBR, Video, CLR = $10^{-5}$, 30 ms max. queuing delay at each layer (MAC and ATM) | I and II | Variable |

that are available to be used by the ABR service. Therefore, all ABR services that are accepted at the MAC layer have no resource limitation at the ATM layer (i.e. their ATM rate is always zero) and there is no new call blocking at the ATM layer for any service. However, calls that cannot be admitted at MAC layer are blocked and cleared from the system. The system capacity, for the simulation scenarios presented in Table 3.7, was set to C = 100 up-link slots.

Two different traffic scenarios were simulated. In traffic scenario I all service classes (0, 1, 2, 3) contribute equally to the total offered load. In traffic scenario II, the contribution of each service class is 20, 40, 20 and 20 per cent respectively, of the total offered load. New calls are generated according to a Poisson distribution; the call duration follows negative exponential distribution with average value 1 (all the other parameters are normalised to this value).

### 3.3.1.6.8 Simulation Results and Discussion
The call blocking probability as a function of the offered load is presented in Figs 3.57 and 3.60 for traffic scenarios I and II respectively. As expected, the service classes that request for a higher number of MAC slots experience higher call blocking rates. The simulation results are in full agreement with those obtained by analysis therefore, the simulation model was used to record very useful statistics that cannot be easily provided by analysis. As shown in Fig. 3.60, the Service Classes 1 and 2 (Traffic Scenario II) experience the same call blocking rate although they represent different ATM service classes (ABR and nrt-VBR). Therefore, if there is no-admission control applied by the network, the traffic from different service classes is regulated only by the MAC layer. However, from the ATM resource utilisation statistics that are plotted in Figs 3.58 and 3.61 for both traffic scenarios, it is clearly shown that different admission control policies are applicable at the ATM layer.

Since the function that calculates the resource utilisation for the ABR service at the ATM layer was unknown, the resource utilisation statistics from all the service classes, excluding the ABR, were collected and presented in Figs 3.59 and 3.62 for the Traffic Scenarios I and II respectively.

The ABR resource availability shown in Fig. 3.63 was calculated by subtracting from the total MAC resources the ones that were used in each traffic scenario. These values are normalised to the overall MAC resources. The Probability Density Function (PDF) of the ATM resource utilisation excluding the ABR services is shown in Fig. 3.64 (Traffic Scenario I, total offered load = 0.72).

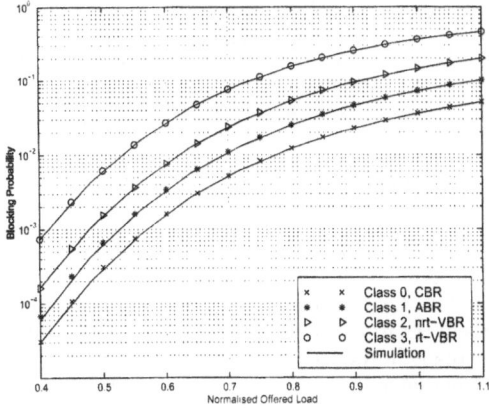

**Figure 3.57.** Call blocking rates for Scenario I.

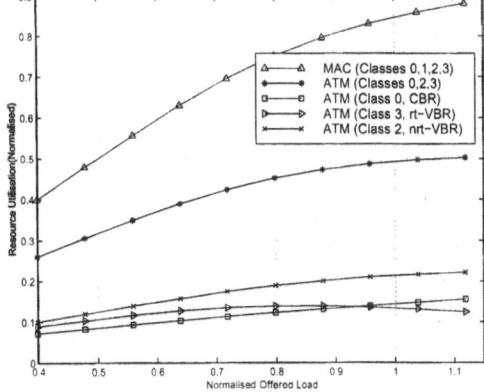

**Figure 3.58.** Resource utilisation for Scenario I.

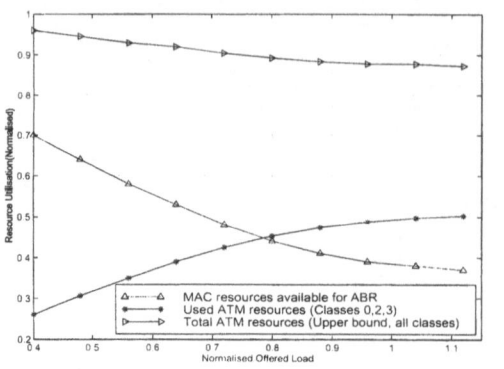

**Figure 3.59.** ABR resource availability for Scenario I.

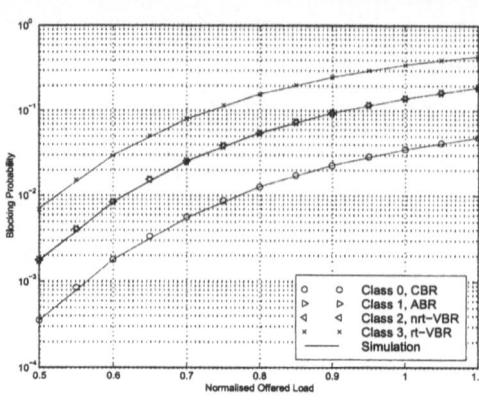

**Figure 3.60.** Call blocking rates for Scenario II.

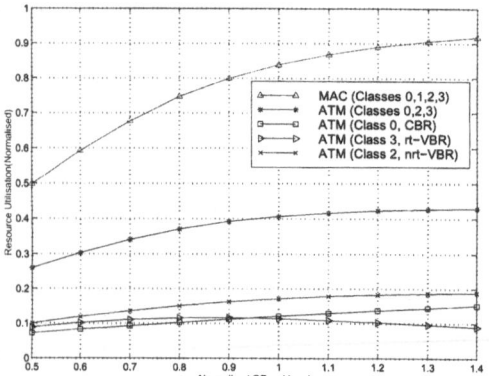

**Figure 3.61.** Resource utilisation for scenario II.

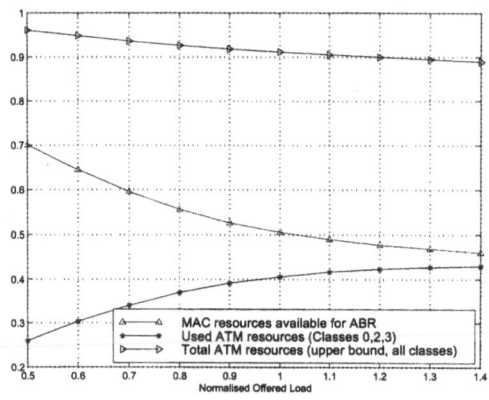

**Figure 3.62.** ABR resource availability for scenario II.

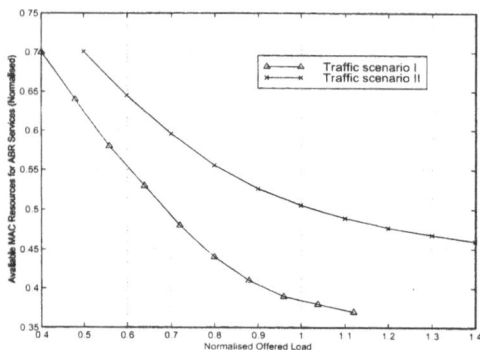

**Figure 3.63.** ABR resource availability at MAC layer.

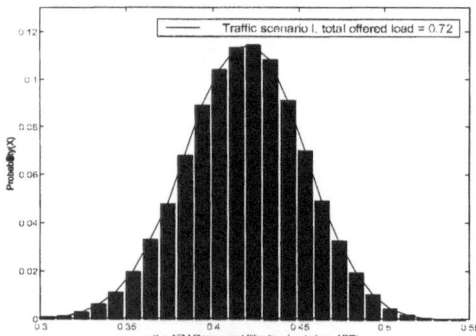

**Figure 3.64.** ATM resource utilisation for classes (0, 2, 3).

From these results, we can easily plot the PDFs of the ABR resource availability at the ATM layer, for different values of the overall ATM resources (see Fig. 3.65). In fact, any connection admission control algorithm that is applied to the ABR service will try to restrict the switch capacity requirements by taking into account the maximum multiplexing gain that can be achieved. In any case, the overall switch capacity should be less than the upper bound values shown in Figs 3.59 and 3.62 for the Traffic Scenarios I and II.

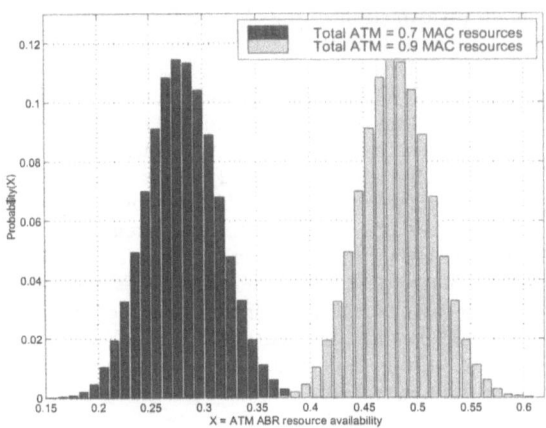

**Figure 3.65.** ABR resource availability.

### 3.3.1.7 Summary and Conclusions

In this section an overview of the protocol architecture for the future broadband satellite networks that accommodate ATM compatible equipment is given. The new satellite multimedia components could be highly integrated into the future broadband networks by incorporating the emerging B-ISDN protocol standards into the satellite specific protocol operations. Two main scenarios that are applicable to the broadband satellite access networks. The first one is based on ATM protocol encapsulation and use of fast packet switching on the satellite segment. The second is a new proposal driven by the definition of a future S-ATM protocol layer which is highly optimised for the satellite segment. There are some strong similarities between the two protocol scenarios, included in the term "ATM compatibility". This term implies the existence of a common fixed size information unit that traverses all the different network interfaces and can carry both control and user data. It relies on a very fast switching fabric at the lower layers of the communication protocols. The higher layer protocols in different planes (i.e. control, user plane) are responsible to establish, maintain, release or transfer user data during a network connection. It is expected that in the next two to five years most of the Ka-band proposals will base their satellite network infrastructure on a new modified version of the ATM protocol layer (S-ATM) which will be satellite specific with no major modifications to the ATM protocol stack. The S-ATM cell header fields will carry essential routing and control information for the satellite segment and different techniques such as the PPD for reliable non-real time data transfer could be used.

One system design issue that still remains open for furthr research is the traffic management and control for the future S-ATM networks. In this report a block diagram of the control functions that are distributed between the space and the ground segment is presented along with the mapping of the ATM layer services to the MAC layer service classes. In addition, a new methodology for evaluating the performance of multi-rate, multi-class services is described. It calculates the call blocking probability per service class at the MAC layer and the ATM resources utilisation. Assuming a MAC layer that supports bandwidth on demand, all the available ATM resources can be allocated to the ABR service without affecting the GoS of the rest of the service classes. Simulation results are presented for two traffic scenarios. Additional work can be found in (77) where the use of a self-similar traffic model for the nrt-VBR service has been considered and more traffic scenarios were evaluated.

### 3.3.2 Service Mapping and GoS Provisioning in Non-GEO Satellite Networks

#### 3.3.2.1 Introduction

The use of multi-beam non-geostationary (non-GEO) satellites with advanced on-board processing capabilities in future multimedia satellite communications networks is a great challenge (86). Non-GEO satellites can provide global coverage with higher elevation angles, satellite diversity and offer lower propagation delays with respect to GEO satellites (87). Smaller, high data rate user terminals can be used with an increased level of mobility/portability. Connection admission control is one of the main areas of concern in any network that needs to provide GoS/QoS guarantees. In non-GEO satellite networks after the call establishment phase, service interruption due to unsuccessful inter spot-beam or inter-satellite handoff could always happen.

Therefore, in this section an investigation of the GoS performance of multi-rate, multi-class services is conducted. In addition, a new technique for radio resource management applicable to dynamic satellite networks for multimedia applications is proposed. It uses an adaptive bandwidth reservation policy to map the application specific requirements for accepting and maintaining a call. According to this, the definition of each service class is based on a range of different performance criteria for the call blocking and handoff failure rates. It is assumed that the inter spot-beam handoff times can be accurately predicted by the network during the call set-up phase. The call admission controller takes into account the performance requirements of the established connections and the affect of accepting a new call on the same spot-beam as well as the handover attempts from the neighbouring spot-beams. Both analytical and simulation techniques were used to evaluated the GoS performance.

The simulation model was developed in OPNET modeller and statistics for the call blocking and call dropping probabilities under various traffic conditions were recorded. The simulation results for a typical Medium Earth Orbit (MEO) and a Low Earth Orbit (LEO) constellation are presented and the performance of the new algorithm is discussed. It is shown that this new approach gives a great flexibility to dynamically map different service classes requirements into the network performance characteristics.

### 3.3.2.2 Satellite Constellation and Network Architecture
The choice of the satellite constellation has a great impact on a broadband satellite-ATM system design. On the one hand systems that use GEO satellites (placed at about 36,000 km of altitude) give an almost static user behaviour resulting in negligible handover probabilities and no need for virtual channel rearrangement. On the other hand, systems using Low Earth Orbiting (LEO) satellites (placed at altitudes ranging from 500 to 2000 Km) require a sophisticated handover algorithm and a methodology for accepting new calls, without affecting the quality of service requirements of all the active connections.

Some of the advantages and drawbacks of using GEO satellites instead of LEO satellites can be found in (87). Medium Earth Orbit (MEO) satellite networks seem to be a good compromise (88) between the high orbital distances of the GEO systems (with no extra complexity for handoff) and the low earth orbital distances (with the very high handover rates due to the satellite movement). Systems that use MEO satellites provide large coverage areas and require much fewer number of satellites to cover the whole earth than any LEO satellite system with global coverage. In addition, each satellite can stay in view for over 1 h before a user must switch to the next satellite.

Figure 3.66 illustrates a dynamic multi-spot beam satellite network architecture and its interconnection to an ATM core network. Each satellite coverage area is serviced by at least one earth station that is connected to the fixed ATM network and plays the role of the satellite network gateway (GTW). Inter-satellite Links (ISLs) are used in order to minimise the number of GTW stations.

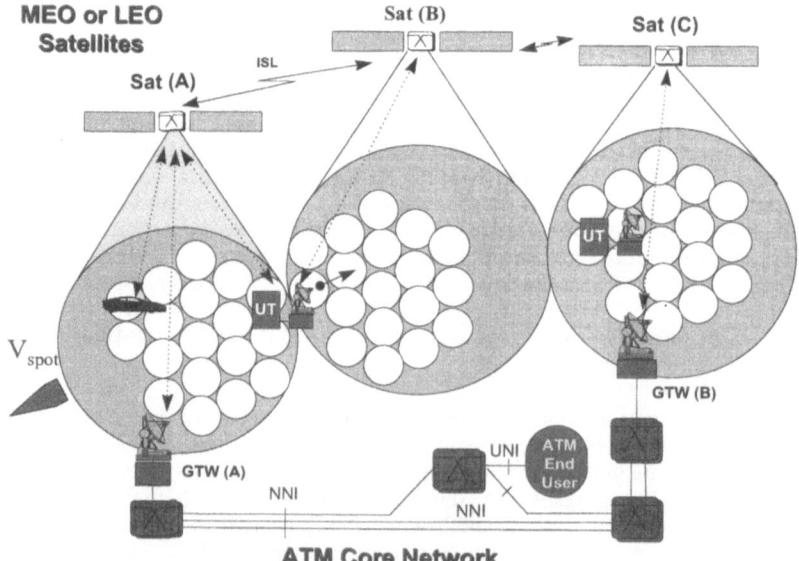

**Figure 3.66.** Non-GEO satellite network architecture.

The number of satellites that are visible any time from any particular user terminal (UT) or GTW stations depends on the satellite constellation design in particular the satellite diversity. It is assumed that during the call set-up phase the UT selects the strongest signal among all the available frequencies that is usually transmitted by the highest satellite in the sky at that particular moment.

### 3.3.2.3 Virtual Connection Tree Concept in Non-GEO Networks

In the last years, a few proposals appeared in the literature suggesting possible ways to overcome the user/terminal mobility problems in wireless ATM networks. A comparison among some of the most recent re-routing and virtual path re-establishment algorithms that can support handoffs in wireless ATM networks can be found in (89). However, the mobility issues related to satellite-ATM networks, are not fully covered yet. In a dynamic satellite-ATM network, the "virtual connection tree" concept (90) can be applied as suggested in (91)(10). This idea was also supported in (92) where a new adaptive routing algorithm applicable in LEO networks with inter-satellite links is presented.

According to the original "virtual connection tree" algorithm, mobile terminals can freely roam around a large area covered by several radio access points and execute handoffs without the involvement of the call processor by simply using a predefined set of virtual circuit numbers for any particular path that activate. A mobile user is admitted to a virtual connection tree during the call establishment phase and look-up tables are created at the intermediate switching points of the tree. These tables contain all the possible paths from the tree root to the access ports that can be reached by the mobile station in that tree. At the root of the tree, the use of a virtual circuit number arriving at a specific input port will give the indication of which path of the virtual connection tree is currently active and update the look-up table in order to correctly route the ATM cells arriving at the reverse direction (from the route to the mobile station).

In a S-ATM network the root of a connection tree can be either a GTW station that serves a particular coverage area or an ATM switch of the fixed network. The leafs will be the input ports that correspond to a single or a group of spot-beams that feed the on-board switch. The virtual trees will be dynamically established and released according to the satellite movement (one or more satellites could be visible from the GTW at any particular moment). Every time a mobile station accesses the satellite station in order to initiate a call, its exact position can be calculated and the next handover time can be predicted with high accuracy. So what becomes important in dynamic satellite systems is that the user handoff times and direction in respect to the moving multi-spot beam pattern can be predicted during the call initiation phase. This give us a great advantage compared to the terrestrial mobile systems since the list of the visited spot-beams can be predefined (satellite diversity and handoff due to link degradation and fading is not addressed here but can be easily included). In cases where a new satellite becomes visible to the GTW station during the call or one that has leaves to an existing virtual connection tree becomes invisible then the call processor gets involved and reconstructs a new virtual connection tree. Using a MEO constellation however, the handoff probability to another virtual connection tree can be minimised by the beam selection algorithm during the call set-up phase (the visibility of a satellite from a fixed point on earth can be for over an hour).

### 3.3.2.4 Mobility Modelling in Non-GEO Networks

The user/terminal mobility in the terrestrial mobile networks is quite different from the fixed/mobile non-GEO satellite networks. In our approach it is assumed that the user terminal positions in respect to the satellite movement can be accurately predicted (93) during the call set-up phase. The user terminals stay within every spot beam time between $0$ and $T_{max}$ which is given by the following formula (94):

$$T_{max} = \frac{2 \cdot R}{V_{spot}} \tag{3.23}$$

If $R$ is the spot beam radius and $V_{spot}$ is the spot-beam speed relative to the ground, the terminal mobility compared to the satellite movement for low mobile or transportable types of terminals can be neglected. In Fig. 3.67 we present two different ways of modelling the user terminal behaviour in respect to the moving pattern of a multi-beam satellite footprint.

#### 3.3.2.4.1 Mobility Model I

The first model (i.e. mobility model I) can be used to represent mobile terminals moving relatively to a LEO polar orbit satellite (15, 16) with orbital velocity of $V_{orb} \cong 26,600$ Km/h, e.g. similar to IRIDIUM system. In this model the UT remains in the spot beam where the call was originated for a time between 0 to $T_{max}$ depending on its position within the spot-beam. $T_{max}$ is given by Eq. 3.23

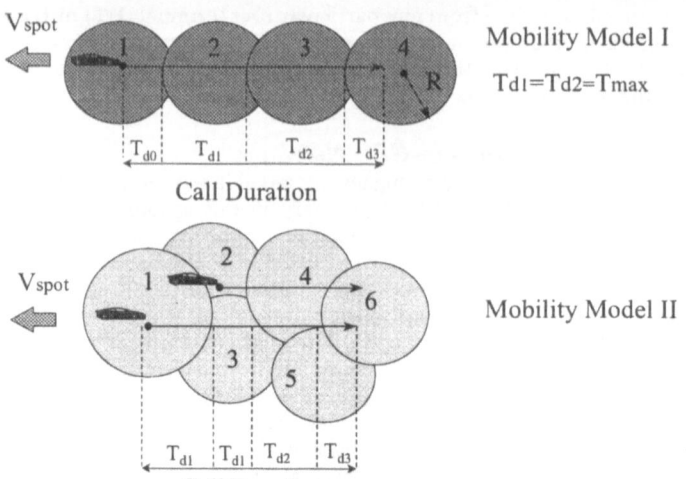

Mobility Model I

$T_{d1} = T_{d2} = T_{max}$

Mobility Model II

**Figure 3.67.** Non-GEO mobility models.

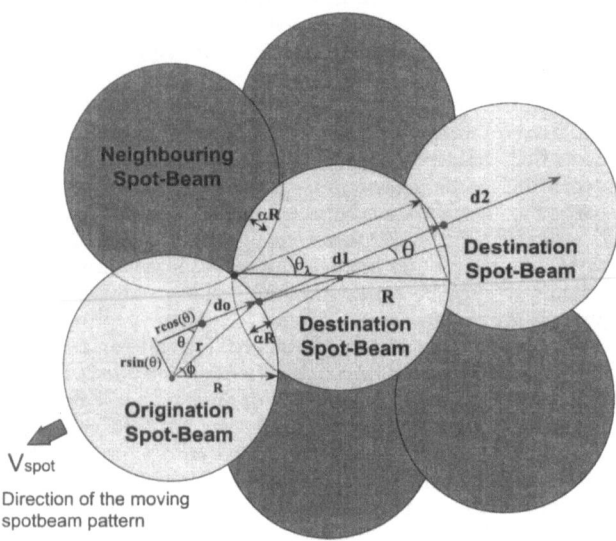

**Figure 3.68.** Geometry of the call origination and destination spot-beams.

and $V_{spot} = V_{orb}$ (all the other component speeds can be neglected due to the very high value of $V_{orb}$). After the first handoff to the adjacent spot beam, a fixed time between inter-beam handoffs is assumed ($T_{d1} = T_{d2} = T_{max}$, as shown Fig. 3.67) until the call termination.

### 3.3.2.4.2 Mobility Model II
In mobility model II the UTs stay within every spot beam they handed-off for a random time between 0 and $T_{max}$. The detail spot-beam geometry for model II is shown in Fig. 3.68. The UT is moving relatively to the satellite footprint at a speed $V_{spot}$ which is assumed to be constant for at least $T_{max}$.

This scenario represents a general scenario of the travelled distances $d_0$ in the spot-beams where calls are originated (origination spot-beam) and $d_1$, $d_2$ ... etc. of the travelled distances in all the following spot-beams. From the spot-beam geometry we can calculate the PDF of the random variable $d$ and the corresponding PDFs of the time $T_d$ that each terminal spends in both the origination and the destination spot-beams (95, 96).

### 3.3.2.4.3 Origination Spot-Beam
The Random Variable (RV) $d_0$ depends on $r$ which is uniformly distributed over the area of a circle, inside the circular pattern of the origination spot-beam. Therefore, the PDFs $f_r(r)$, $f_\phi(\phi)$, $f_\theta(\theta)$ of $r$, $\phi$ and $\theta$ are given respectively by:

$$
f_r(r) = \begin{cases} \dfrac{2\pi r}{\pi R^2} = \dfrac{2r}{R^2} & 0 \leqslant r \leqslant R \\ 0 & \text{elsewhere} \end{cases} \tag{3.24}
$$

$$
f_\phi(\phi) = \begin{cases} \dfrac{1}{2\pi} & 0 \leqslant \phi \leqslant 2\pi \\ 0 & \text{elsewhere} \end{cases} \tag{3.25}
$$

$$
f_\theta(\theta) = \begin{cases} \dfrac{1}{\pi} & 0 \leqslant \theta \leqslant \pi \\ 0 & \text{elsewhere} \end{cases} \tag{3.26}
$$

$$
d_0 = \sqrt{R^2 - (r\sin(\theta))^2} - (r\cos(\theta)) \tag{3.27}
$$

Since the value of $d_0$ depends only on $r$ and $\theta$ following the analysis presented in (95) we can represent the PDF of $d_0$ by the following formulae:

$$
f_{d_0}(d) = \begin{cases} \dfrac{2}{\pi R^2}\sqrt{R^2 - \left(\dfrac{d}{2}\right)^2} & 0 \leqslant d \leqslant 2R \\ 0 & \text{elsewhere} \end{cases} \tag{3.28}
$$

and from (6) and $Td_0 = d_0/V_{\text{spot}}$, assuming a constant $V_{\text{spot}}$ for that period of time, we can derive the following PDF:

$$
f_{Td_0}(t) = \begin{cases} \dfrac{4}{\pi \cdot T_{\max}}\sqrt{1 - \left(\dfrac{t}{T_{\max}}\right)^2} & 0 \leqslant t \leqslant T_{\max} \\ 0 & \text{elsewhere} \end{cases} \tag{3.29}
$$

#### 3.3.2.4.4 Destination Spot-Beams

In the destination spot-beams, the RV $d$ depends on the penetration angle $\theta$ and it is not uniformly distributed over spot-beam area any more. If we assume that there is a certain degree of cell overlapping $a$ (i.e. $a = 0$ for no cell overlapping, $a = 2$ for full cell overlapping) we can express the travelled distance $d$ as:

$$
d = \begin{cases} (2 - \alpha)R, & 0 \leqslant |\theta| < \theta_L \\ 2R\cos\theta, & \theta_L \leqslant |\theta| < \pi/2 \end{cases} \quad 0 \leqslant a \leqslant 2 \tag{3.30}
$$

$$
f_\theta(\theta) = \begin{cases} (1/\pi), & (-\pi/2) \leqslant \theta \leqslant (\pi/2) \\ 0 & \text{elsewhere} \end{cases} \tag{3.31}
$$

where $\theta_L$ is given by $a\cos(1 - a/2)$. By calculating first the PDF of the RV $d$, the PDF of $Td$ is finally given by the following:

$$
f_{Td}(t) = \begin{cases} \left(\dfrac{t}{T_{\max}}\right)\Big/ T_{\max} \cdot \sqrt{1 - \left(\dfrac{t}{T_{\max}}\right)^2}, & 0 \leqslant t < (1 - a/2)T_{\max} \\ \sin(a\cos(1 - a/2)), & t = (1 - a/2)T_{\max+} \\ 0 & \text{elsewhere} \end{cases} \tag{3.32}
$$

### 3.3.2.5 Radio Resource Management Overview

In non-GEO networks the GoS is usually affected by the service interruption probability due to unsuccessful handoffs in addition to the new call blocking probability. The "trunk" reservation method applied in a Fixed Channel Allocation (FCA) scheme, reserves a number of radio channels only for handoffs, so that the forced call termination probability (or call dropping probability) can be reduced to an acceptable level that is usually lower than the call blocking probability. Another method, is the exploitation of priority based Dynamic Channel Allocation (DCA) schemes that although more complex, in some cases perform better than the simple FCA. The combination of different channel allocation and handoff policy schemes have been presented in the literature mainly for voice communications such as: FCA with Handoff-Priority (97) and DCA-Handoff Queuing (15). The DCA scheme provides flexibility and adaptability to the traffic conditions at the expense of an increased complexity. However, DCA are reported to be less efficient (due to their complexity) than the FCA under uniform high load conditions. The proposed DCA method (16) claims to overcome these problems for LEO mobile satellite systems.

Our new approach, the Adaptive Bandwidth Reservation Scheme (ABRS) (98), can be applicable to both FCA and DCA. A metric called "mobility reservation status" was first introduced in (91, 10) that is updated every time a new call is accepted, released or a handoff to another spot-beam is performed. This metric provides information about the bandwidth requirements of all the active connections in a specific spot-beam in addition to the "possible" bandwidth requirements of mobile terminals currently connected to the neighbouring spot-beams. A new call is accepted in a spot beam if there are enough channels available for that spot beam and the mobility reservation status of the adjacent spot-beams has not exceeded a predetermined threshold ($\theta_n$). Handoffs from one spot-beam to another are successful, if there are channels available in the target spot-beam and the mobility reservation status of that beam is less than another threshold ($\theta_h$). This work was extended in (99) by considering additional parameters for the reservation mechanism and the different service classes performance criteria. The satellite orbit parameters and the service specific requirements as well as the mean call duration directly affect the number of spot-beams that are involved in the reservation matrix calculations.

### 3.3.2.6 "Trunk" Reservation for Multi-Rate Services

As explained in a previous section the call blocking probabilities of multi-rate services can be calculated by an accurate and very fast recursive solution. It is shown that the higher the bit rate the higher the call blocking probability. "Trunk" reservation is the method that is used to balance the call blocking probabilities of multi rate services. According to this, each service class is admitted into the system only up to a certain bandwidth threshold ($\vartheta_i$). The formula that approximates the call blocking probabilities (100, 101) has a very similar form to the one presented in (78). It is assumed that each service class ($i$) of constant rate ($c_i$) is characterised by a Poisson arrival process with mean call arrival rate ($\lambda_i$) and mean call holding time ($1/\mu_i$) following a general distribution. Then, if ($I$) is the number of service classes in a shared link of capacity ($C_{IN}$) the approximated un-normalised probability $P_{un}(m)$ and the normalised probability $P(m)$ of ($m$) occupied bandwidth units are given by:

$$P_{un}(m) = \begin{cases} 1 & : m = 0 \\ 0 & : m < 0 \\ \dfrac{1}{m}\sum_{i=1}^{I} P_{un}(m - m_i) \cdot m_i \cdot \dfrac{\lambda_i}{\mu_i} & 0 < m \leq M \end{cases} \qquad P(m) = P_{un}(m)\left(\sum_{m=0}^{M} P_m(m)\right)^{-1} \qquad (3.33)$$

where $M = C_{IN}/dc$, $m_i = c_i/dc$ and ($dc$) represents the greatest common divisor of ($c_i$) and:

$$m_i(m) = \begin{cases} m_i & m \cdot dc \leq \vartheta_i \\ 0 & m \cdot dc > \vartheta_i \end{cases} \qquad (3.34)$$

The approximated call blocking probability per class ($i$) can be calculated from:

$$P_{bi} = \sum_{m=\min\{M - m_i, (\vartheta_i/dc)\}+1}^{M} P(m) \qquad (3.35)$$

According to (102), a very simple and general rule applies when trying to balance the call blocking probabilities of multi-rate services: *For any subset of traffic classes $\omega \subseteq \Im = \{1,2,...I\}$ the call blocking probabilities are equalised if a bandwidth threshold ($\vartheta_i$) = M − max{$m_i$} is applied $\forall i \in \omega$.* However, it should be noted that call blocking equalisation imposes a penalty on the link utilisation.

### 3.3.2.6.1 "Trunk" Reservation in Non-GEO Networks

The "trunk" reservation method also appeared in the early terrestrial cellular-based and the satellite PCNs proposals. A certain number of "guard"[10] or "trunk" channels are used so that the call dropping probability due to unsuccessful handoffs is much less than the call blocking probability. For

---

[10] These are the channels that are reserved for handoff requests only

**Table 3.8.** Non-GEO multi-rate mixed traffic service parameters

| Service class | Service rate ($m_i$) | Offered traffic (%) |
|---|---|---|
| Class-$0_n$ (new call) | 1 | 22.22 |
| Class-$1_n$ (new call) | 2 | 22.22 |
| Class-$2_n$ (new call) | 4 | 22.22 |
| Class-$0_h$ (handoff) | 1 | 11.11 |
| Class-$1_h$ (handoff) | 2 | 11.11 |
| Class-$2_h$ (handoff) | 4 | 11.11 |

Spot-beam capacity M = 50, $m_0$ = 1, $m_1$ = 2, $m_2$ = 4 , "Trunk" channels only of handoff classes = $m_h$ = 0, 1, 2

single service systems Erlang-B can be used to calculate the offered GoS (it is expressed as a function of both the call blocking and forced call termination probability). In our study, we concentrate on a broadband non-GEO multi-rate, multi-service network environment where not much attention has been given in the existing literature. Looking at the analytical solutions provided by Kaufman (see (78)) and Roberts (Eq. 3.33) the multi-class service separation is based only on the service rate ($m_i$) and the results depend only on the offered traffic load ($\lambda_i/\mu_i$) and the system capacity ($M$). As a result, service classes with the same bandwidth requirements ($m_i$) but different offered load are not distinguished.

Therefore we adopted a slightly different methodology to investigate the call blocking performance as well as the call dropping due to unsuccessful handoffs in a broadband multi-service non-GEO network. Each traffic class is subdivided into the one that represents the new calls with mean arrival rate ($\lambda n_i$) and offered load ($\lambda n_i/\mu'_i$) and the one that represents the handoff requests with mean handoff rate ($\lambda h_i$) and offered load ($\lambda h_i/\mu'_i$). The total arrival rate per traffic class is still given by $\lambda_i = \lambda n_i + \lambda h_i$. However, the ratio ($1/\mu'_i$) represents the mean channel holding time[11] in a given spot-beam which should be noted that in general it is different from the mean call holing time ($1/\mu_i$). In order to investigate the rules that regulate the blocking probabilities of multi-rate services using "trunk" reservation in a non-GEO network with multi-spot beam payload configuration a mixed traffic scenario is assumed. If we define $\Im_n = \{1_n, 2_n, \dots I_n\}$ and $\Im_h = \{1_h, 2_h, \dots I_h\}$ the new call and handoff services subclasses $\forall\ i \in \Im$ so that $\Im = \Im_n \cup \Im_h$, then both subclasses will experience the same blocking rates although they might correspond to different offered traffic loads (i.e. $\lambda n_i/\mu'_i \neq \lambda h_i/\mu'_i$) assuming that no trunk reservation takes place (i.e. $\vartheta_{in} = M \mid i_n \in \Im_n$, $\vartheta_{ih} = M \mid i_h \in \Im_h$).

In the following figures, we present the blocking probabilities of a hypothetical mixed traffic scenario shown in Table 3.8, using "trunk" reservation and different rules for regulating the ratios among them. In these results it is assumed that all the main service-classes (i.e. $i$ = 0, 1, 2) contribute equally to the total offered load and a fixed ratio $\lambda n_i/\mu'_i = 2 \times \lambda h_i/\mu'_i$ between the new call and handoff arrival traffic load. In Fig. 3.69, it is shown that when "trunk" reservation is not used the blocking probability of each traffic subclass depends only on the offered load and the service rate ($m_i$). Therefore, new calls and handoff requests of each traffic class experience the same blocking rate.

A balance of all traffic classes and subclasses can be achieved if "trunk" reservation is applied with $\vartheta_{in} = \vartheta_{ih} = \max\{m_i\} = m_2\ \forall\ i \in \Im$. When comparing this plot (straight line) with the "no trunk" cases in Figure 3.69, it is shown that for class 0 (the one with the lowest service rate) the new call blocking probability ($Pb_n$) and the handoff failure probability ($Pb_h$) increases whereas for Class 2 (the one with the highest service rate) decreases. Furthermore, Class 1 maintains almost the same new call blocking.

In fact, that is slightly higher for total offered load values less than 0.6 and then becomes lower as the offered load increases. In mix traffic scenarios where there is a great difference between the min $\{m_i\}$ and max $\{m_i\}$, e.g. higher that 10, an oscillation of the blocking probabilities as a function of the offered load has been observed by (101, 103).

As shown in Figs 3.70 and 3.71 "trunk" reservation can also be used to balance both the new call blocking and handoff failure probabilities and at the same time provide a desired ratio between them. The bandwidth thresholds should be set equal to $\vartheta_{in} = M - m_2 - m_h$ and $\vartheta_{ih} = M - m_2\ \forall\ i \in \Im$. The ratio ($Pb_h/Pb_n$) depends on the number of "trunk" channels ($m_h$) that are reserved for the handoff classes only. In the plotted graphs, the straight line between the $Pb_{ih}$ and $Pb_{in}$ data points represents the common call blocking probability of all subclasses if the same bandwidth threshold $\vartheta_{in} = \vartheta_{ih} = \max\{m_i\} = m_2\ \forall\ i \in \Im$ is used.

---

[11] The mean channel holding time ($1/\mu'_i$) depends on the mean call holding time ($1/\mu_i$), the constellation dynamic and the user/terminal mobility

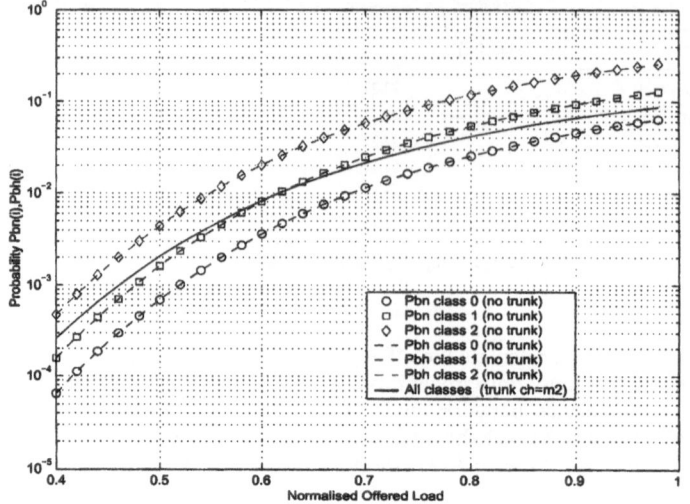

**Figure 3.69.** Balancing the call blocking probabilities using "trunk" reservation (no trunk: $\vartheta_{in} = \vartheta_{ih} = M$, with "trunk": $\vartheta_{in} = \vartheta_{ih} = m_2$, $m_h = 0$).

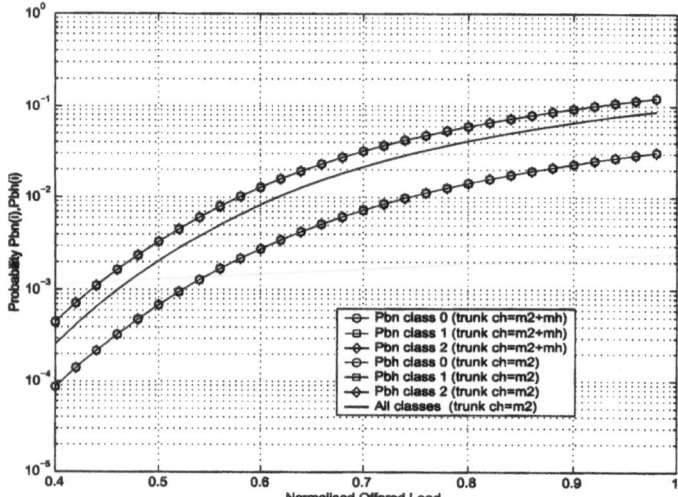

**Figure 3.70.** Balancing both new call blocking and handoff failure probabilities using "trunk" reservation ($\vartheta_{in} = M - m_2 - m_h$ and $\vartheta_{ih} = M - m_2$, $m_h = 1$).

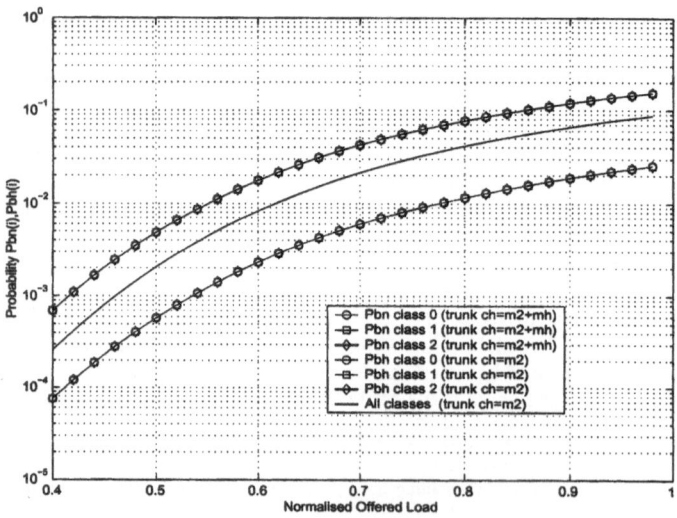

**Figure 3.71.** Balancing both new call blocking and handoff failure probabilities using "trunk" reservation ($\vartheta_{in} = M - m_2 - m_h$ and $\vartheta_{ih} = M - m_2$, $m_h = 2$).

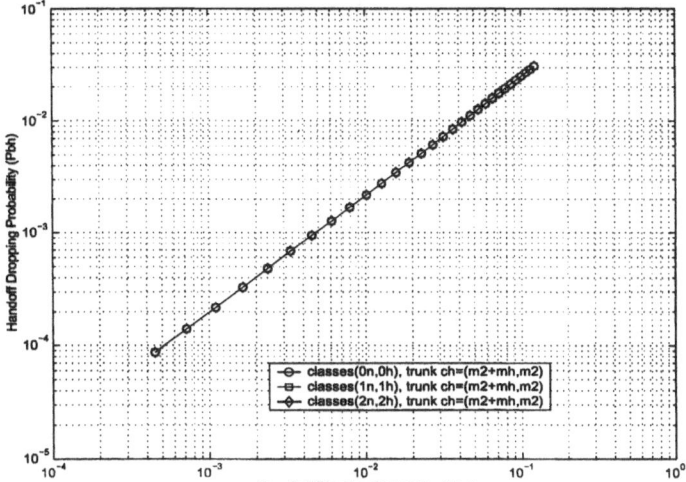

**Figure 3.72.** New call blocking versus handoff failure probabilities using "trunk" reservation (all classes $\vartheta_{in} = M - m_2 - m_h$ and $\vartheta_{ih} = M - m_2, m_h = 1$).

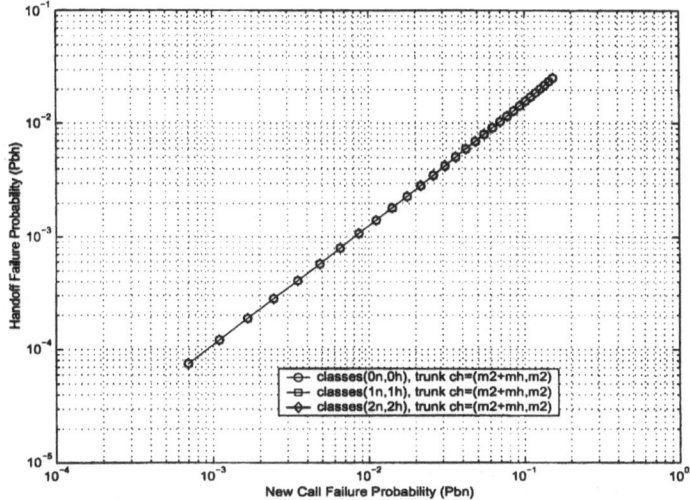

**Figure 3.73.** New call blocking versus handoff failure probabilities using "trunk" reservation ($\vartheta_{in} = M - m_2 - m_h$ and $\vartheta_{ih} = M - m_2, m_h = 2$).

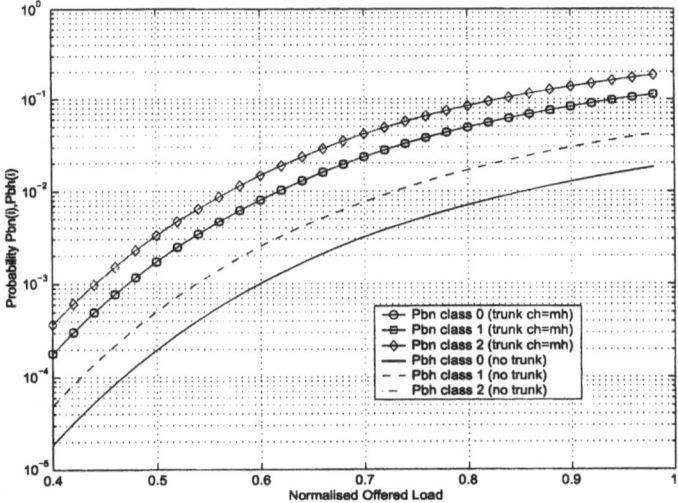

**Figure 3.74.** "Trunk" reservation only for handoff calls ($\vartheta_{in} = M - m_h$ and $\vartheta_{ih} = M, m_h = 2$).

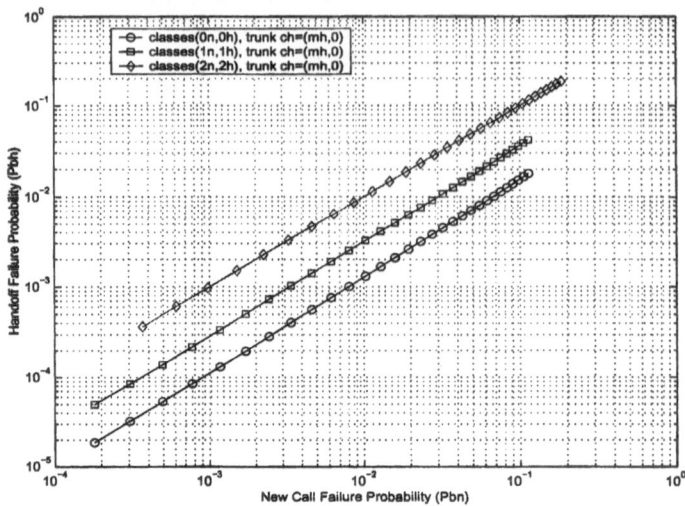

**Figure 3.75.** New call blocking versus handoff failure probabilities using "trunk" reservation only for handoff calls ($\vartheta_{in} = M - m_h$ and $\vartheta_{ih} = M$, $m_h = 2$).

In Figs 3.72 and 3.73, two different ratios of $Pb_h/Pb_n \cong 5$ and 10 are plotted for $m_h = 1$ and 2 respectively (i.e. the first points on the left and represent a total offered load of 0.4). However, as the offered load increases the ratio $Pb_h/Pb_n$ becomes 4 and 6.67 for $m_h = 1$ and 2 respectively (i.e. the top right points which represent a total offered load close to 1). This change is due to the non-linear call blocking and offered load relation.

As noticed in most of the previous graphs, there is an associated link utilisation penalty for the lower bit rate service classes when "trunk" reservation is applied. It seems that this is unavoidable when a fair link utilisation among the supported service classes is needed. However, this is not always the case especially in a non-GEO broadband multimedia network. Therefore, in Figs 3.74 and 3.75 we apply "trunk" reservation only for the handoff calls (i.e. $\vartheta_{in} = M - m_h$ and $\vartheta_{ih} = M$, $m_h = 2$) and try to achieve the best $Pb_h/Pb_n$ performance without considering the balance among $Pb_{in}$ and $Pb_{ih}$ subclasses. As a result, the best $Pb_h/Pb_n$ is acheived by class 0 (i.e. $\cong 10$ in Fig. 3.75) followed by Classes 1 and 2 with 3.3 and 1 respectively. That indicates that any value of $m_h \lesssim m_2$ does not affect the $Pb_{2h}/Pb_{2n}$ ratio but it creates a balance between $Pb_{0n}$, $Pb_{1n}$ since $m_h = m_1 = \max\{m_0, m_1\}$.

### 3.3.2.7  The Adaptive Bandwidth Reservation Scheme (ABRS)

In existing proposals for personal communications systems, mobile services require a bandwidth reservation mechanism that provides certain guarantees for the blocking and dropping probabilities. As shown in a previous paragraph, in a non-GEO multi-rate, multi-service network environment "trunk" reservation could be used in a static way with some extensions to the rules for regulating the call blocking and call dropping probabilities. Some typical real time services such as telephony, video telephony and video-conference are usually offered much higher call blocking rates than call dropping rates (ratios of 100 over 1 is a typical example) whereas for non-time sensitive applications a service interruption might be tolerated.

Nevertheless, when looking at the way the future wireless multimedia services are developed, a more dynamic approach should be adopted. Therefore, a new adaptive bandwidth reservation scheme has been developed in order to cope with the future demands. This new method introduces a great flexibility to support various service classes with user/application defined performance criteria. The designing targets are listed below:

- Provide a flexible and dynamic radio resource management scheme that best satisfies the application performance criteria.
- Incorporate all the useful information concerning the terminal locations at different points in time that a multi spot-beam satellite network can provide.
- Accept the calls for which most probably the requested service class requirements are met.
- Block the calls that the probability of meeting the service class criteria is low.
- During inter spot-beam handoffs assign priorities according to the application performance characteristics.

**Table 3.9.** Wireless service classification

| Service class (i) | $\theta_n$ | $\theta_h$ | ATM class |
|---|---|---|---|
| Class 0 | LOW | HIGH | CBR, rt-VBR |
| Class 1 | HIGH | LOW | nrt-VBR, ABR |
| Class 2 | LOW | LOW | UBR |
| Class 3 | HIGH | HIGH | CBR*, rt-VBR* |

*Distress calls or calls with high priority

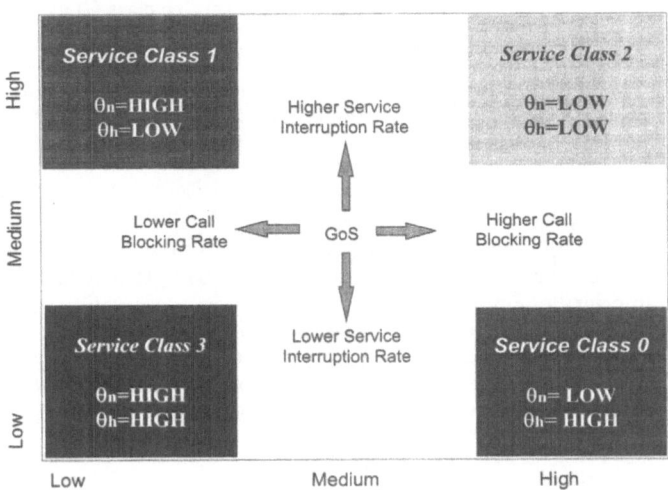

**Figure 3.76.** GoS expectation map for future wireless multimedia services.

In addition, a new scheme for grouping the future radio multimedia services is proposed according to the user service expectations and the network tariff policy. For this scheme, two mandatory and one optional parameters are required by the network for the radio resource management decisions. The mandatory ones are the expected call blocking and handoff failure rates which are expressed as the user/application defined threshold values for accepting and maintaining a call (i.e. $\theta_n$ and $\theta_h$ respectively). The optional one is the expected call duration (a different range of values can be given by the application). When the expected call duration is not given, a default parameter can be used by the network depending on the application (i.e. 3 or 5 min for telephony, 10 min for video-telephony, etc.). In such a way, new calls and handoff requests can be treated differently by the radio resource manager using the two threshold values $\theta_n$ and $\theta_h$ respectively.

### 3.3.2.7.1 Service classification and GoS

An example of a possible classification of the wireless services according to the user defined performance requirements is given in Table 3.9. At the application level, only two values (HIGH and LOW) are given for the threshold values and therefore up to four different service classes are created. Class 0 represents all the typical real time applications such as CBR and rt-VBR voice/video where the user expectation for a call to be dropped due to the frequent handoffs is less than the call to be blocked. On the contrary, Class 1 represents all those applications where the user expectation for accepting the call is higher than maintaining the call. For example, it includes all applications that need to have immediate access to the network no matter if the call is forced to be dropped in the following spot-beams due to the system overload. This service class can be quite attractive for both the user and the network when it is combined with a traffic adaptive tariff policy.

Class 2 represents all user applications that give no priority to a new call compared to a handoff call and uses a LOW threshold for both. Finally, Class 3 accommodates applications with the highest priority in the system for both call blocking and call dropping rates. Class 3 can be used for distress calls, services with the highest tariff policy or for calls that a book-in-advance policy applies.

In Fig. 3.76, we present the GoS performance expectation map. The user performance criteria for accepting a new call and maintaining an ongoing call are shown on the horizontal and the vertical axes respectively. At the application level, a short list of two (HIGH and LOW) or even more (including MEDIUM, etc.) priority levels are specified.

On the top-right and bottom-right side of Fig. 3.76, service Classes 2 and 0 are shown respectively, whereas service classes 1 and 3 are on the top-left and bottom-left side of the figure. These four distinct service classes represent the four extreme points on the user expectation map. In practice, any point between those four points could be used in the system to dynamically assign different performance requirements. In such a way, the network performance under a certain load and available resources can be dynamically change, using different threshold values. These thresholds represent the rate in which the user's requirements are met. It is then a network specific task to provide the absolute values for $\theta_n$ and $\theta_h$ as well as the ratios between the call blocking and handoff failure probability. As a result we can define the expected GoS of a service class ($i$) as:

$$GoS_i = \theta_{in} \cdot Pb_{in} + \theta_{ih} \cdot Pb_{ih} \tag{3.36}$$

where ($Pb_{in}$) is call blocking probability and ($Pb_{ih}$) the handoff failure probability using $\theta_{in}$ and $\theta_{ih}$ as the waiting factors respectively. It should be noted that in the above formula the handoff failure probability is used instead of the forced call termination probability (i.e. used in traditional telephony networks).

### 3.3.2.8 Wireless Call Admission Algorithm

In Fig. 3.77, the logic diagram of the call admission algorithm is presented. Every new call request is associated with a set of call parameters that are used by the system to dynamically reserve network resources and assign priorities among different connections.

Let us assume that a new call request of type ($i$) is received by the network in spot-beam ($k$). A list of the predicted spot-beam residual times ($Td_0$, $Td_1$, $Td_2$, ...$Td_{s-1}$) is calculated based on the constellation dynamics and the expected mean call duration ($1/\mu_i$) of that type if this is provided. Therefore, the size ($S_i$) of this list is variable and can be represented by the first integer that satisfies the following formula:

$$\sum_{0}^{Si-1} Td_k \geq \frac{1}{\mu_i} \tag{3.37}$$

In cases where the service mean call holding time is unknown, a default value for $Si$ is used.

The call admission algorithm is completed in a few steps:

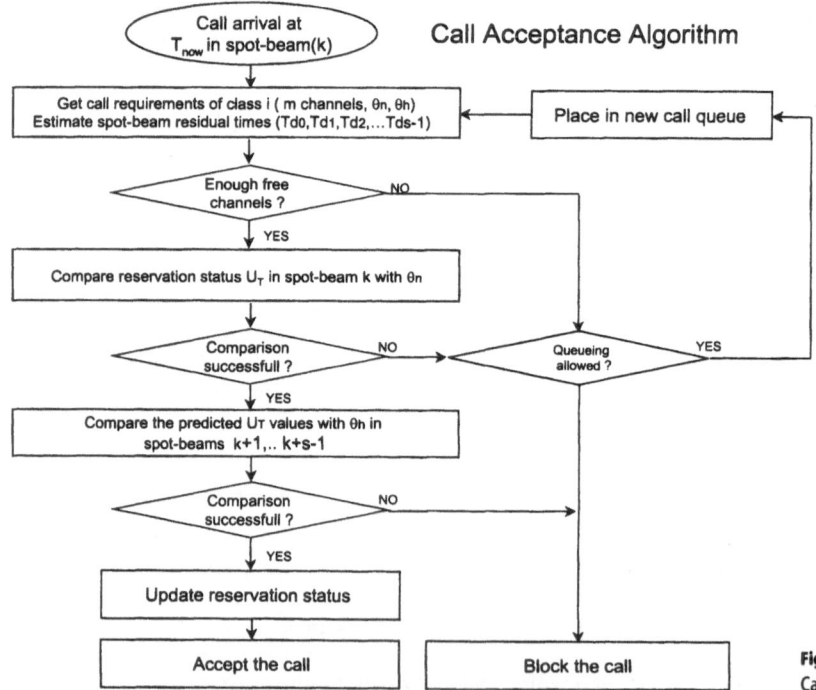

**Figure 3.77.**
Call acceptance algorithm.

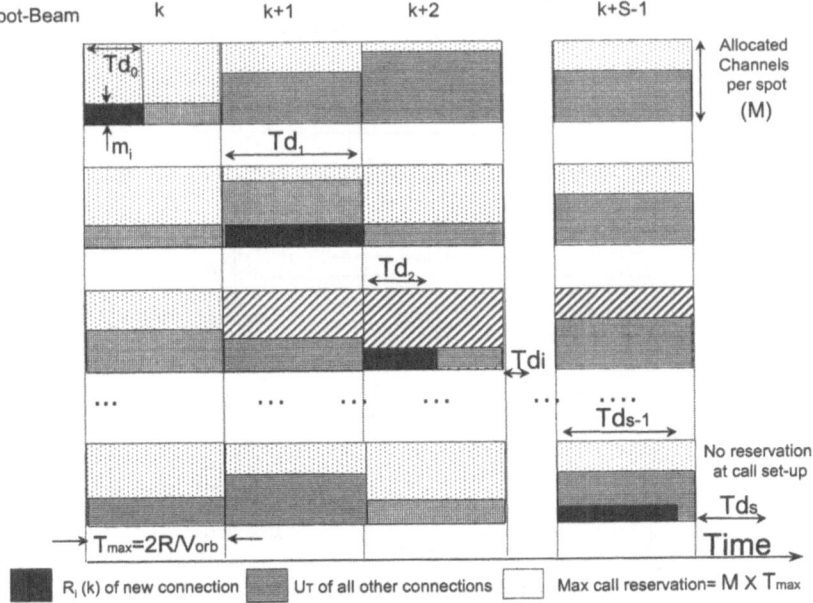

**Figure 3.78.** Resource reservation table updates.

1. First the number of the free radio channels in the spot-beam $k$ where the call is originated is checked

2. If there are enough free radio channels for a new call to be accepted, then the bandwidth reservations $U_T$ for the call origination beam ($k$) and the handoff destination beams ($k + 1 \ldots k + S_i - 1$) is checked against the service class specific threshold values $\theta_{in}$ and $\theta_{ih}$ respectively (see example in Table 3.9).

The admission controller keeps a record of the bandwidth requirements of all the existing calls in each beam that might be affected by the new connection. The call is accepted if the following criteria are met:

$$U'_T(k) = U_T(k) + R_i(0) \leq \theta_{in} \qquad (3.38)$$

$$U'_T(k + j) = U_T(k + j) + R_i(j) \leq \theta_{ih} \qquad (3.39)$$

where:

$$R_i(j) = Td_j \cdot m_i, \quad j = 0, 1, \ldots S_i - 1 \qquad (3.40)$$

$R_i(j)$ represents the bandwidth reservation for the call type ($i$) in beam ($j$) and ($mi$) is the requested number of channels (normalised bandwidth units).

3. If condition 2 is satisfied, then the bandwidth reservation table of the spot-beams that are included in the list is updated by replacing the values of $U_T(k)$ with $U'_T(k)$.

If any of the call admission criteria is not satisfied, then the call is blocked and it is removed from the system. If new call queuing is allowed, then the new call request is placed in a priority queue where $\theta_{in}$ is used to place the call request according to its service class. However, new call queuing was out of the scope of the performance evaluation.

Figure 3.78 illustrates the bandwidth reservation mechanism during the call set-up phase. $T_{max}$ represents the maximum time a call can affect a particular spot-beam and therefore the maximum value of $R_i(k)$ is $M \times T_{max}$.

### 3.3.2.8.1 Handoff Policy

The logic diagram for the handoff algorithm is given in Fig. 3.79. Provided that there are enough channels in the target spot-beam ($j$) and the number of traveled spot-beams is less than $S_i$, the handoff is successful if the following condition is true:

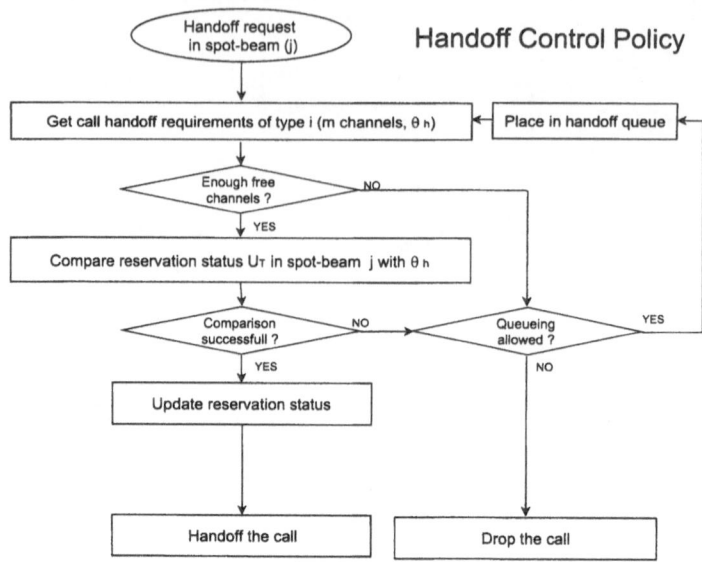

**Figure 3.79.** Handoff control algorithm.

$$U_T(j) \leq \theta_{ih} \tag{3.41}$$

If the above condition is not satisfied, the call is dropped in cases where queuing is not allowed, otherwise the request is placed in a priority queue. In any case, the reserved bandwidth in the previous spot-beam is released. When the terminal has reached the limit $S_i$ of the visited number of spot-beams for which all the bandwidth reservations were made two possible solutions are envisaged. Either to allow each call to retain the same performance criteria without any extra reservation, or to engage the call admission controller to perform new bandwidth reservations. The first approach is the simplest one and provided that the parameters chosen for the admission algorithm are highly optimised, it can give satisfactory results. However, if the second approach is used, the new reservations will have to be made for a much smaller number of spot-beams.

In cases where handoff queuing is allowed, the value ($\theta_{ih}$) of each service class will be used for the positioning of the request in the queue. However, new call or handoff request queuing was out of the scope of our investigation since we concentrate only on the parameters that affect the performance of the dynamic bandwidth reservation mechanism.

### 3.3.2.8.2 Call Termination
At the call termination phase if the number of the visited spot-beams is less than $S$, then the release of the reserved resources in all the remaining spot-beams takes place and the call is cleared form the system. Otherwise, the call is cleared without any further action.

### 3.3.2.9 Analytical Approximation
In this section, an analytical approach that approximates the GoS performance using the adaptive bandwidth reservation scheme is presented. Each service class ($i$) is characterised by its new call blocking probability $Pb_{in}$ and the handoff failure probability $Pb_{ih}$. If we assume uniformly distributed traffic in all the spot beams and a steady state system, we can then calculate these probabilities $Pb_{in}$ and $Pb_{ih}$ as a function of the probability (Pr) that reservation status $UT(k)$ in any spot beam ($k$) exceeds certain threshold values. According to the call admission algorithm steps 1 and 2 (shown in Section 3.3.2.8), and the conditions shown in Eqs 3.38 and 3.39, the new call and handoff failure probabilities of service type ($i$) are given by the :

$$PB_{in} = 1 - (1 - PB_i) \times \Pr(U_T < \theta_{in}) \times \prod_{1}^{Si} \Pr(U_T < \theta_{ih}) \tag{3.42}$$

$$Pb_{ih} = 1 - (1 - Pb_i) \times \Pr(U_T < \theta_{ih}) \tag{3.43}$$

$Pb_i$ represents the call blocking probability due to the lack of radio channels. Since the probability $\Pr(U_T < \theta)$ can be given by the Cumulative Distribution function (CDF) of the random variable $U_T$ we first need to calculate the PDF of $U_T$ i.e. $f_{UT}(U_T)$. Having in mind that the reservation status ($UT$) at each spot beam consists of the reservations ($U_{Tn}$) made from the new calls that have arrived in that beam and the reservations ($U_{Th}$) made from calls that have arrived in a different spot-beam, then the random variable $U_T = U_{Tn} + U_{Th}$ can be calculated by the convolution of the PDFs of the random variables $U_{Tn}$, $U_{Th}$. In addition, the random variables $U_{Tn}$, $U_{Th}$ are expressed as a function of the RVs $Td_0$ and $Td$ for which the PDFs are given by Eqs 3.29 and 3.32.

$$U_{Tn} = \sum_{i=1}^{I} \sum_{j=1}^{N_{in}} R_{in} = \sum_{i=0}^{I} \sum_{j=1}^{N_{in}} Td_0 \cdot m_i \tag{3.44}$$

$$U_{Th} = \sum_{i=1}^{I} \sum_{j=1}^{N_{ih}} R_{ih} = \sum_{i=0}^{I} \sum_{j=1}^{N_{ih}} Td_0 \cdot m_i \tag{3.45}$$

where $N_{in}$, $N_{ih}$ are the mean number of new calls of type ($i$) and the mean number of handoff calls of that type respectively. Therefore by using the following formula:

$$N_{in} = \frac{\lambda_{in}}{\mu_i'} (1 - Pb_{in}), \quad N_{ih} = \frac{\lambda_{ih}}{\mu_i'} (1 - Pb_{ih}) \tag{3.46}$$

and the PDFs of $R_{in}$ and $R_{ih}$ from:

$$f_{Rin}(R_{in}) = \frac{1}{m_i} f_{T_{d0}}(T_{d0}), \quad f_{Rih}(R_{ih}) = \frac{1}{m_i} f_{T_d}(T_d) \tag{3.47}$$

we can express the PDFs $f_{UTn}(U_{Tn})$, $f_{UTh}(U_{Th})$ as the convolution of the PDFs of the RVs $R_{in}$ and $R_{ih}$ respectively.

$$f_{U_{Tn}}(U_{Tn}) = \overbrace{f_{R_{0n}}(R_{0n}) \otimes f_{R_{0n}}(R_{0n}) \otimes \dots f_{R_{0n}}(R_{0n})}^{N_{0n}} \otimes \dots \overbrace{f_{R_{1n}}(R_{1n}) \otimes f_{R_{1n}}(R_{1n}) \otimes \dots f_{R_{1n}}(R_{1n})}^{N_{1n}} \tag{3.48}$$

$$f_{U_{Th}}(U_{Th}) = \overbrace{f_{R_{0h}}(R_{0h}) \otimes f_{R_{0h}}(R_{0h}) \otimes \dots f_{R_{0h}}(R_{0h})}^{N_{0h}} \otimes \dots \overbrace{f_{R_{1h}}(R_{1h}) \otimes f_{R_{1h}}(R_{1h}) \otimes \dots f_{R_{1h}}(R_{1h})}^{N_{1h}} \tag{3.49}$$

So finally we can write:

$$f_{U_T}(U_T) = f_{U_{Tn}}(U_{Tn}) \otimes f_{U_{Th}}(U_{Th}) \tag{3.50}$$

and by defining as $F_{U_T}(U_T)$ the CDF of the RV $U_T$ Eqs 3.42 and 3.43 can be written as:

$$Pb_{in} = 1 - (1 - Pb_i) \times F_{U_T}(\theta_{in}) \times \prod_{1}^{Si} F_{U_T}(\theta_{ih}) \tag{3.51}$$

$$Pb_{ih} = 1 - (1 - Pb_i) \times F_{U_T}(\theta_{ih}) \tag{3.52}$$

The above analytical approach assumes that the offered load per traffic class ($\lambda_{in}/\mu_i' + \lambda_{ih}/\mu_i'$) is known. $Pb_{in}$ and $Pb_{ih}$ are expresses in terms of $N_{in}$, $N_{ih}$ and depend on $Pb_{in}$. Therefore, this solution needs some iterations in order to have convergence in a given solution.

### 3.3.2.9.1 Numerical Examples
In this section some representative analytical results of the expected new call blocking and handoff failure probabilities are presented. The scope of this investigation is to demonstrate the impact of all the involved system parameters on the GoS performance of the new adaptive bandwidth reservation scheme. In Table 3.10, the parameters of two different satellite payload configurations that were investigated are shown. They represent a typical MEO and a typical LEO system.

The PDFs of the times $Td_0$, $Td$ that a $U_T$ spends in the call origination and destination spot beams are plotted from Eqs 3.29 and 3.32 respectively.

As shown in Figs 3.81 and 3.83 the function $f_{Td}(T_d)$ have a peak value of 0.6 at $T_d = 376$ and $T_d = 89$ respectively whereas the PDFs $f_{Td}(T_{d0})$ at the call origination spot-beams have a totally different shape as shown in Fig. 3.80 and Fig. 3.82.

**Table 3.10.** MEO and LEO constellation parameters

| Constellation parameters | MEO | LEO |
|---|---|---|
| Orbit height (Km) | 10355 | 1410 |
| Elevation angle (deg) | 40 | 40 |
| Spot-beam tears | 7 . | 3 |
| Orbit time (sec) | 21531.86 | 6836.91 |
| Orbit speed (Km/h) | 17578.20 | 25766.07 |
| Total number of spot beams | 169 | 37 |
| Max speed relevant to UT, $V_{spot}$ (Km/h) | 5132.63 | 16164.47 |
| Max satellite visibility to UT | 65 min 50.3 s | 7 min 3.3 s |
| Spot-beam diameter (Km) | 668.37 | 496.25 |
| Spot-beam overlapping ($\alpha$) | 0.4 | 0.4 |
| $Tmax = 2 \times R/V_{spot}$ (sec) | 468.79 | 110.52 |

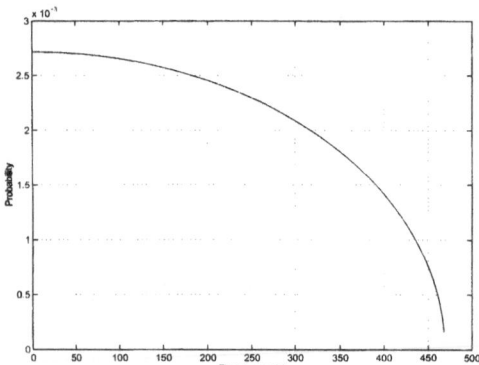

**Figure 3.80.** PDF of time $Td_0$ spend in the call origination beam for a MEO system ( see Table 3.10).

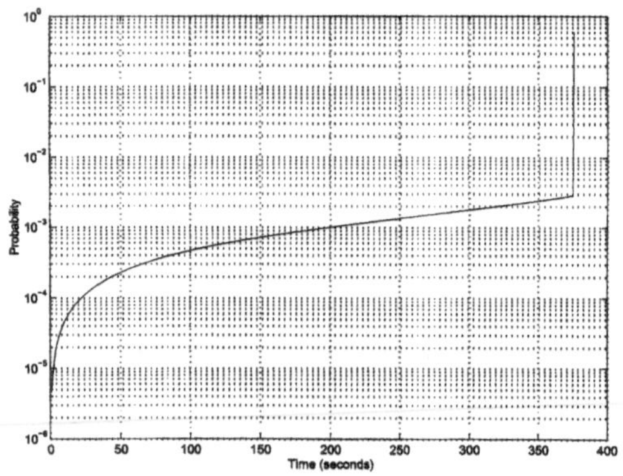

**Figure 3.81.** PDF of time $Td_i$ spend in the call destination beams for a MEO system (see Table 3.10).

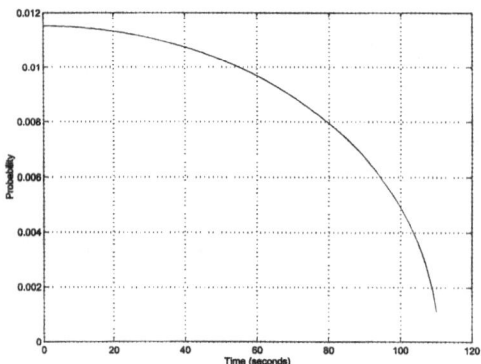

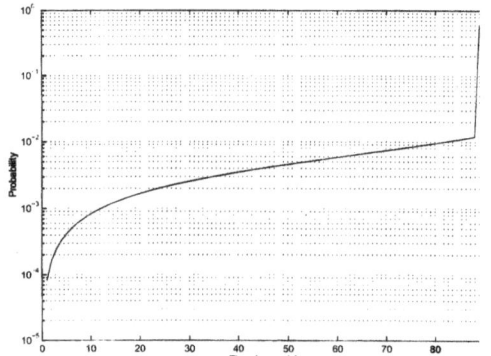

**Figure 3.82.** PDF of time $Td_0$ spend in the call origination beam for a LEO system (see Table 3.10).

**Figure 3.83.** PDF of time $Td_i$ spend in the call destination beams for a LEO system (see Table 3.10).

**Table 3.11.** MEO and LEO service and traffic parameters

| Service class | $\theta_n$ | $\theta_h$ | Offered traffic (%) MEO | Offered traffic (%) LEO |
|---|---|---|---|---|
| Class-$0_n$ (new call) | LOW | N/A | 16.667 | 8.333 |
| Class-$1_n$ (new call) | HIGH | N/A | 16.667 | 8.333 |
| Class-$2_n$ (new call) | LOW | N/A | 16.667 | 8.333 |
| Class-$3_n$ (new call) | HIGH | N/A | 16.667 | 8.333 |
| Class-$0_h$ (handoff) | N/A | HIGH | 8.333 | 16.667 |
| Class-$1_h$ (handoff) | N/A | LOW | 8.333 | 16.667 |
| Class-$2_h$ (handoff) | N/A | LOW | 8.333 | 16.667 |
| Class-$2_h$ (handoff) | N/A | HIGH | 8.333 | 16.667 |

Spot-beam capacity M = 50, $m_0 = m_1 = m_2 = m_3 = 1$, "Trunk" channels only of handoff classes = $m_h = 0, 1, 2$
HIGH = 1, 0 < LOW ⩽ 1.

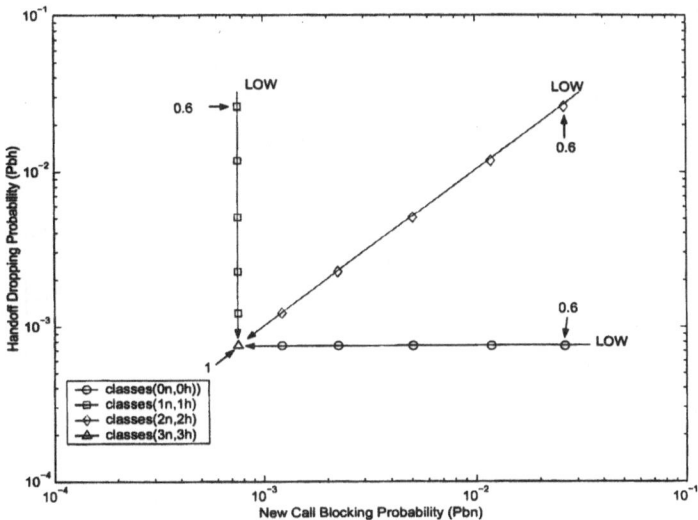

**Figure 3.84.** Impact of the LOW threshold variations on the GoS performance (MEO, S = 1).

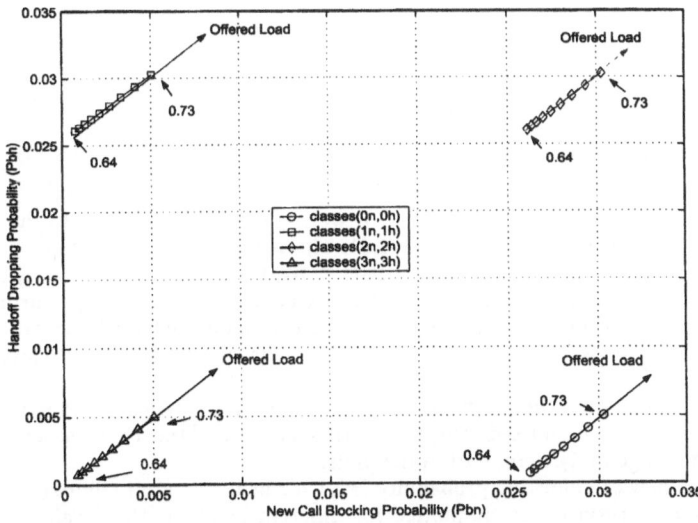

**Figure 3.85.** Multi-class GoS performance as a function of the offered load (MEO, S = 1, LOW threshold = 0.6 ).

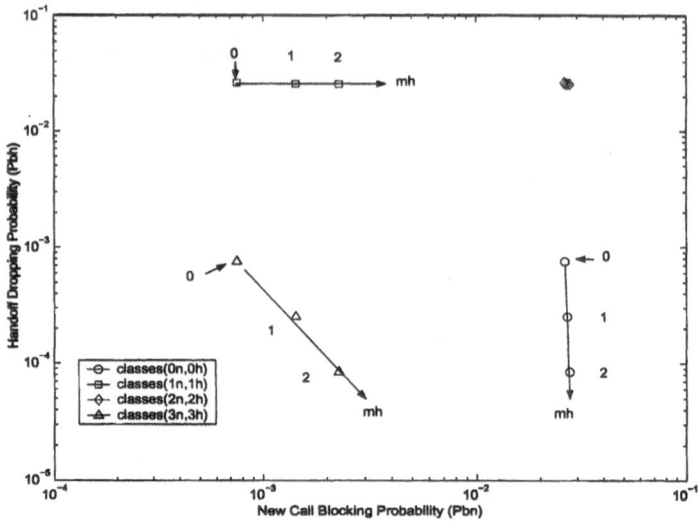

**Figure 3.86.** GoS performance of the adaptive bandwidth reservation scheme combined with "trunk" reservation for handoff calls (MEO, S = 1, LOW threshold = 0.6, $m_h$ = 0,1,2).

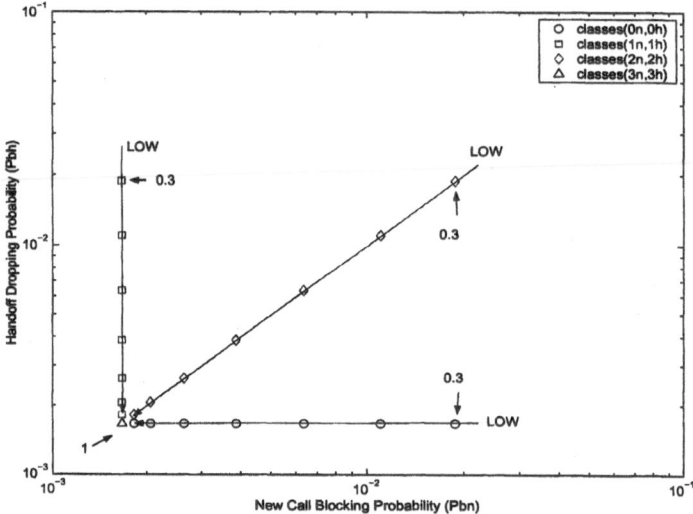

**Figure 3.87.** Impact of the LOW threshold variations on the GoS performance (LEO, S = 1).

Four different service classes (i.e. $i = 0, 1, 2, 3$) were considered according to the service classification criteria shown in Fig. 3.76 and Table 3.11. The required GoS criteria per service class are $\theta_{in}$ = {LOW, HIGH, LOW, HIGH} and $\theta_{ih}$ = {HIGH, LOW, LOW, HIGH} for $i = 0, 1, 2, 3$. In order to clearly demonstrate the impact of various system parameters on the GoS performance, the same number of radio channels for all service classes is assumed in the analytical calculations (i.e. $m_i = 1$). The total normalised offered load was 0.64 E and it is distributed among all classes as shown in Table 3.11.

The following figures show the results of one iteration of the analytical solution using as an initial point for the value $Pb_{in} = Pb_i$. A fixed value for the HIGH threshold (i.e. 1) and several LOW threshold values within the range (0, 1) were used in the plots.

In Fig. 3.84, the new call blocking probability ($Pb_h$) versus the handoff failure (or dropping) probability ($Pb_n$) for each service class is plotted for different LOW threshold values. By setting both thresholds to (HIGH, LOW) = (1, 1) all services meet at the same point in the GoS map. As expected, by reducing the LOW threshold value to 0.6 all service classes that use it as an admission criterion

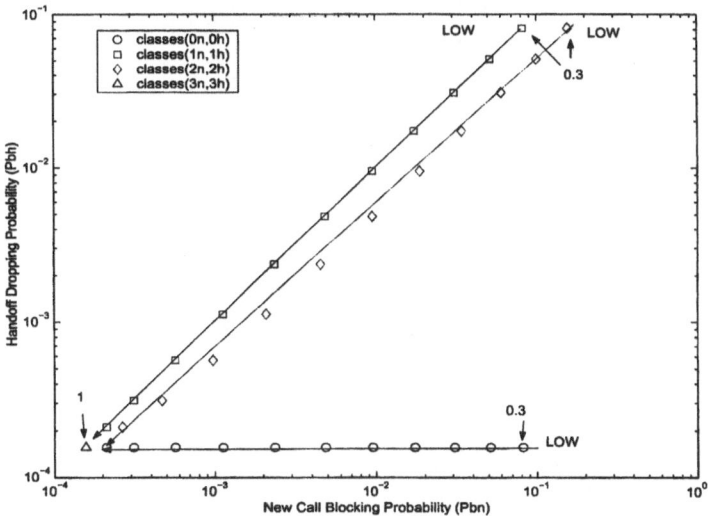

**Figure 3.88.** Impact of the LOW threshold variations on the GoS performance (LEO, $S = 2$).

are affected (i.e. Classes 1, 2, 3). Therefore, class 0 with admission criteria $(\theta_{0n}, \theta_{0h}) = $ (LOW, HIGH) moves horizontally towards higher $Pb_n$, Class 1 with $(\theta_{1n}, \theta_{1h}) = $ (HIGH, LOW) moves vertically towards higher $Pb_h$ and Class 2 with $(\theta_{2n}, \theta_{2h}) = $ (LOW, LOW) moves diagonally towards both higher $Pb_n$ and $Pb_h$. Finally, the GoS performance of class 3 is not affected by the LOW threshold variations since $(\theta_{3n}, \theta_{3h}) = $ (HIGH, HIGH).

In Fig. 3.85 the impact of the offered load on the GoS performance is shown. The results represent a MEO satellite configuration scenario with $S_i = 1, \forall\, i \in \{0, 1, 2, 3\}$, LOW = 0.6 and total offered load (normalised) from 0.64 to 0.73. As the offered load increases, all service classes are affected and the GoS performance moves towards both higher $Pb_n$ and $Pb_h$ values.

In Fig. 3.86, another possible way to affect the performance of all service classes including service Class 3 is presented. By using "trunk" reservation (for handoff calls only) in addition to the dynamic bandwidth reservation scheme, we can achieve (if this is required) a better $Pb_h/Pb_n$ ratio at a given offered load and LOW threshold value. The combination of both techniques improves the ratio $Pb_h/Pb_n$ only for the service classes that give HIGH priority to handoffs (e.g. Classes 0, 3). Consequently $Pb_h/Pb_n$ of Class 1 will be reduced and for Class 2 remains the same. In the presented results the number of "trunk" channels reserved for the handoff subclasses only varied from $m_h = 0$ to 2.

For the LEO satellite constellation scenario, the ratio between new call and handoff call arrival rate was much higher (e.g. 2) when compared to the MEO satellite constellation (e.g. 0.5) Therefore, the contribution of all handoff subclasses to the total offered load is double the one that is offered by the new call subclasses (see Table 3.11). In Figs 3.87 and 3.88 the GoS performance of the same set of service classes is examined under a LEO satellite configuration scenario using $S_i = 1$ and $S_i = 2$ respectively.[12]

When compared the two different admission policies the one that is more conservative in terms or bandwidth reservations ($S = 1$) results in much lower $Pb_h/Pb_n$ ratios than the one with $S \geqslant 2$ for the same LOW threshold value (e.g. 0.3). This is mainly due to the high percentage of the handoff calls into the system. However, as shown in Fig. 3.88, there is a penalty on the GoS performance for service Class 1, for admission policies that use $S > 1$. The new call blocking probability for Class 1 becomes very close to the one of Class 2 even though they use different $\theta_n$ values.

### 3.3.2.10 Simulation Modelling Approach

In addition to the analytical approach presented in the previous sections, both the ABRS scheme and "trunk" reservation were also evaluated by simulation means. The provision of handoff queuing increases the system performance under certain traffic conditions (e.g. non-uniform), however queuing (either for the new calls or for the handoff requests) was not considered in the simulation

---

[12] $S_i = 1$ involves CAC only at the call origination spot-beam and $S_i = 2$ at the call origination beam and one call destination spot-beam.

**Table 3.12.** Simulation model parameters

| Simulation parameters | Value |
|---|---|
| User population | 1200 |
| Service class ($i$) | 0,1,2,3 |
| Population per service class | Variable |
| Requested channels per call ($m_i$)* | 1,2,4 |
| Mean call rate $\lambda_i$ (calls/h/user)* | 0.3 |
| Mean call duration ($1/\mu_i$) (min)* | 1,3 |
| No of simulated spot-beams | 5 |
| Channels per spot-beam ($M$) | 10 |
| LOW priority threshold ($U_T(k)$ %) | Variable |
| HIGH priority threshold ($U_T(k)$ %) | 110% |

*See the simulated mixed traffic scenario parameters in Table 3.13

**Table 3.13.** Simulation mixed traffic parameters

| Traffic scenario | Traffic parameters | Service class ($i$) | User population | | | Service rate ($m$) | | | Offered load ($E$) |
|---|---|---|---|---|---|---|---|---|---|
| | | Class 0 | 300 | | | 1 | | | 4.5 |
| I | $1/\mu_i = 3$ min | Class 1 | 300 | | | 1 | | | 4.5 |
| | | Class 2 | 300 | | | 1 | | | 4.5 |
| | | Class 3 | 300 | | | 1 | | | 4.5 |
| | | | Total = 1200 | | | | | | Total = 18 |
| II | $1/\mu_0 = 3$ min | Class 0 | 200 | 200 | 200 | 1 | 2 | 4 | 21 |
| | $1/\mu_1 = 1$ min | Class 1 | 100 | 100 | 100 | 1 | 2 | 4 | 3.5 |
| | $1/\mu_2 = 1$ min | Class 2 | 100 | 100 | 100 | 1 | 2 | 4 | 3.5 |
| | | | Total = 1200 | | | | | | Total = 28 |

model since it is out of the scope of this work. The simulation model was implemented in OPNET modeler by MIL3. Both MEO and LEO satellite payload configurations were simulated according to the parameters shown in Table 3.10 and the mobility Model II presented in Section 3.3.2.4.2.

All the remaining simulation parameters are shown in Tables 3.12 and 3.13. A total number of five spot-beams with 1200 uniformly distributed user terminals were simulated. One dimensional model was used for the satellite spot-beam pattern movement, with a cycling approach for the users that exit the coverage area without completing their call.

The requested traffic channels ($m_i$) per service class represent the normalised bandwidth units of each connection at the MAC layer including any multiplexing gain could be achieved when VBR or ABR traffic sources are involved at the terminal side. This effective rate is assumed to be constant during the call and could represent multiples of a basic service rate such as: 32, $2 \times 32$, $4 \times 32$ kbps. The call inter arrival times and call holding times follow negative exponential distributions. For the inter spot-beam handoffs at both the call origination and destination spot-beams the distributions given by equations (Eq. 3.29) and (Eq. 3.32) were used.

Two different threshold levels (i.e. HIGH and LOW shown in Table 3.12) control the new calls (through $\theta_n$) and the handoff requests (through $\theta_h$) as shown in Table 3.9. The HIGH threshold was kept constant at 110 per cent of the bandwidth reservations $U_T(k)$ and the LOW threshold varied from 60 to 110 per cent. It should be noted that at the actual upper bound of $U_T(k)$ was one of the unknown simulation parameters since it depends on the bandwidth reservations that are initiated form other spot-beams beams. As a result, values for the LOW and HIGH thresholds that exceed the maximum reservation per call per spot-beam (i.e. $M \times T_{max}$) were allowed in order to obtain the initial system configuration parameters. A typical value of the maximum $U_T(k)$ that was used during our simulations was $1.1 \times (M \times T_{max})$. Values of $UT(k)$ higher than $1.4 \times (M \times T_{max})$ were not recorded in the simulated scenarios.

In Table 3.13, the traffic scenarios that were simulated are presented. In Scenario I, all traffic classes ($i$) request for 1 radio channel with mean call duration of 3 min and each traffic class contributes equally to the total offer load. Scenario II represents a mixture of different user traffic classes with various number of requested radio channels (i.e. 1, 2 and 4) and user population. The mean call holding time ($1/\mu_i$) was 3 min for Class 0 and 1 min for Classes 1 and 2.

### 3.3.2.11 Simulation Results and Discussion

In addition to the new call blocking probability and handoff failure probability the simulation model was used to record a number of other useful statistic that cannot be easily obtained by analysis. Hence, the simulation results of the call dropping probability (i.e. the forced call termination due to unsuccessful handoff) versus the call blocking probabilities are shown in the following figures. The point where all the service classes have exactly the same performance is when both LOW and HIGH thresholds are equal to the recorded maximum value of the reservation status. At this point it is obvious that no priority is given to any service class and therefore for the same traffic load all curves cross at the same point. As the LOW threshold value decreases, it starts to affect the priorities given to the new calls over the handoff requests for the different classes.

In Figs 3.89 and 3.90, the service separation according to the performance requirements given in Table 3.9 is clearly shown. Therefore, the four different points of each service class in the plotted graphs, represent the class position in the performance expectation map (see Fig. 3.76).

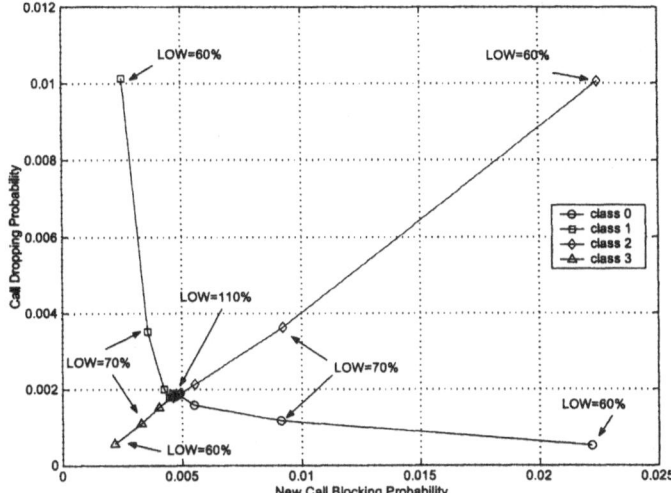

**Figure 3.89.** Simulated MEO traffic scenario I.

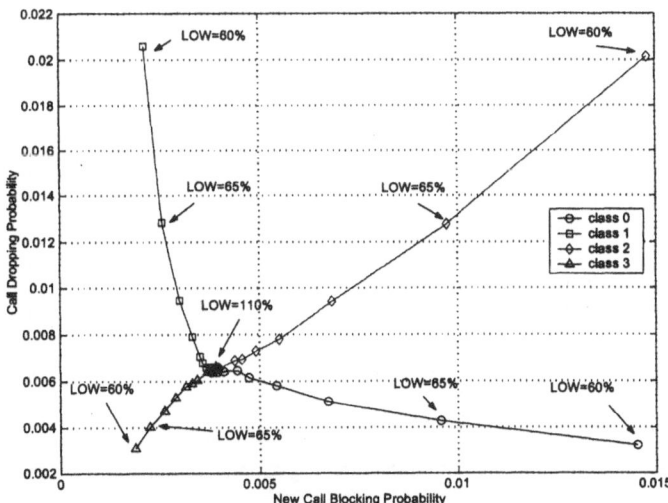

**Figure 3.90.** Simulated LEO traffic Scenario I with constant S = 2.

However, as explained in section 3.3.2.9.1 where the analytical plots are presented (see Fig. 3.87 and Fig. 3.88) the admission policy has a great impact on the GoS performance. The difference between the two graphs is in the number of spots-beams ($S$) that are involved in the bandwidth reservations. This is also illustrated in the simulation results shown in Figs 3.90 and 3.91 for traffic Scenario I and LEO satellite constellation. For example, in Fig. 3.90 the ratio between the call blocking over the call dropping probability for Class 0 and LOW = 60 per cent is around five, whereas in Fig. 3.91 for LOW = 90 per cent the same ratio is around 30.

That is due to the fact that a fixed assignment $S = 2$ in a LEO constellation results in low levels of bandwidth reservation for handoffs (the recorded mean number of handoffs for the successfully completed calls was 2.4). However, as the level of bandwidth reservation increases (i.e. Fig. 3.91 with variable $S$) the Classes 0 and 3 become more dominant and the performance curve of Class 1 becomes closer to the one of Class 2. Nevertheless, an overall balance of the call dropping over the call blocking rates for all service classes is still maintained. Therefore, depending on the two threshold values and the level of bandwidth reservations a dynamic range of performance ratios can be achieved.

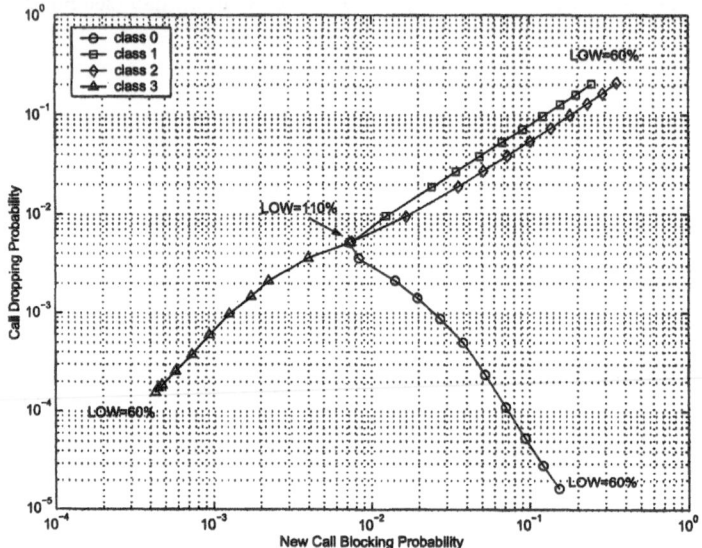

**Figure 3.91.** Simulated LEO traffic Scenario I with variable S.

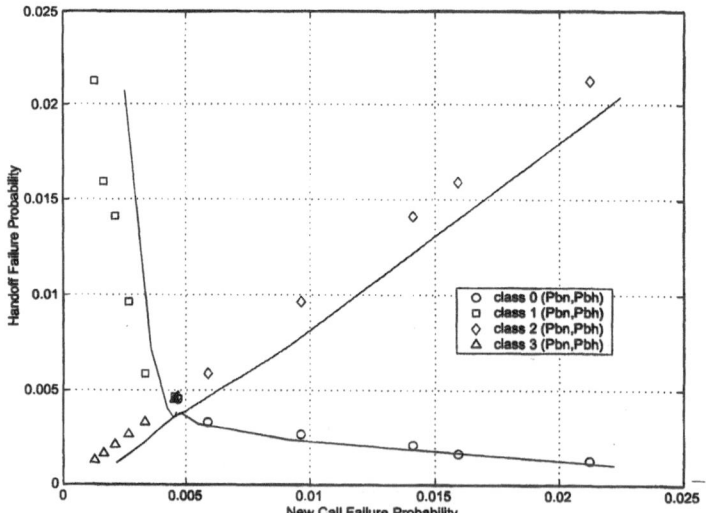

**Figure 3.92.** Comparison between the simulation and the analytical results (MEO).

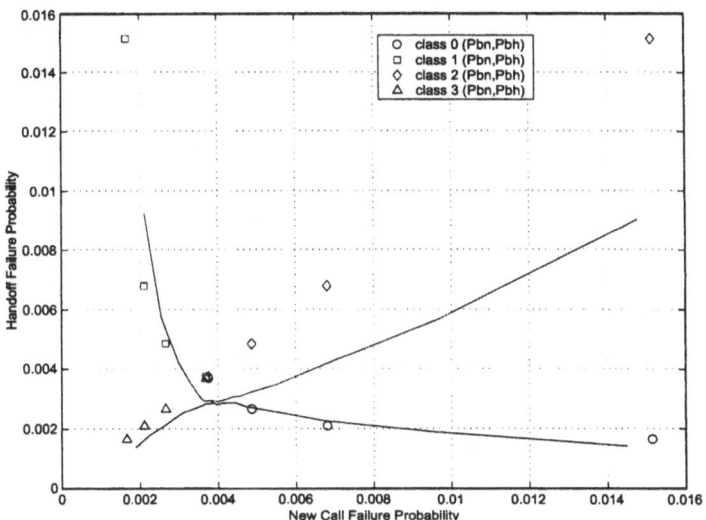

**Figure 3.93.** Comparison between the simulation and the analytical results (LEO, S = 2).

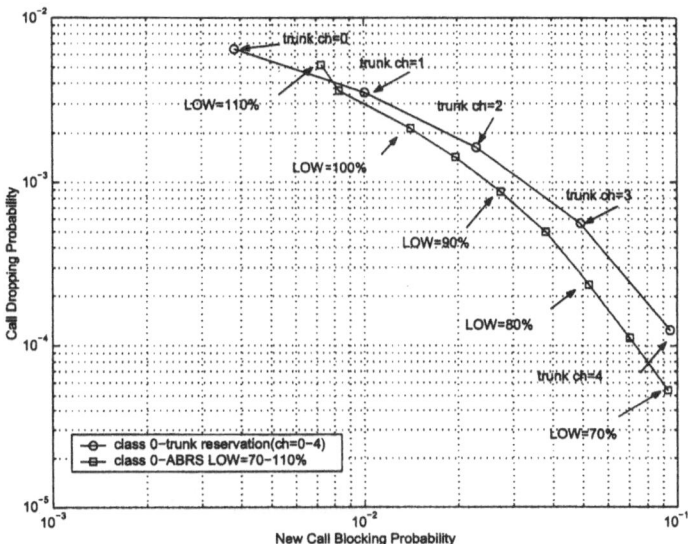

**Figure 3.94.** Comparison with the "trunk" reservation LEO.

In the following figures a comparison between the results obtained by analysis and simulation is presented. The continuous lines in Figs 3.92 and 3.93 represent the simulation results and the points were calculated according to the analytical approximation presented in the previous section.

By selecting the proper parameters for the analytical model it is shown that the results fit quite well with the simulation data. A slight mismatch between the presented graphs cannot be avoided due to the effect of the *ceiling* function in the analytical implementation and the adjustments in the offered load. For the analytical results the spot-beam capacity was 50 channels and the normalised offered load was 0.66–0.71. This adjustment in the analysis was necessary, due to the fact that the analytical model becomes accurate when the number of the users per traffic class in the system is high enough to represent a realistic traffic scenario. In cases where the traffic load is very low, the average number of users becomes less that one and as a result the required number of convolutions is always the same (e.g. one per traffic class). Therefore, when the offered load is not high enough it was observed that the LOW threshold has little affect on the service class performance.

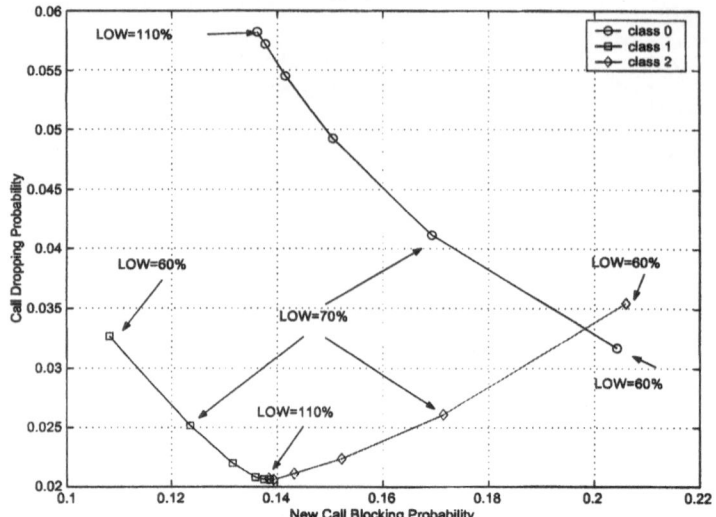

**Figure 3.95.** MEO Traffic Scenario II.

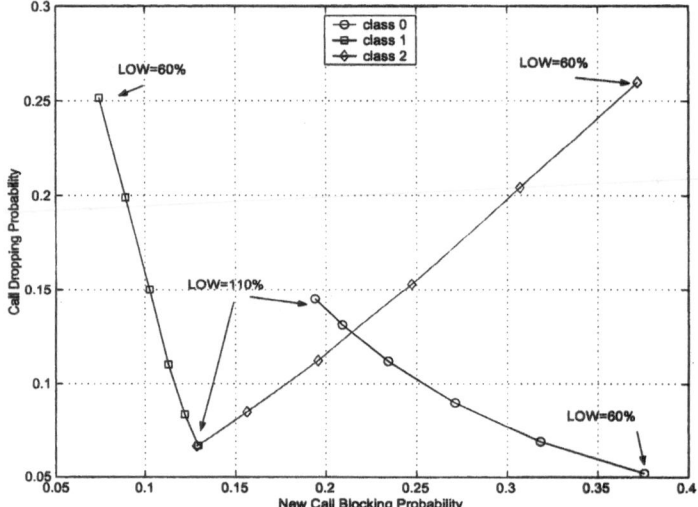

**Figure 3.96.** LEO traffic Scenario II.

A comparison between the performance obtained by ABRS and the "trunk" reservation method is shown in Fig. 3.94 although the two methods could be seen as complementary to each other. As expected, in a uniformly distributed traffic scenario with a single service class both methods perform in a similar way. That is because on average, they both introduce certain levels of admission thresholds to satisfy the needs of traffic with different priorities. Nevertheless, an absolute gain in terms of resource utilisation for service Class 0 can be achieved by ABRS for as the ratio between the call blocking over the call dropping rates increases and was recorded in the LEO constellation scenario. Thus, depending on the network design requirements both methods can be combined as shown in by Section 3.3.2.9.1.

Finally, in cases where the total offered load is not uniformly distributed among the service class, the GoS performance curves are shifted according to the offered load per traffic class. As shown in Figs 3.95 and 3.96 for the simulated traffic Scenario II (see Table 3.13), the same analogies for the LOW and the HIGH threshold values still apply.

### 3.3.2.12 Conclusions

In future broadband satellite networks, where dynamic satellite networks are expected to play an essential role for providing global coverage, traditional methods used mainly for voice communications are not flexible enough to satisfy the needs of various multimedia applications. In this section, an investigation of the bandwidth reservation techniques and their performance that could be used in a non-GEO multi-rate, multi-service class environment has been conducted. The adopted methodology and the analytical results of applying new rules in maintaining the desired GoS performance at the access radio part is presented. A new approach for creating service classes according to the user defined performance criteria is developed and its performance has been evaluated both analytically and by simulation means. This approach, can be used to dynamically regulate the system's call blocking and handoff failure probabilities for different services classes. A new radio resource management scheme, the ABRS is proposed by incorporating two different threshold values for handling the new calls and the handoff requests. The bandwidth reservation mechanism is simple enough that can be executed in real time for a large population of users. It also provides the flexibility to map in a dynamic way the service GoS requirements into the non-GEO network performance characteristics. In addition, due to the dynamic bandwidth reservation each service class maintains a GoS profile that has been accepted or negotiated during the call set-up phase and can be easily modified within the call. The great advantage of the ABRS is that it does not have the limitations of the fixed predefined way of handling calls in a wireless network that "trunk" reservation has. It incorporates a service class separation in such a way that can be adjusted to any multi-class traffic scenario that needs to be supported in the future.

### 3.3.3 ISL Network Routing and Dimensioning

Future broadband LEO satellite communication systems will increasingly rely on an intersatellite link (ISL) trunk network with time-variant topology. Within the network planning process, the routing and dimensioning tasks are in general closely coupled and do essentially influence both installation and operation costs of a system.

The routing of complex global traffic flows over dynamic network topologies has already been addressed in former work to a level of detail that allows to provide necessary input for the dimensioning task. In (104), a new concept for connection-oriented ATM-based routing for periodically time-variant ISL networks has been developed, called Discrete-Time Dynamic Virtual Topology Routing (DT-DVTR), (Fig. 3.97). DT-DVTR works completely off-line, i.e. prior to system operation. In a first step, a virtual topology is set up for all successive discrete time steps $s = 1 \dots S$ of the system

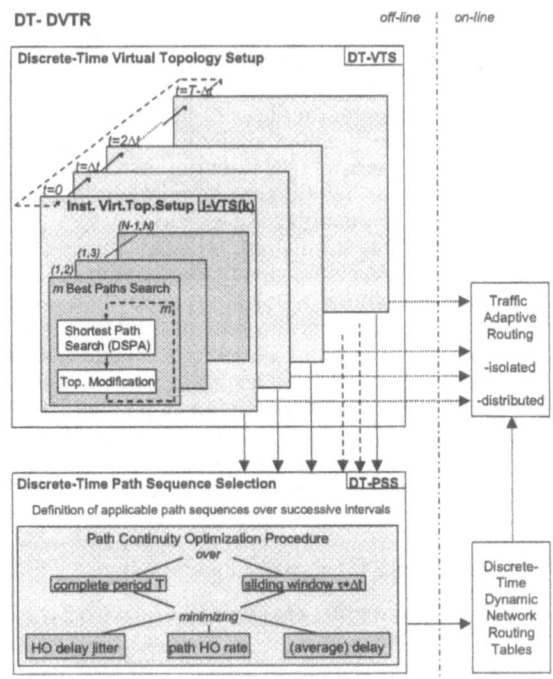

**Figure 3.97.** Discrete-Time Dynamic Virtual Topology Routing (DT-DVTR): concept and implementation.

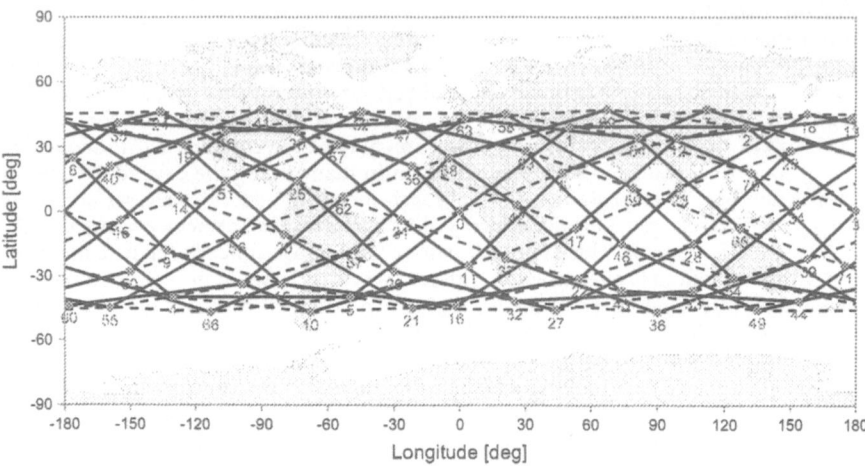

**Figure 3.98.** Considered permanent ISL topology for M-Star; solid lines: intra-plane ISLs; dashed lines: inter-plane ISLs.

period, providing instantaneous sets of alternative paths between all origin- destination (OD) pairs. In the second step, path sequences over a series of time intervals are chosen from that according to certain optimisation procedures. A VPC-based implementation of DT-DVTR, together with a sophisticated path sequence selection, has shown a good performance with respect to path switching offset even in an environment with switched ISLs, as encountered in polar constellations like Iridium.

However, especially in the light of broadband multimedia networks, the seam between contrarotating orbits and the on/off-switching of inter-plane ISLs seem to be too severe drawbacks for the connection-oriented operation. This has stimulated the investigation of inclined Walker constellations employing ISLs (105). It has been shown that such constellations generally provide the possibility to set up a number of inter-plane ISLs that can be maintained *permanently* with acceptable pointing, acquisition and tracking (PAT). Obviously, this is a highly desirable feature in the light of real-time connection-oriented services, as problems due to path switching can be avoided.

M-Star has been one of the first commercial system proposals aiming at the promising combination of Walker orbits and ISLs. A proper design of the ISL topology to be implemented is a first important step to guarantee efficient networking in the operational system. In (2) we have presented in detail a pragmatic approach to the ISL topology design in Walker constellations, deriving a reference topology for M-Star as an example. Figure 3.98 illustrates the final outcome. The following dimensioning approaches have been performed on top of this shown topology.

For the capacity dimensioning of ISL networks one may use approaches, algorithms and tools known from terrestrial (ATM) networks to a certain extent, but has to take into account specific additional constraints like the time-variance of the topology and also potentially different target functions adapted to the LEO ISL scenario.

With respect to an appropriate target function for the ISL network dimensioning, there is an important difference compared to the terrestrial case, which is due to the dynamic topology encountered. With the complete satellite constellation periodically orbiting the Earth and thus the source traffic demand, each satellite and each ISL will face the same worst case requirements (in terms of traffic load) one day and has hence to be dimensioned and built with respect to this unique value; as a consequence, all satellites in the constellation, including especially the ISL equipment, will be identical. Capacity requirements on an ISL translate into bandwidth and thus RF power requirements, whereas capacity requirements for a satellite node mainly drive processing power and buffer sizes. Altogether, this translates into DC power requirements as well as into size and weight of single onboard components and finally of the whole satellite, the latter being a major cost factor of the satellite constellation.

Based on these considerations it is straightforward to formulate the two most appropriate target functions for the ISL network dimensioning:

- Minimise the worst case link (WCL) capacity, which is the maximum capacity required on any link at any time.
- Minimise the worst case node (WCN) capacity (being correspondingly defined).

Of course, some combined metric is possible as well. In this paper, we consider the WCL target exemplarily.

The routing/distribution of given demand pair traffics on the available VPCs is either performed heuristically, according to fixed rules, or it can be treated as an optimisation (minmax) problem which is formulated and solved using linear programming (LP) techniques. Assuming that a limited set of alternative VPCs may be used for splitting the traffic between a specific pair of end nodes, the main optimisation parameters are then the splitting factors. The required capacity of a single link at a given step is determined by simply summing up all VPC capacities crossing it.

For the numerical studies, the following three basic routing/splitting approaches have been used:

1. *Equal Sharing (ES)*: This simple but pragmatic approach is based on observations made in earlier research (2) and used mainly for reference purposes. ES means that each OD traffic will be equally distributed on the $k$ best paths, $k$ being fixed for all OD pairs in the topology and over time.

2. *Full Optimisation (FO)*: In contrast to the heuristic approaches presented above, one may consider a dedicated optimisation of a given target function, which is in our case the minimisation of the overall WCL capacity. As already indicated, the major part of this optimisation consists of $S$ dimensioning subtasks, namely minimising the WCL capacity for all steps $s$ independently, and the overall WCL capacity is then taken being the maximum of all minimised per-step WCL capacities.

3. *Bounded Optimisation(BO)*: With FO, we have not introduced any specific constraint on the share of the total traffic that one path is allowed to carry. As a consequence, a single path may convey the complete offered OD traffic alone, whereas other paths may remain empty. Although FO certainly leads to the maximum possible WCL load reduction per step, there are some reasons – e.g. consequences for operation in failure situations, potentially enormous load variations on single links from step to step, etc. – to introduce an additional linear constraint in form of an upper bound $\alpha$ for the normalised share of the OD traffic one path is allowed to carry, with $\alpha = 0 \ldots 1$.

The results presented in the following have been derived for a homogeneous traffic scenario in order to isolate and study the influence of the topology on the dimensioning performance (106).

Figure 3.99 compares the performance of ES, FO and BO with respect to the WCL traffic load reduction vs. the number of alternative OD pairs for routing/splitting, $k$. The first impressive comparison is between the pure ES and FO curves, where FO achieves an improvement of 45 per cent compared to ES in the saturation zone. With $k = 5$ it is already possible to use the full optimisation potential. Concerning BO, as expected, the additional constraint reduces the optimisation potential with respect to the WCL target value. However, with moderate values of $\alpha$ the results come pretty close to the FO ones in the saturation; for $\alpha = 0.6$, the WCL load is less than 10 per cent higher than for FO.

In order to understand *how* the various approaches operate on the topology, it is helpful to study the load variation of single ISLs over the steps. One observes that in the considered M-Star ISL topology the inter-plane ISLs over mid-latitudes are always the critical, i.e. most loaded links. WCL reduction consequently works by shifting traffic away from those links, as illustrated in Fig. 3.100. Again, the superior performance of FO compared to ES becomes obvious; already with a moderate $k = 5$ FO achieves the full optimisation potential.

Finally, Fig. 3.101 presents another impressive view on this behaviour, displaying the ISL load distributions over all ISLs at a certain step. It becomes obvious how traffic from higher loaded links is

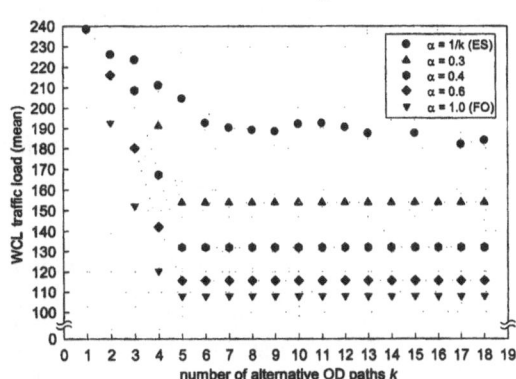

**Figure 3.99.** Mean WCL traffic load versus $k$: performance comparison for different values of $\alpha$ in the range between the two extreme cases, ES and FO.

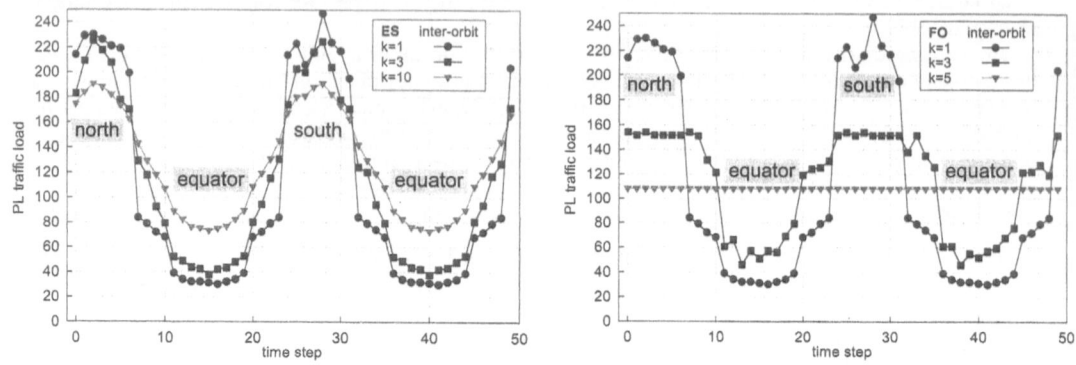

**Figure 3.100.** Traffic load shaping on an inter-plane ISL through the (left) ES, and (right) FO approaches with varying *k*.

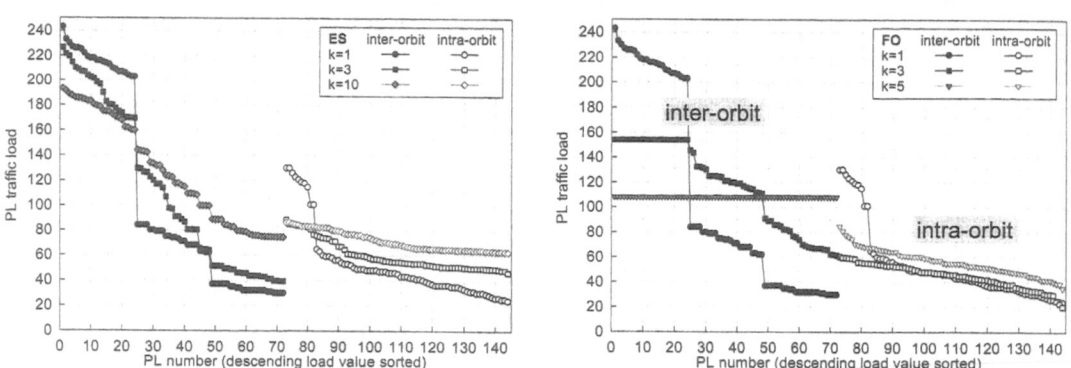

**Figure 3.101** ISL traffic load distribution in the network at step 0 for (left) ES and (right) FO with selected *k*.

shifted to lower loaded ones, and a complete balance is achieved for the inter-plane ISLs when FO with $k = 5$ is used. In reality, of course, a certain traffic is not just shifted from one link to another but OD traffic is shifted to other paths – sometimes to paths with more hops, so that we can expect a growing ISL load in the network with increasing $k$. This is one of the prices to be paid for WCL load reduction/minimisation, and here again FO performs better: whereas ES with a reasonably high $k = 10$ leads to an increase of more than 20 per cent, the effect is nearly negligible for the FO counterpart.

Last, but not least, it is necessary to emphasise that the most important trade-off from a user or QoS perspective is of course between the WCL load minimisation target (which is mainly important from a network performance and system cost viewpoint) and the correlated degradation in terms of path delay encountered. One basically observes from all numerical studies that with increasing optimisation freedom (from ES over BO to FO) not only the WCL load target value, but also the average path delay can be reduced for a given $k$. For appropriate parameter settings, the increase of path delay can be kept fairly low around 5–10 per cent with respect to the reference case of routing each OD traffic completely over the shortest path alone.

# 3.4 Packet-Based Inter-Satellite Link Routing

## 3.4.1 Introduction

Traditionally satellites have been extensively used for broadcast services and telephone trunk links, both taking advantage of satellites positioned in geostationary orbit (GEO). Recently telecommunication industry is facing a challenge to provide a variety of new, broadband multimedia services for users equipped with fixed and mobile terminals. Requirements for higher capacity and lower propagation delay made non-geostationary satellite constellations increasingly interesting, especially with advances in technology which enabled the implementation of intersatellite links (ISLs).

### 3.4.1.1 Performance Evaluation of Adaptive Routing in ISL Networks

Many satellite communication systems have been proposed in the last few years both, for the provision of personal communication services and Internet-in-the-sky. Several of these proposals (e.g. Iridium, Teledesic, Celestri) have already assumed the use of ISLs suitable for traffic interconnection in the satellite segment of the network, but they still lack efficient routing algorithms, adaptive to inherent dynamics of topology and traffic load. Current solutions are mainly reusing algorithms developed and optimised for the use in terrestrial networks with static topology, thus having only limited capability to grasp the characteristics of non-geostationary satellite system. Dynamic topology of satellite networks and variable traffic load in satellite coverage areas, due to the motion of satellites in their orbits, pose stringent requirements to routing algorithms.

Having this in mind we have developed a simulation model of a satellite communication system which allows to test and analyse various algorithms suitable for adaptive routing in ISL networks. This simulation model consists of (i) a traffic flow model, which takes into account geographic distribution of traffic sources and time variation of traffic intensity, and of (ii) an ISL network model.

### 3.4.1.2 Satellite System Dynamics

We have limited the scope of our interest in the first phase to traffic adaptive routing. Thus, in order to avoid the necessity to perform also topology adaptive routing required in polar constellations, we have chosen inclined (rosette) satellite constellation, based on LEO Celestri constellation (107). The selected constellation has been adapted so as to fit the needs of our study; each satellite was assigned four permanent intersatellite links to the nearest satellites.

The parameters of the constellation are summarised in Table 3.14, while the constellation with ISLs is shown in Fig. 3.102.

In order to make the simulation model sufficiently realistic, it has to take into account the temporal dynamics of a non-geostationary satellite system, caused by the revolution of satellites around the earth. Except for satellite or link failures, ISL topology dynamics is deterministic and periodic, and can be easily computed in advance. Similar to other authors (e.g. 108, 109 and 110), we are considering topology dynamics through a set of time-discrete snapshots of satellite positions. The position of satellites at a given snapshot is identified by respective longitude and latitude values of the subsatellite point on the surface of the earth, calculated using equations from (111). Time steps have to be chosen so as to guarantee a reasonably smooth sequence of snapshots. In our Celestri-like LEO

**Table 3.14.** Parameters of the selected satellite constellation

|  | Celestri-like constellation |
| --- | --- |
| Orbit altitude | 1400 km |
| Orbit period | 114 min |
| Orbit inclination $\beta$ | 48° |
| Number of satellites | 63 |
| Number of orbits | 7 |
| Number of ISLs | 2 intra-orbit + 2 inter-orbit |
| Minimum elevation angle $\varepsilon_{min}$ | 16° |

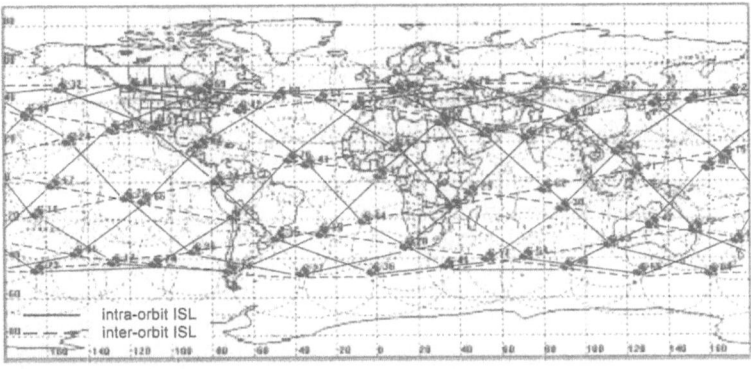

**Figure 3.102.** Snapshot of Celestri-like constellation with four ISLs per satellite.

constellation they have been set to one minute, which is the time in which a satellite moves for approximately 7 per cent of its footprint diameter.

Besides the dynamic topology non-geostationary satellite systems are subject also to the dynamic variation of traffic load with changing satellite location and with time of the day. Given source traffic demand, a proper model for user activity, and a serving satellite selection scheme (considering multiple visibility), the resulting traffic load dynamics is also deterministic and can be computed in advance before the start of the simulation.

Due to periodic changes of topology and traffic load, the whole satellite communication system also exhibits periodicity. Thus in order to take into account all possible combinations of network topology and traffic load it is enough to simulate only one system period, defined in (112) as the smallest common integer multiple of the orbit period and traffic intensity period (usually 24 hours).

### 3.4.1.3 Traffic Flow Model

Besides the selection of the traffic source model, most appropriate for the type of services to be provided by the system, a suitable traffic model for non-geostationary satellite system has to consider also the geographic distribution of traffic sources and daily variation of their traffic intensity. Uniform models for both geographic distribution of traffic sources and for daily variation of traffic intensity are useful for studying the inherent characteristics of the selected satellite constellation. For the evaluation of the system performance in dynamic conditions, however, one has to use a more precise model for geographic distribution of relative traffic intensity, which can be used to generate cumulative traffic on each satellite, proportionally to the traffic intensity in its coverage area. In this study we have only considered uniform model for geographic distribution of traffic sources.

Destinations of traffic flows can be defined using the source-destination traffic flow tables between the earth's six continental regions. After the definition of destination region, the traffic is divided among the satellites proportionally to their coverage in that region. Different tables have to be defined for different types of services, because telephony traffic for example certainly exhibits different behaviour than www traffic. Telephony traffic mainly remains within the originating country, and international traffic is mainly terminated within the originating continent. WWW traffic, on the other hand, is expected to have a significantly larger international share with pronounced hot spots (in terms of amount of sources and/or destinations) as they can be already observed today within or being terminated in North America and some specific European countries.

Time variation of traffic load in a non-geostationary satellite system is first of all caused by two phenomena on the user activity side, namely (i) daily variation of traffic load due to local time of the day, and (ii) geographical variation of just this daily load behaviour according to geographical time zones. Daily variation can be taken into account with an appropriate daily profile curve, while for geographical time zones a simplified model can be considered, which increments the local hour each $360/24 = 15$ degrees longitude. In addition to user activity dynamics, which is a somewhat general feature of any (global) communications system, another contribution and specific challenge for LEO satellite systems is the rapidly changing satellite visibility, and consequently active users' coverage, on ground.

### 3.4.1.4 ISL Network Model

The ISL network simulator for performance evaluation of routing algorithms should not be restricted to a certain communication protocol or connection mode. Furthermore, it has to consider the specific characteristics of a packet or cell switching communication system, especially those having considerable impact on performance of different routing algorithms. Consequently, the simulation model has been built on the packet level, which increases the system complexity and computational effort, but on the other hand it allows studying adaptive routing algorithms considering the actual status of the network. Such network status is usually described in a suitable performance parameter, such as delay or congestion, where the corresponding metric can be calculated/evaluated by a node either directly from the incoming/outgoing packets and/or queues, or from dedicated status information being exchanged with neighbouring nodes. The ISL network simulator has been realised in BONeS Designer, a powerful event driven software simulation tool developed for the simulation of communication systems. In order to use this tool for its main purpose, which is, as mentioned above, the simulation of communication systems, satellite system specific dynamics (ISL topology dynamics and traffic load variation) has been implemented separately, calculating all deterministic changes in advance. Two input files have been prepared for the main simulator based on topology snapshots. One file contains matrices of propagation delay between satellites, simply calculated from the position of satellites at a given snapshot, and the other file contains matrices of normalised traffic load between all possible source-destination satellite pairs.

In the ISL network simulator a limited functionality has been implemented of the first four OSI layers: physical layer, data link layer, network layer and transport layer.

In the transport layer the traffic generator is implemented, which uses imported matrices with normalised traffic load between all satellite pairs. Thus, according to the traffic intensity of its coverage area, each satellite generates packets with all the necessary data for routing and for analysis. The transport layer also terminates packets which have reached the destination, and generates reply packets according to the selected symmetry-shaping model, which enables studying the performance of routing algorithms under both symmetric or asymmetric traffic load.

The most important task of the network layer is the computation of routing tables and routing of packets towards their destination. The next node on the shortest path is determined from the currently valid routing table, which is calculated in the network layer with the selected routing algorithm, which receives the information about the network status (link cost) from the data link layer. Currently only best effort routing has been implemented using a centralised version of the Dijkstra shortest path algorithm.

In the data link layer packets are actually sent to and received from ISL. Before being sent to the appropriate intersatellite link, packets are put in a FIFO queue which enables a distinction of different packet priorities. Data link layer also gathers information about the link cost, required for calculation of routing table in networking layer. A link cost metric is composed of traffic load and the propagation delay on the link ($D$propagation). Their relative impact on the link cost (LC) is linearly regulated with a traffic weight factor (TWF) and a propagation delay weight factor (PDWF), respectively, as shown in Eq. 3.53. In fact, the impact of the traffic load on link cost is considered via the delay in the satellite output queue ($D$queuing).

$$LC = PDWF \cdot D_{\text{propagation}} + TWF \cdot D_{\text{queuing}} \tag{3.53}$$

On the physical layer only the propagation delay between satellite pairs connected with ISLs is taken into account. Actually, the propagation delays are computed for all snapshots in advance and imported in the simulator as a propagation delay matrix.

### 3.4.1.5 Simulation Parameters and Results

The main motive for the development of the ISL network simulator was to study adaptive routing in the highly dynamic non-geostationary satellite systems. The simulator enables us to evaluate the effect of using different assumptions regarding the distribution of traffic sources and destinations, the user daily activity profile, the source-destination traffic flow tables, the traffic source model, symmetry shaping, and the parameters and weight factors for the link cost calculation.

The ISL network simulator has many simulation parameters to study the performance of the simulated routing algorithm under various network conditions. The most important simulation parameters, and their values used for obtaining the simulation results presented below, are listed in Table 3.15.

The simulator can provide various results, such as average delay experienced by packets in the network, average number of hops between origination and destination satellite node, total number of sent and received packets, and normalised link load. The representative simulation results reported in the following for the selected Celestri-like LEO constellation have been obtained using:

- Uniform distribution of traffic sources and destinations over the surface of the earth.
- A uniform source-destination traffic flow table.
- A constant daily profile curve.
- Poisson traffic source generator.
- Symmetry shaping for an unidirectional traffic flow (RF = 0) and a symmetric bi-directional traffic flow (RF = 1).
- Dijkstra shortest path algorithm for central calculation of routing tables.
- The duration of the simulation run equal to one orbit period.

First, average packet delay and average number of hops in the Celestri-like network are presented for different values of RF and TWF in Figs 3.103 and 3.104, respectively. In order to make a fair comparison between the results for the simulation runs with (RF = 1) and without (RF = 0) replying packets, twice as many packets have been generated in case of RF equal to zero, thus guaranteeing the similar traffic load in the network except for the symmetry. The results show that the traffic symmetry under assumed conditions does not have any significant influence on either average delay or average number of hops. The increase of relative impact of traffic load on the link cost for the

**Table 3.15.** Simulation parameters

|  | Celestri-like constellation |
| --- | --- |
| Reply factor | {0, 1} |
| Traffic weight factor | {0, 1, 10} |
| Propagation delay weight factor | 1 |
| Mean packet length | 1000 bit |
| Maximum packet length | 10 000 bit |
| Link capacity | 100 000 bit/s |
| Total network source traffic | 100/200 packets/s |
| Simulation duration | 6840 s |
| Snapshot duration | 60 s |
| Number of snapshots | 114 |
| Refresh cost matrix window size | 30 s |
| Refresh routing rate | 30 s |

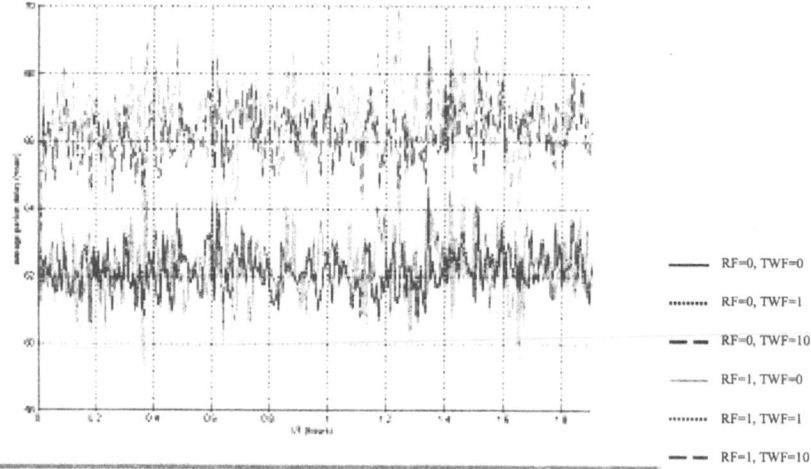

**Figure 3.103.** Average packet delay in the Celestri-like network for asymmetric (RF = 0) and symmetric RF = 1) traffic load.

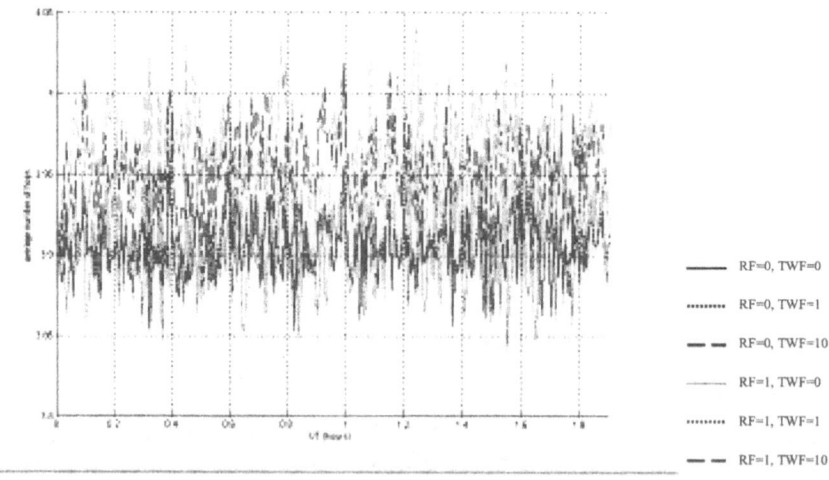

**Figure 3.104.** Average number of hops in the Celestri-like network for asymmetric (RF = 0) and symmetric RF = 1) traffic load.

**Table 3.16.** Queuing delay statistics for all ISLs

|  |  | Mean | Standard | Minimum | Maximum |
|---|---|---|---|---|---|
| RF = 0 | TWF = 0 | 0.431 ms | 0.445 ms | 0 ms | 5.349 ms |
|  | TWF = 1 | 0.431 ms | 0.446 ms | 0 ms | 5.349 ms |
|  | TWF = 10 | 0.4854 ms | 0.523 ms | 0 ms | 18.22 ms |
| RF = 1 | TWF = 0 | 0.457 ms | 0.464 ms | 0 ms | 5.526 ms |
|  | TWF = 1 | 0.458 ms | 0.466 ms | 0 ms | 5.526 ms |
|  | TWF = 10 | 0.484 ms | 0.556 ms | 0 ms | 17.0 ms |

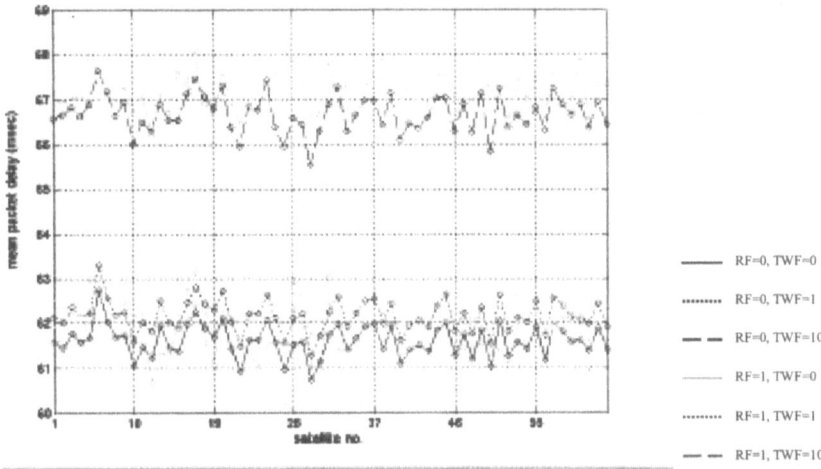

**Figure 3.105.** Mean delay of received packets on satellites in one orbit period for asymmetric (RF = 0) and symmetric (RF = 1) traffic load.

calculation of routing tables, however, has slightly more noticeable effect to results for average delay and average number of hops. The differences for different values of TWF factor are still negligible, which is a consequence of the high connectivity of the ISL network, providing several routes between a pair of nodes with the same number of hops, and thus similar propagation delay.

The impact of TWF to the link cost assuming symmetric or asymmetric traffic load can be estimated using Eq. 3.53, taking into account typical val es for propagation delay in the selected constellation between 10.3 ms and 19.7 ms, and the statistics for the queuing delay summarised in Table 3.16.

Apparently, the traffic load in case of TWF = 10 can have a higher impact on the calculation of routing tables than propagation delay, thus enabling routing via longer but less burdened links. This is confirmed by the results in Fig. 3.105 showing the mean delay of packets received in one orbit period by each of 63 satellites in the network. Packets received by a particular satellite in case of TWF = 10 exhibit longer delays than in case of TWF equal 0 or 1 regardless of traffic symmetry.

Furthermore, the variation of queuing delay as reported in Table 3.16 indicates that by using fixed weight factors the ratio between traffic load and the propagation delay in the link cost metric is changing during the simulation run. In order to keep this ratio within certain limits weight factors should be adaptive to topology and traffic load variations.

Figures 3.106, 3.107 and 3.108 depict satellites with the highest normalised link load for symmetric traffic load (RF = 1) and for different values of traffic weight factor in dependence of time and of the latitude of the subsatellite point. Link load is calculated from the throughput of the ISL in certain time-step, normalised with the link capacity. Similar results have been obtained also for asymmetric traffic load (RF = 0). Table 3.17 summarises results for the highest normalised link load identifying the ISL, the time of the occurrence of the peak, and the location of the corresponding satellite.

The results for TWF = 0 in Fig. 3.106 show the fluctuation of normalised link load due to updating of routing tables considering only propagation delay. A significantly higher fluctuation is experienced in case of a high impact of traffic load on the link cost. Results in Fig. 3.108 for TWF = 10 show unstable network conditions, since routing tables are changing significantly after every update interval

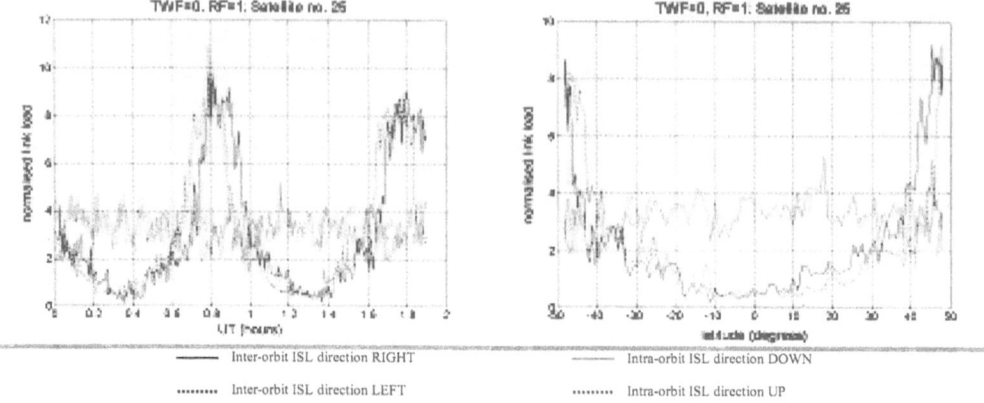

**Figure 3.106.** Satellite with the highest normalised link load for TWF = 0 and RF = 1.

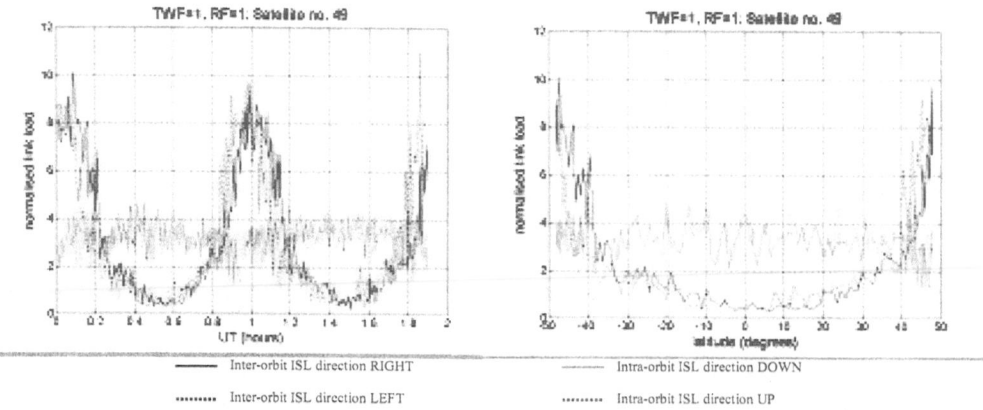

**Figure 3.107.** Satellite with the highest normalised link load for TWF = 1 and RF = 1.

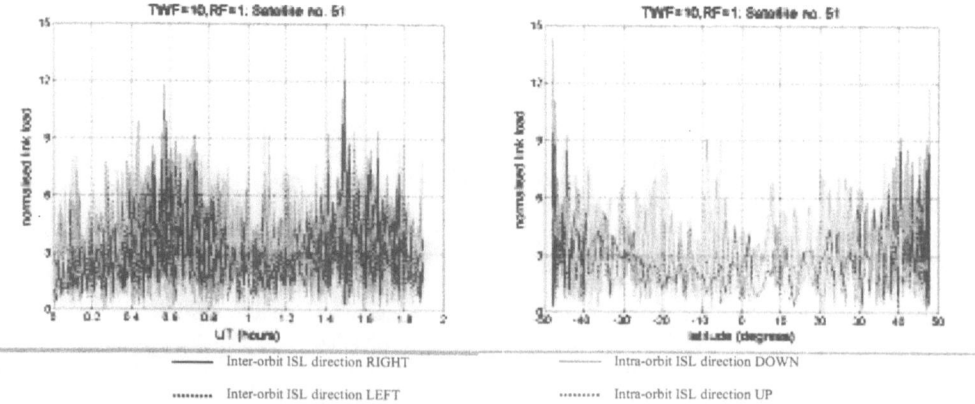

**Figure 3.108.** Satellite with the highest normalised link load for TWF = 10 and RF = 1.

**Table 3.17.** The highest normalised link load for different values of TWF in case of symmetric traffic load (RF = 1).

| TWF | Value | The highest normalised link load | | |
| | | ISL | Universal time | Latitude |
|---|---|---|---|---|
| 0 | 10.887 | inter-orbit left (Sat 25 to Sat 17) | 0h 48min | 47.93° |
| 1 | 11.054 | inter-orbit left (Sat 49 to Sat 41) | 1h 52min | −46.07° |
| 10 | 14.383 | inter-orbit right (Sat 51 to Sat 59) | 1h 30min | −47.85° |

equalling 30 s. Figures 3.106, 3.107 and 3.108 also indicate that regardless of simulation parameters the highest normalised link load is always achieved on the inter-orbit links (left or right ISLs). The periodicity, distinctively demonstrated first of all for inter-orbit links in case of TWF = 0 and TWF = 1, indicates the impact of satellite location on routing tables. Results for low impact on link cost of traffic load relative to propagation delay (TWF equal 0 and 1) clearly indicate that most of the traffic tends to be routed between orbits at higher latitudes, where the distance between the neighbouring satellites decreases. Results for TWF = 10, on the other hand, show higher link utilisation also for intra-orbit links and at the latitudes around equator, due to lower impact of propagation delay on the calculation of routing tables.

Furthermore, as shown in Figs 3.106, 3.107 and 3.108, in case of uniform distribution of traffic sources and uniform source-destination traffic flow tables, the link utilisation is periodic with half of the orbit period, or with 180° latitude.

### 3.4.1.6 Conclusions

This contribution addresses the problems of routing in the ISL segment of non-geostationary satellite systems, characterised by dynamic network topology and fluctuating traffic load experienced by a particular satellite. As such, these systems are well suited for the implementation of adaptive routing algorithms. In order to estimate and compare the performance of different routing algorithms we have realised a simulation model of a satellite communication system with intersatellite links. The model consists of the ISL network model realised on the packet level and a model for generation of traffic flow between satellites, which considers the motion of satellites, the geographic distribution of traffic sources, and the daily variation of traffic intensity. Currently we have implemented a Celestri-like LEO satellite constellation in the ISL network model, using a centralised version of the Dijkstra shortest path algorithm for calculation of routing table updates.

The simulation results obtained for different assumptions regarding traffic asymmetry and the impact of traffic flow on link costs are presented, considering uniform geographic distribution of traffic sources, a constant daily profile curve, a uniform source-destination traffic flow table, and a Poisson traffic source generator. The motivation for using such idealised (uniform) assumptions in a first step lies in the fact, that only such investigations reveal the inherent characteristics of the selected satellite constellation clearly enough. This is of great importance especially for further envisaged studies, enabling us to distinguish between intrinsic effects of the selected constellation and effects introduced with more realistic assumptions regarding the traffic model. Also, the comparison of these results with future results obtained with more realistic traffic models is expected to show the level of detail that needs to be considered in the traffic model for comparison of different adaptive routing algorithms, thus also indicating which effect causing traffic variation is the most important for traffic adaptive routing.

## 3.5 MAC Protocols for Satellite Multimedia Networks

### 3.5.1 PRMA Protocols in Low Earth Orbit Mobile Satellite Systems

Future global-coverage mobile networks will integrate a terrestrial cellular component with a *Mobile Satellite System* (MSS) (113, 49, 11). This research activity has focused on *Low Earth Orbit* – MSSs (LEO-MSSs) that will be able to provide mobile users with multimedia traffic worldwide (114).

Two solutions have been proposed for the implementation of a LEO-MSS able to cover all the earth (115): (*i*) *satellite-fixed cells* (adopted by the IRIDIUM system), where cells are fixed with respect to satellites and move as regards the earth; (*ii*) *earth-fixed cells* (used by the TELEDESIC system), where cells are fixed on the earth and satellite antenna spot-beams are steered so that they point to the same area on the earth as long as possible. The following study assumes an earth-fixed cell system.

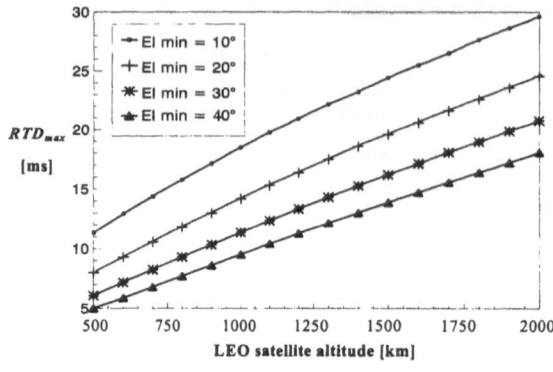

**Figure 3.109.** Range of possible $RTD_{max}$ values for different LEO satellite altitudes.

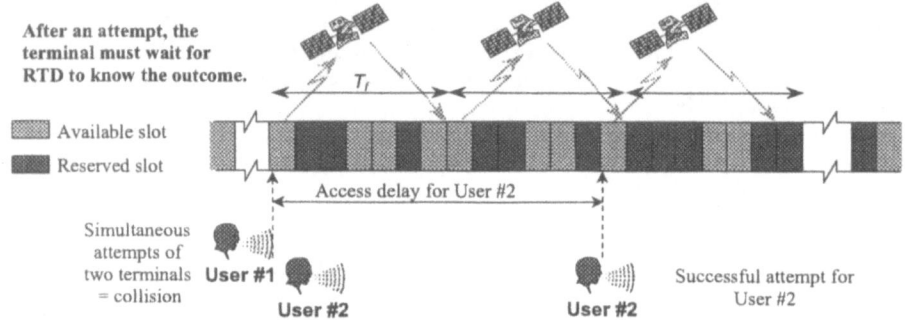

**Figure 3.110.** Description of the access phase in the PRMA protocol in LEO-MSSs.

Therefore, we neglect the impact of terminal mobility on the performance of the protocols considered in this work.

Satellites are both power and frequency limited. Hence, suitable *Medium Access Control* (MAC) protocols must be designed in order to share the LEO satellite resources among the highest number of simultaneous transmissions, each with specific *Quality of Service* (QoS) requirements. In particular, we consider the *Packet Reservation Multiple Access* (PRMA) scheme to support *Variable Bit Rate – Real Time* (VBR-RT) voice traffic and *Available Bit Rate* (ABR) data traffic in LEO-MSSs (116).

The classical PRMA scheme has been proposed for terrestrial microcellular systems and it is based on Time Division Multiple Access (TDMA) and combines a random access with a slot reservation scheme (117, 118, 119 , 120, 121, 122, 123). The efficiency of PRMA relies on managing voice sources with speech activity detection: only during a talkspurt, a voice source has reserved one slot per frame to transmit its packets; when there is a silent pause, this slot can be destined to another active source. A feedback channel broadcast by the cell controller informs the terminals about the state of each slot (i.e. *idle* or *reserved*) in a frame. As soon as a new talkspurt is revealed, the terminal attempts to transmit a packet in the first idle slot (*contending state*), according to a *permission probability* scheme. If more terminals attempt to send their packets on the same slot, there is a collision (we neglect the capture effect (123)): these terminals must reschedule new attempts. As soon as the transmission attempt of a terminal is successful on a slot, the terminal attains the reservation of this slot in subsequent frames.

The main limiting factor for the use of PRMA in LEO-MSSs is the high Round Trip propagation Delay (RTD) that prevents the mobile terminals on the earth to know immediately the outcome of their transmission attempts. In LEO-MSSs, RTD values vary from 5 ms to 30 ms, depending on the satellite constellation altitude and the minimum elevation angle for mobile terminals (see Fig. 3.109).

If a collision occurs during the random access phase, the terminal knows that it must reschedule a new transmission only after RTD (see Fig. 3.110). This delay is particularly significant for the real-time voice service. In order to relax this drawback, we have also proposed a modified PRMA protocol (4, 124, 125) named PRMA scheme with Hindering States (PRMA-HS).

Let us consider two types of User Terminals (UTs): that is Voice Terminals (VTs) and Data Terminals (DTs). This differentiation is functional rather than physical, since we could have that a

**Table 3.18.** System parameter values

| Parameter | Definition | Value |
|---|---|---|
| Rc | Channel bit-rate | 765 kbit/s |
| Rs | Source bit-rate | 32 kbit/s |
| H | Packet header length | 64 bit |

VT and a DT are integrated in the same terminal. We assume that $M_v$ VTs and $M_d$ DTs share the access to the same PRMA carrier in a cell. VTs and DTs use different permission probabilities to access the shared channel to take into account their different service priorities. Let $p_v$ denote the permission probability for VTs and $p_d$ the permission probability for DTs. In general, we must have that $p_v > p_d$.

Every terminal has a buffer to store packets. A VT discards the first packet from its buffer, when it experiences a delay to be transmitted greater than a maximum value, $D_{max}$; typically, $D_{max} = 32$ ms (123). When a packet is discarded, the VT tries to obtain a reservation with the next packet. Let $P_{drop}$ denote the packet dropping probability for a VT; it is required $P_{drop} \leq 1$ per cent for an acceptable speech quality. As for DTs, the service is characterised by the average message delay, $T_{msg}$, that is the time from the message arrival instant to the DT buffer to the instant when this message is completely sent.

### 3.5.1.1 Protocols Description

Let $T_f$ denote the frame duration, $Rc$ the transmission bit-rate in the channel, $Rs$ the voice source bit- rate. The number of slots per frame and the packet length are dimensioned on the basis of VT characteristics: an active VT generates a packet in a frame; each packet is composed of $L_{pkt} = R_s T_f$ information bits and $H$ header bits. The same packet length is used for DTs. The number of slots per frame is $N = \lfloor R_c T_f / (L_{pkt} + H) \rfloor$, where $\lfloor x \rfloor$ denotes the greatest integer number lower than or equal to $x$, and the slot duration is $T_s = T_f / N$. The values assumed in this work for parameters $R_c$, $R_s$ and $H$ are shown in Table 3.18.

For each VT we consider a Slow Speech Activity Detector (SSAD) (123) that distinguishes between talkspurts and silent pauses within a conversation. The durations of talking phases and silent ones are exponentially distributed with mean values $t_1 = 1$ s and $t_2 = 1.35$ s, respectively. The voice activity factor $\psi_v$ is given by $t_1/(t_1 + t_2) \approx 0.425$; the voice throughput is $\eta_v = M_v \psi_v (1 - P_{drop})/N$ pkts/slot.

We assume that DTs adopt an *exhaustive discipline* to manage the transmissions of the messages stored in their buffers (5): a DT releases its slot reservation only when its buffer is idle. The service envisaged for DTs is loss-less. We assume that all the packets of the same message arrive simultaneously and synchronised with slot times. The input traffic to each DT is modelled as follows: each DT receives messages independently of other DTs and according to a Poisson arrival process with mean rate $\lambda$ msgs/s. The message length in packets, $l_d$, is geometrically distributed with expected value $L_d = 1/\gamma_{fd}$:

$$\text{Prob. \{message length } l_d = n \text{ pkts}\} = \gamma_{fd} (1 - \gamma_{fd})^{n-1}, n = 1, 2, \ldots \qquad (3.54)$$

The mean message length (only information part) in bits, $L_{d,bit}$, is related to $L_d$ as:

$$L_{d,bit} = L_d L_{pkt} \text{ (bits/msg)} \qquad (3.55)$$

According to (126), this model is suitable to represent e-mail traffic.

Let $M_{v,max}$ denote the maximum number of VTs which can be served by a PRMA carrier (supporting also DT traffic) with $P_{drop} \leq 1$ per cent. Moreover, $\lfloor (R_c - \lambda L_{d,bit} M_d)/R_s \rfloor$ represents the maximum number of slots that can be used by VTs on an ideal TDMA carrier (i.e. without packet overhead). We define the VT *multiplexing gain*, $\mu_v$, expressed in conversations/voice channel as the ratio between $M_{v,max}$ and $\lfloor (R_c - \lambda L_{d,bit} M_d)/R_s \rfloor$. If $\mu_v > 1$ conversations/voice channel, PRMA allows a more efficient VT management than the ideal TDMA scheme.

For a conservative performance evaluation of both PRMA and PRMA-HS protocols in LEO-MSSs, we assume RTD equal to its maximum value for a given satellite constellation, $RTD_{max}$. The satellite acknowledges the packet used for the reservation as soon as it correctly decodes its header. We assume $T_f = nRTD_{max} + \varepsilon$, with $n \geq 1$ and $\varepsilon$, the packet header transmission time equal to $H/R_c$. Hence, a UT knows the outcome of its transmission attempt on a slot before the same slot in the next frame. Since $\varepsilon << RTD_{max}$, we have $T_f \approx nRTD_{max}$. We consider for $n$ only those integer values that divide $N$; then, $RTD_{max}$ contains an integer number of slots, i.e. $N/n$ slots.

With the classical PRMA scheme, assuming a typical LEO system where $RTD_{max}$ is equal to

$D_{max}/2 = 16$ ms, a VT could perform at most two access attempts before dropping the first packet. To remove this constraint, the PRMA-HS protocol allows that a UT may perform new access attempts also while it waits for receiving the outcome of a previous attempt. The following two cases are possible: (i) the UT makes a successful transmission attempt on a slot, the UT leaves the contending state and enters a block of $N/n$ states (i.e. *hindering states*) that model the delay $RTD_{max}$ to receive the positive acknowledgement from the satellite; during this time the UT can still try to transmit on available slots, but these attempts are useless and may hinder the accesses of other UTs. (ii) The UT attempt has been unsuccessful; hence, the UT remains in the contending state, but it can retry to transmit the same packet on successive idle slots without waiting for the acknowledgement from the satellite. We have obtained that the high attempt rate of PRMA-HS allows a capacity advantage with respect to PRMA that overcomes the hindering contentions described at point (i). Note that PRMA-HS coincides with the PRMA scheme proposed in (123), if RTD $<< T_s$. If more transmission attempts are successful for the same UT, the satellite acknowledges only a reservation of one slot per frame. A special end-of-reservation flag is set by the UT in the header to notify the satellite the release of a reservation in the next frame.

The behaviour of each VT is described by the state diagram shown in Fig. 3.111. An active VT stays in the SIL state as long as it has no speech packet to transmit. As soon as the first packet of a talk-spurt is generated the VT enters the CON state. It remains in the CON state until its attempt is successful; in this case, the VT leaves the CON state and enters the block from $HIN_{N-1}$ to $HIN_{N-N/n}$; these states model the delay to know the positive outcome of a transmission attempt (i.e. $RTD_{max}$ = $N/n$ slots). While the VT is in $HIN_i$ states, it may continue to attempt transmissions on available slots even if it has already obtained a reservation, because the positive acknowledgment is received by the VT only after the $RTD_{max}$ time. From the VT standpoint, CON and $HIN_i$ states are indistinguishable. This justifies the introduction of the *global contending state* (GCON) in Fig. 3.111. When the VT receives a positive acknowledgement, the VT leaves the hindering states and must wait for $N-N/n-1$ slots to transmit the subsequent packet of the talkspurt or to release the access to the channel if the talkspurt is ended. In order to take into account this time, a special block of reservation states ($RES_i'$), which is different from the main chain of reservation states ($RES_i$), has been considered. The overall time spent in the hindering states, $HIN_i$, plus the time spent in the block of reservation states, $RES_i'$, must be equal to $N$ slots. Whenever a VT leaves the $RES_i'$ states, with still voice packets to transmit, it enters the loop from $RES_{N-1}$ to $RES_0$. The behaviour of the VT when it is in the block of $RES_i$ states is the same as in the PRMA protocol. In the VT state diagram we have also considered the backward transition from the CON state to the SIL one to take into account that a talkspurt may end before obtaining a reservation.

The transitional probabilities that appear in Fig. 3.111 are defined as follows: $\sigma_v$ is the probability that a silent gap ends within $T_s$; $\gamma_v$ is the probability that a talkspurt ends during $T_s$; $\gamma_{fv}$ is the probability that a talkspurt ends within $T_f$; $u_v$ is the probability that an attempt is successful; $a_v$ is the probability that a slot is used for an attempt.

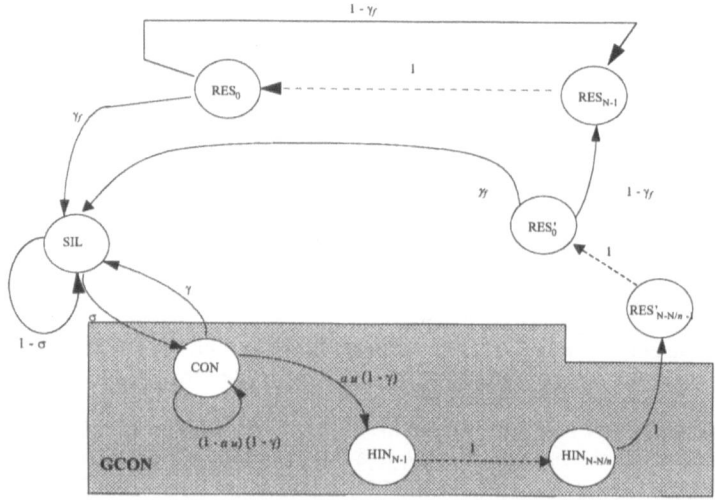

**Figure 3.111.** State diagram of a voice terminal with the PRMA-HS protocol.

Let us discuss about the stability of these protocols. Since a packet dropping mechanism is considered for VTs, the VT buffers can not be unstable. Whereas, in the case of the data sub-system we have that the data throughput $r_d$ must be equal to the mean input traffic in pkts/slot in order to guarantee the DT buffer stability: $r_d = \lambda T_s L_d M_d$ pkts/slot. Hence, we have the following stability condition to be fulfilled by the data traffic:

$$\lambda T_s L_d M_d + \eta_v < 1 \quad \left[\frac{\text{pkts}}{\text{slot}}\right] \tag{3.56}$$

### 3.5.1.2 Results

We have considered very long simulation runs in order to achieve an accurate and reliable estimations. In particular, we have selected a duration of $50 \times 10^{6}$ slots which allows 5 per cent confidence interval for the estimations of $P_{\text{drop}}$ and $T_{\text{msg}}$. The results shown hare been obtained by considering optimised choices of the permission probabilities and the frame duration (4, 5, 124, 125, 127). Finally, all the cases fulfill the condition described by Eq. 3.56 for the stability of the system.

Let us first examine voice-only cases. Fig. 3.112 shows the PRMA multiplexing gain $\mu v$ as a function of $T_f$ with $p_v = 0.4$ for both the terrestrial and the satellite scenario (4, 124, 125). The system parameter values in Table 3.18 have been assumed. Therefore, an increase in $T_f$ causes longer packets. The behaviour of $\mu_v$ in Fig. 3.112 can be explained as follows:

- A low value of $T_f$ implies to reduce the slot duration by assuming fixed values of $R_c$ and $R_s$. Therefore, a VT needs a large number of reserved slots to transmit its talkspurt. This entails a reduced number of slots available for transmission attempts by new active VTs. In this case, the multiplexing capabilities of the PRMA protocol are significantly reduced and $P_{\text{drop}}$ increases.
- An excessive value of $T_f$ causes a worse system performance, because it leads to reduce the maximum number of attempts available for a contending VT before dropping a packet. If $T_f > 40$ ms, the multiplexing gain $\mu v$ suddenly reduces and undergoes 1 conversation/channel, i.e. the PRMA protocol loses its advantages as regards TDMA.

In Fig. 3.112, we can note a significant difference between the terrestrial case and the satellite one for high values of $T_f$: in the satellite case we have a more evident reduction of $\mu_v$ when $T_f$ exceeds 40 ms. This difference is due to the high value of RTD which significantly reduces the number of possible attempts within $D_{\text{max}}$. However, in both the terrestrial case and the satellite one, $\mu_v$ has a quite flat maximum for $T_f$ around 15 ms. Moreover, it is important to note that, according to Fig. 3.112, the PRMA protocol is suitable for application in LEO systems (i.e. $\mu_v > 1$ conversations/ channel), but not in Medium Earth Orbit – MSSs (MEO-MSSs), where $RTD_{\text{max}} > 70$ ms.

Figure 3.113 presents the comparison between PRMA and PRMA-HS in terms of $P_{\text{drop}}$ in LEO-MSSs for both $n = 3$ (i.e. $RTD_{\text{max}} = 5$ ms) and $n = 1$ (i.e. $RTD_{\text{max}} = 15$ ms) with $p = 0.4$ and $T_f = 15$ ms (4, 125).

In this figure we note that the performance difference between PRMA and PRMA-HS is slight for $n = 3$ and significant for $n = 1$. Hence, we may state that the PRMA performance strongly depends on $n$: an increase in RTD leads to a worse behaviour. Whereas, PRMA-HS is less sensitive to variations of RTD. This is an interesting result that makes this protocol quite insensitive to the variations

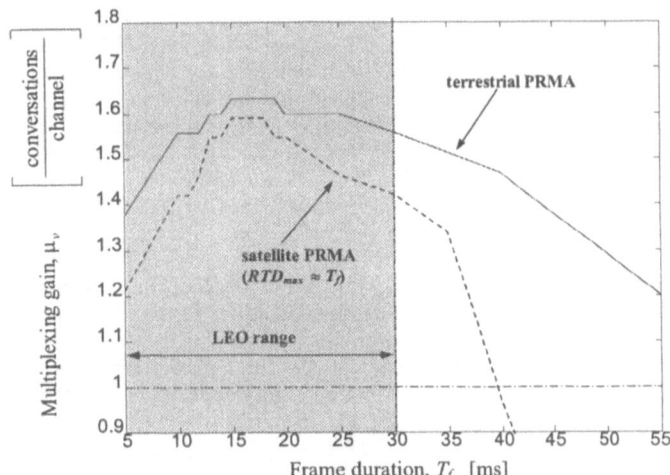

**Figure 3.112.** Parameter $\mu_v$ as a function of $T_f$ with $p = 0.4$ for both the terrestrial and the satellite case.

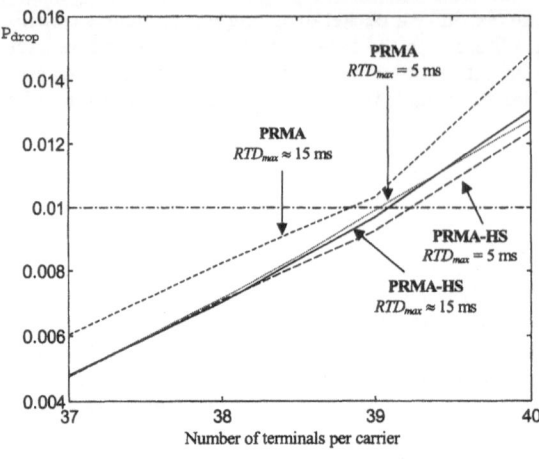

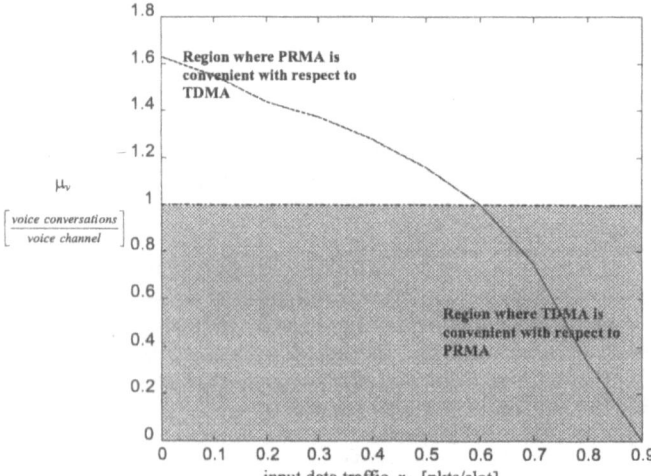

**Figure 3.113.** Comparison in terms of $P_{drop}$ between the PRMA-HS protocol and the original PRMA scheme in LEO-MSSs with $p = 0.4$, $T_f = 15$ ms, $n = 3$ (i.e. $RTD_{max}$ = 5 ms) and $n = 1$ (i.e. $RTD_{max} \approx 15$ ms).

**Figure 3.114.** Behaviour of $\mu_v$ as a function of $r_d$ for $L_{d,bit}$ = 25 kbit/msg, 20 DTs/carrier, $p_v = 0.6$, $p_d = 0.2$ and $T_f$ = 16 ms.

of RTD experienced in LEO systems during call lifetime, mainly due to the motion of LEO satellites. Moreover, the PRMA-HS performance obtained for $n = 1$ (i.e. $T_f \approx RTD_{max}$) can be considered as a conservative evaluation of the PRMA-HS performance for any $RTD_{max}$ value less than $T_f$.

Let us focus now on voice and data transmissions with PRMA (5). Figure 3.114 shows the behaviour of $\mu_v$ as a function of $r_d$ for $L_{d,bit}$ = 25 kbit/msg, 20 DTs/carrier, $p_v = 0.6$, $p_d = 0.2$ and $T_f$ = 16 ms (note that these results have been obtained for slightly different parameter values with respect to Table 3.18; in particular, $R_c$ = 720 kbit/s). We can note that $\mu_v$ decreases as $r_d$ increases. This behaviour can be justified as follows: $\mu_v$ depends on the number of slots devoted per frame to support voice transmissions; when $rd$ increases we have a reduced number of slots for voice transmissions and, therefore, a low $\mu_v$ value. For $r_d \leqslant 0.6$ data pkts/slot, we have $\mu_v \geqslant 1$. In these conditions, PRMA allows a more efficient management of VTs than TDMA. Moreover, when $r_d = 0$ (i.e. no data traffic) we obtain $\mu_v \approx 1.6$ voice conversations/voice channel, which is very close to the $\mu_v$ value of terrestrial microcellular systems with only voice sources and optimised parameter values (123). This result highlights that the high RTD value in LEO-MSSs as regards that experienced in microcellular systems does not change the multiplexing capabilities of the PRMA protocol.

In Fig. 3.115 we present simulation results for PRMA in a terrestrial cellular system and in a LEO satellite scenario as a function of the input data traffic, $r_d$ (i.e. the mean number of data packets arrived per slot), in order to evaluate the impact of RTD on the performance of the PRMA protocol in terms of both $P_{drop}$ and $T_{msg}$ (5). In performing these simulations we have assumed an *exhaustive discipline* for the service of DTs, $T_f = RTD_{max}$ = 16 ms, 16 VTs/carrier, 16 DTs/carrier, $p_v = 0.35$, $p_d = 0.15$, $L_d = 20$ pkts/msg and the parameter values shown in Table 3.18.

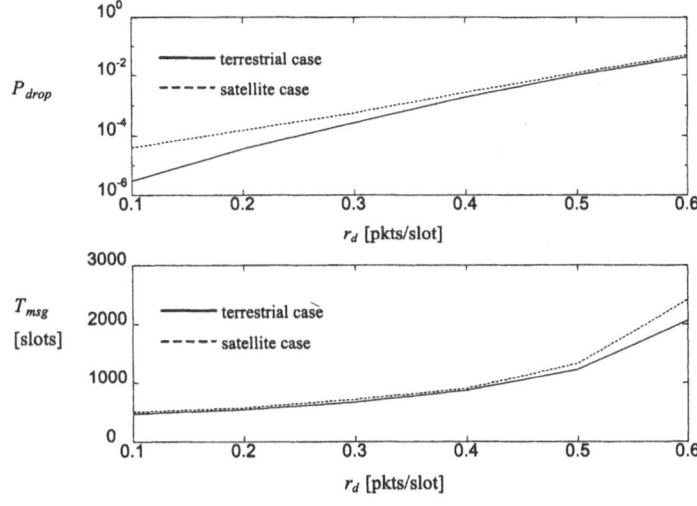

**Figure 3.115.** Comparison between the performance of the terrestrial PRMA (*continuous line*) and the satellite PRMA (*dashed line*).

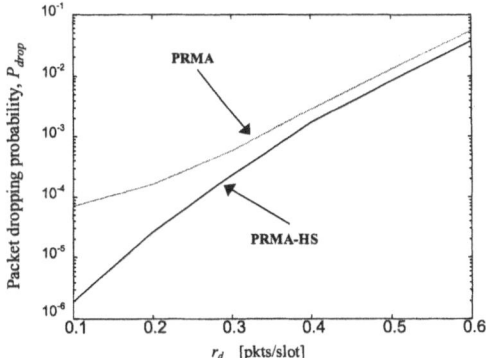

**Figure 3.116.** Behaviour of $P_{drop}$ for PRMA and PRMA-HS as a function of the input data traffic $r_d$, for $T_f = 15$ ms, $n = 1$, $p_v = 0.4$, $p_d = 0.2$, 16 VTs/carrier, 16 DTs/carrier.

**Figure 3.117.** Behaviour of $T_{msg}$ for PRMA and PRMA-HS as a function of the input data traffic $r_d$, for $T_f = 15$ ms, $n = 1$, $p_v = 0.4$, $p_d = 0.2$, 16 VTs/carrier, 16 DTs/carrier.

We have obtained that the differences between terrestrial and satellite cases are small: the maximum $r_d$ value (constraint $P_{drop} \leqslant 1$ per cent) is around 0.5 pkts/slot in both the terrestrial case and the LEO one. Correspondingly, there is a difference of few slots (practically, RTD) in $T_{msg}$ between these two cases. Hence, the PRMA protocol manages voice and data traffics in LEO-MSSs with a quality of service very close to that obtained in terrestrial systems.

Finally, we examine the PRMA-HS performance in supporting voice and data transmissions (5). Figs 3.116 and 3.117 respectively show $P_{drop}$ for voice packets and the mean message delay (in slots), $T_{msg}$, as functions of the input data traffic $r_d$. We have considered 16 VTs/carrier and 16 DTs/carrier with the system parameter values given in Table 3.18 and $L_d = 20$ pkts/msg. From Figs 3.116 and 3.117 it is evident that the PRMA-HS protocol permits a higher value of $r_d$ under the constraint $P_{drop} \leqslant 1$ per cent and a lower $T_{msg}$. Hence, PRMA-HS attains a higher resource utilisation than PRMA.

Finally, we have also evaluated the performance of the PRMA scheme in the presence of a mixed traffic, where for the data traffic we have used a different model with respect to that described in Section 2. In (127) a new data source has been defined which permits to model interactive traffic (Internet-like). In particular, a Modulated Markov Process (MMP) generator has been considered, which has two message sub-generators with mean arrival rates $\lambda_A$ and $\lambda_B$. Each sub-generator produces messages geometrically distributed with means $L_A$ and $L_B$. The mean sojourn times in the two states are $T_A$ and $T_B$, respectively. By properly selecting the values for these six parameters, the MMP generator can reproduce the effects of a bursty traffic source. In particular, we have assumed the MAC parameters in Table 3.18 with the slight change in $R_c$, that is $R_c = 720$ kbit/s. Then, we have

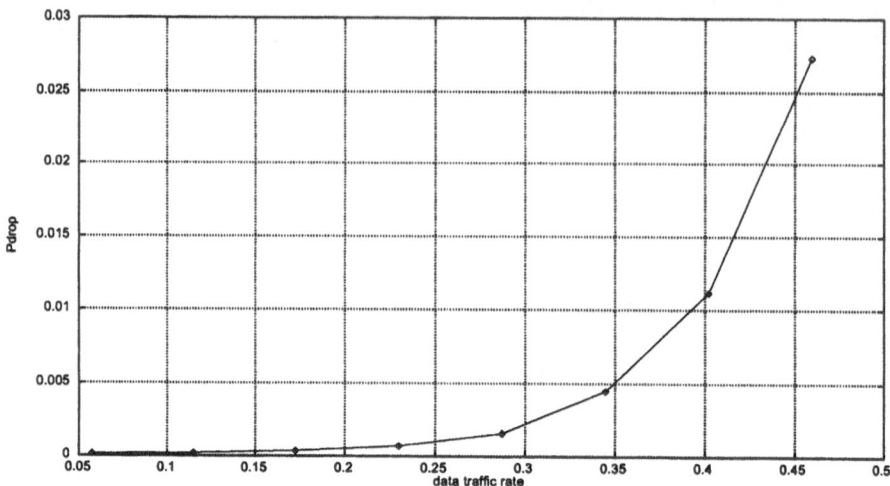

**Figure 3.118.** Behaviour of $P_{drop}$ as a function of the input data traffic $r_d$ ($T_f$ = 16 ms, $M_v$ = 20 VTs/carrier, 20 slots/frame).

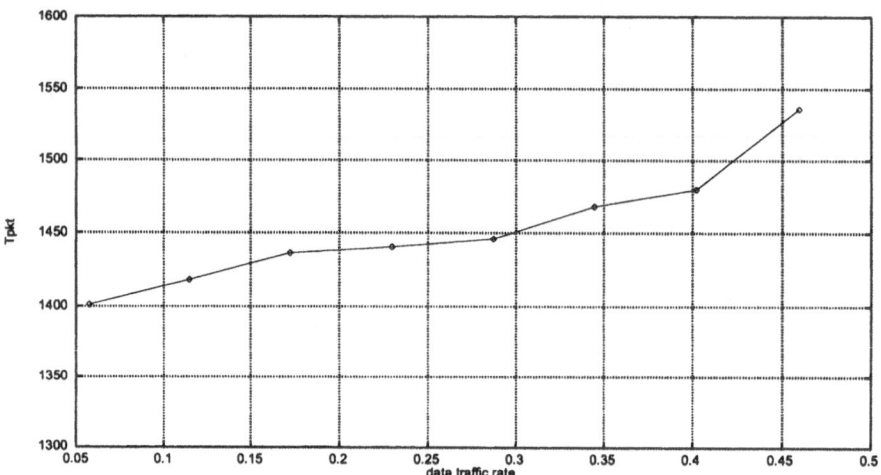

**Figure 3.119.** Behaviour of $T_{pkt}$ in slots as a function of the input data traffic $r_d$ ($T_f$ = 16 ms, $M_v$ = 20 VTs/carrier, 20 slots/frame).

selected: $T_A$ = 6000 slots, $T_B$ = 15 000 slots, $\lambda_A^{-1}$ = 1200 slots, $\lambda_B^{-1}$ = 173.68 slots, $L_A$ = 50 pkts/msg, $L_B$ = 4 pkts/msg. The burstiness degree of a traffic source, $B$, is given by the peak-to-mean traffic ratio. The data traffic source described in Section 3.5.1.1 is characterised by $B$ = 1, whereas, the two-state data traffic source considered here has $B$ = max $\{\lambda_A L_A, \lambda_B L_B\}/(\lambda_A L_A P_A + \lambda_B L_B P_B)$ , where $P_A$ = $T_A/(T_A + T_B)$ and $P_B$ = $T_B/(T_A + T_B)$. With the values selected for this source we have $B \approx 1.5$, that is a more bursty data traffic source than that considered in Section 2. Results in Figs 3.118 and 3.119 refer to $T_f$ = 16 ms and 20 VTs/carrier.

According to Fig. 3.118, we have that the bursty data traffic causes a worse $P_{drop}$ performance. Moreover, the impact on the packet transmission delay, $Tpkt$, is significant.

### 3.5.2 Advanced PRMA for Real-time VBR Traffic Category

The advanced PRMA (A-PRMA) Fig. 3.120 is similar to PRMA, only the transmission does not depend on the permission probability as in the original PRMA. It is also based on frames consisting of $N$ time slots each. Let us have a vector with $N$ components. The components are zeros and user identity numbers, where zero indicates an unreserved time slot, and an identity number specifies a reserved slot. Every user unit $X$ produces from this vector two new vectors with identity numbers of time slots: (1) a vector of reserved slots for user $X$, and (2) a vector of unreserved slots as shown in

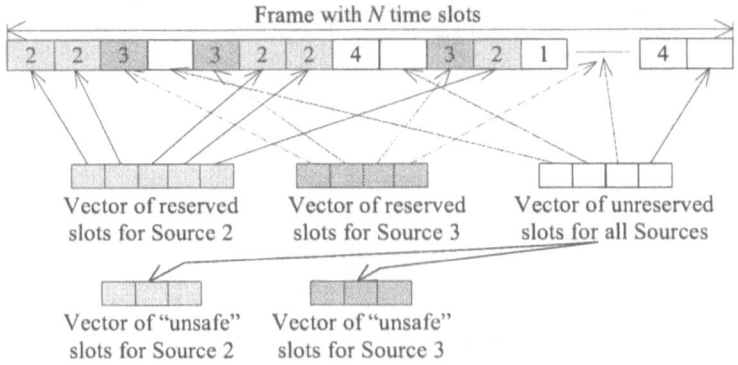

**Figure 3.120.** Creation of reserved and unreserved vectors within each source in A-PRMA.

Fig. 3.120. Then, if the user source has more packets in the buffer than reserved slots, it randomly selects the difference number of slots from the vector of unreserved slots. These slots are then put into vector of "unsafe" (unreserved) slots, while these are not reserved yet. This procedure takes place after each time slot, and therefore the contents of the last two vectors is highly dynamic. This technique is still contention based. The user source then always transmits a packet, when the actual time slot number and the time slot number in the reserved or the unsafe vector are identical.

The supported elementary channel bit rate is $R_e = 16$ kbit/s. The 26.5 ms frame contains 125 identical time slots. The payload of one slot has been assumed to be an ATM cell including the header, i.e. $53 \cdot 8 = 424$ uncoded bits. This corresponds to an uplink carrier information bit rate of 2 Mbit/s.

The rt-VBR source is modelled using an autoregressive model with parameters valid for a typical video CODEC according to (128). Throughout the simulations, a mean bit rate of 128 kbit/s has been used, and the reference value for the standard deviation $\sigma$ has been 57 kbit/s; this latter parameter has been varied in a certain range for specific investigations. In the reference case, the VBR source simulations have revealed a peak rate of approximately 300 kbit/s corresponding to a 0.43 burstiness value.

The A-PRMA protocol including the used source traffic models has been specified in SDL, and the simulations have been performed using the SDL tool SDT.

The comparison of the original and advanced PRMA with respect to the cell loss probability is shown in Fig. 3.121a. The superior performance of A-PRMA is obvious. It is also interesting that the difference for original PRMA protocols with different permission probabilities is very small, and it is almost independent from the delay in contrast to A-PRMA, as can be seen in Fig. 3.121b. The curves in the latter show the cell loss probability for different number of VBR user sources with different (one-way) delays. The results are compared with the TDMA curve, where each user source has fixedly allocated $1/N$ of the carrier capacity, where $N$ is the number of user sources. One obtains that up to the average load of approximately 0.5 the TDMA protocol is more efficient. Beyond this load value A-PRMA clearly outperforms the fixed allocation scheme. However, the performance gain decreases with growing delay.

The average throughput of A-PRMA with respect to the average load shows the typical behaviour: the curve increases almost linearly over the load range from 0 to 0.8, since all VBR sources have the same mean bit rate; it then turns into saturation reaching a throughput of between 0.8 and 0.9.

Figure 3.122. displays the relation between throughput and cell loss probability. The graph confirms that with increasing throughput also the cell loss probability raises, since higher throughput reflects more competing user sources, consequently causing more collisions.

Figure 3.123 finally shows the comparison of the cell loss probability for a system with (one-way) delay 12.6 ms and different standard deviations $\sigma$ of source bit rates. The reference curves are again for the fixed allocation TDMA scheme. It can be clearly obtained that for both the reference scheme and A-PRMA the cell loss probability raises with the average load and standard deviation. Comparing the three pairs of schemes, one also finds that the A-PRMA tends to outperform the corresponding TDMA at already lower load values with a higher source bit rate dynamics, i.e. higher standard deviation.

For multimedia networks the quality of service (QoS) is essential. The multimedia applications bit rates are normally variable. For efficient use of transmission resources the statistical multiplexing should be used. At the call or connection level the connection admission control (CAC) algorithm

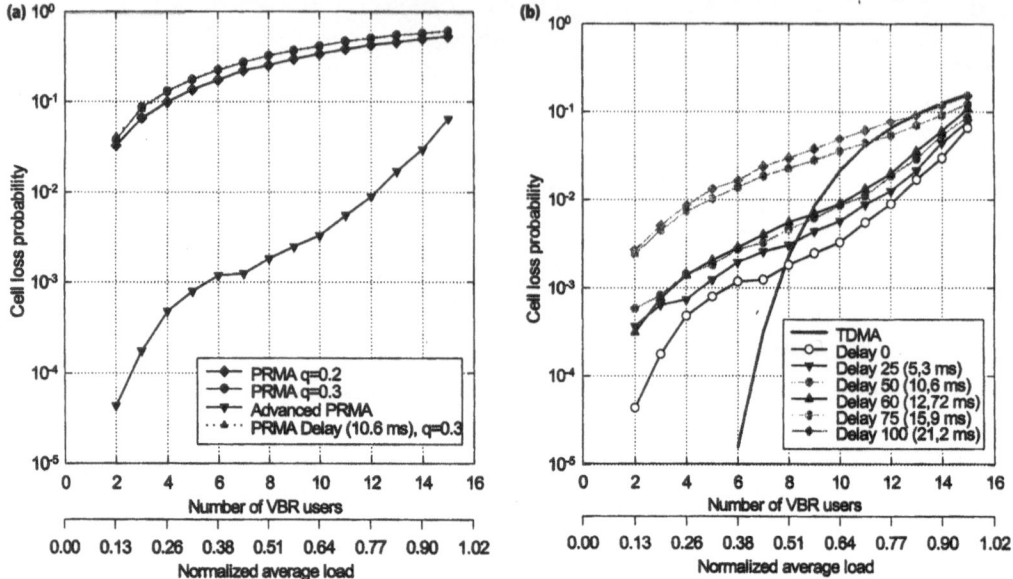

**Figure 3.121.** Cell loss probability performance of advanced PRMA: **a** comparison of original and advanced PRMA; **b** dependency on load and (one-way) propagation delay values (in slots and ms).

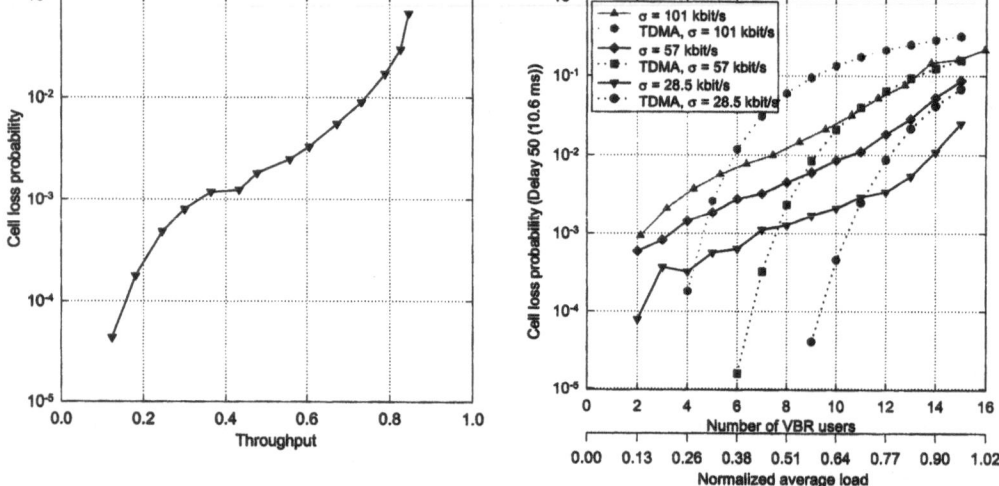

**Figure 3.122.** Cell loss probability vs. throughput of advanced PRMA.

**Figure 3.123.** Cell loss probability performance of advanced PRMA for different bit rate dynamics of rt-VBRMAC with QoS for Satellite Multimedia Networks.

decides if a new connection can be established or not. The approach can be similar to the one for terrestrial fixed ATM networks. At the cell level an entity called "scheduler" redirects the cells or packets to the correct outgoing links. In the case of wireless or radio access the scheduler has to also manage the uplink access (terminal to access point, which is base station or satellite).

The MAC protocol for the satellite broadband multiservice systems must achieve the following goals:

- Support of different kinds of traffic categories such as constant bit rate (CBR), real-time variable bit rate (rt-VBR), non-real-time variable bit rate (nrt-VBR), unspecified bit rate (UBR) and available bit rate (ABR).

- Guarantee QoS, which means that the MAC protocol will guarantee the connection parameters negotiated at the connection setup for the time of the connection. The real-time services are very sensitive to cell transfer delay (CTD) and cell delay variation (CDV) QoS parameters, whereas the data services are very sensitive to cell loss ratio (CLR) QoS parameter.
- Fairness, i.e. the MAC protocol must serve the terminals of the same priority class with equal probability.
- Efficiency. i.e. the MAC protocol must minimise the network bandwidth usage while guaranteeing QoS,
- Small signaling overhead of the MAC protocol functions, i.e. the information flow between the access point and terminal should be as small as possible.

The PRMA, PRMA-HS and A-PRMA do not support all of these requests. Especially the QoS which is a problem. The entities which are mainly responsible for guaranting the QoS parameters in fixed networks are the usage parameter control (UPC) and the traffic scheduler. Therefore, we will first look into packet scheduling in fixed networks and the differences for scheduling in radio networks.

### 3.5.2.1 Scheduling of Packets in Fixed Networks
In fixed networks the scheduling is associated only with the outgoing links as shown in Fig. 3.124. The scheduling function, which redirects the cells according the pre-negotiated QoS parameters to outgoing links, is realised within the ATM switch. In this context, it is assumed that the access links from the sources and the input buffers are dimensioned in such a way that they do not impose any constraints on the traffic, i.e. cell rate.

The scheduling can also be seen as the mechanism that determines which queue is given the opportunity to transmit a cell. In general, the queues can be organised as per-group queuing or per- virtual channel/virtual path (per-VC/VP) queuing (129). With per-group queuing a number of connections share the same queue in a first-in–first-out (FIFO) arrangement. In this queuing structure, the connections can be categorised according the service category, service class or conformance definition into groups. On the other hand, with the per-VC/VP queuing the cells of each VC or VP are queued independently.

A very simple queue structure scheduling technique is the priority based scheduling, which assigns a priority to each queue and serves them in order of priority. Under this scheme, the QoS support is limited. Better techniques are those based on fair share scheduling, where for each queue is guaranteed that it gets a share of link bandwidth according to a defined weight. The weight is the criterion for connections with equal priority but different traffic and QoS parameters. In this way, they guarantee a certain minimum rate allocated among the queues and are further classified as rate allocation (work conserving) and rate-controlled (non-work conserving) schedulers. With a work conserving scheduling a server is never idle when there is a packet to send. On the other hand, with a non-work conserving scheduling the server may be idle even when there are packets waiting to be sent (130).

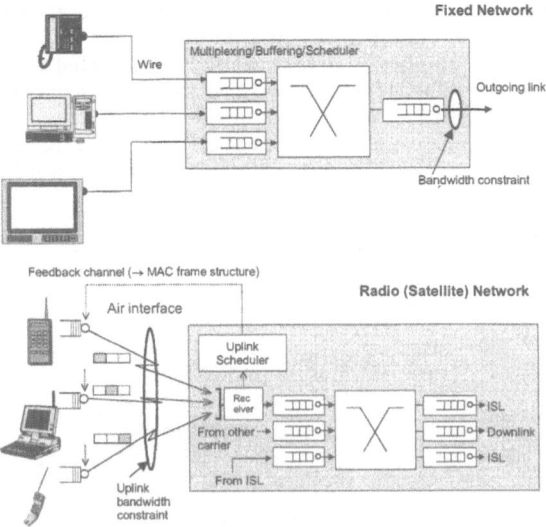

**Figure 3.124.** Scheduling in fixed and satellite networks.

The entities that control and guarantee that the traffic parameters are in accordance with the pre-negotiated at the connection setup are called usage parameter control (UPC) and network parameter control (NPC) and are not part of scheduling (131). They have similar functions but at different interfaces: UPC is done at UNI, whereas NPC is done at the NNI. Their functions include monitoring cell streams, checking the conformity between the actual cell stream and the nominal cell stream (the traffic descriptor values) and taking necessary actions when unconformity is detected. The actions that can be taken are discard of cells immediately or tagging of them for discard at a congestion point when the network is congested. The cell loss priority (CLP) bit is used for tagging.

### 3.5.2.2 Scheduling of Packets in Radio (Satellite) Networks

In radio networks, the constraint is on the total bandwidth available to all users before the access point, i.e. satellite or base station. Because of that, for the exploitation of the statistical multiplexing in a TDMA system, a scheduler is needed to organise the frame structure in the satellite uplink channels. The information on the slot assignments, as decided by scheduler, is broadcast to the user terminals via a feedback channel as shown in Fig. 3.124.

The connection QoS parameters are negotiated during the connection setup. The entity which decides if a new connection can be accepted or not runs a connection admission control (CAC) algorithm. This entity can be placed on the on-board processing satellite or in the ground network control station. However, the parameters of the accepted connection must be transmitted to the scheduler which uses them for the process of slots allocation.

As already mentioned the big difference between the scheduler in fixed networks and the scheduler in radio networks is that the latter one does not schedule the traffic based on arrived cells in queues but on arrived requests for the bandwidth capacity (number of time slots in case of TDMA-based MAC).

The first criterion for the scheduling decision is the priority of the service category. As shown in Table 3.19 the highest priority is assigned to CBR service category which is then followed by other service categories.

The UPC for the uplink could be performed in the terminal or within the scheduler entity on on-board the satellite. The downlink does not need UPC because this traffic has already gone through the uplink UPC and the traffic is conformed to traffic descriptors. The UPC functions implemented in the terminal can only use discard principles since the traffic over the air interface has also to be conformed to traffic descriptors. On the other hand, the scheduler implemented UPC functions could also use the tagging principle if there is enough capacity. In this case the scheduler could allocate additional slots if there are non-conforming requests and when the cells would reach the access point, the CLP bit of cells in non-conforming slots could be set to lower priority like in fixed networks.

### 3.5.2.3 MF-TDMA-based Multiple Access Technique

The MF-TDMA access scheme has been proposed for different GEO and LEO systems. The MF-TDMA frame is normally divided into two areas: the first one is intended for synchronisation and signalling information transmission, whereas in the second one the data is transmitted. The requests for bandwidth allocations can be transmitted via out-of-band request slots, which are part of the first area. However, the in-band (implicit) requests can be sent together with data packets.

The first task of the scheduler is to calculate the number of slots which will be allocated to the specific connection. This has been presented for hierarchical round robin (HRR) scheduling in (132). Then, the allocated number of slots has to be assigned to actual slots in MF-TDMA frame. This process is much more complex and needs also the time component, such as virtual arrival times.

For this first approach, we decided to implement a very simple strategy which is based on priority

**Table 3.19.** ATM traffic categories priorities

| Priority number | Service category |
|---|---|
| 5 | CBR |
| 4 | Rt-VBR |
| 3 | Nrt-VBR |
| 2 | ABR |
| 1 | UBR |

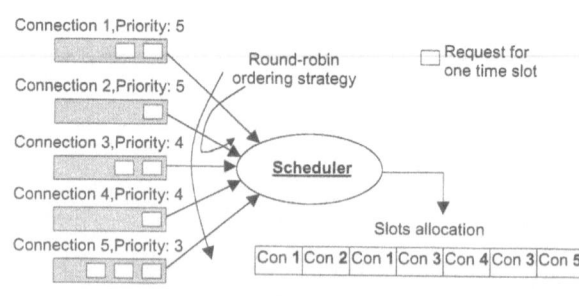

**Figure 3.125.** Scheduler operation.

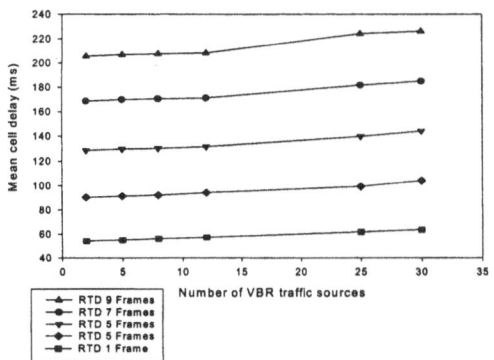

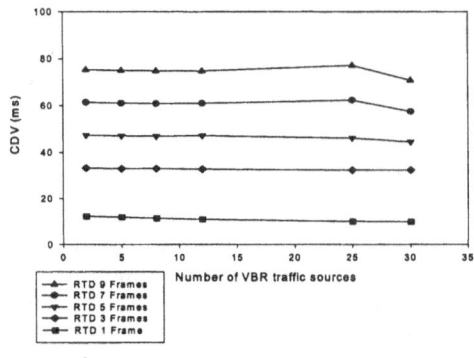

**Figure 3.126.** Mean delay of cells for various number of VBR traffic sources and channel delay.

**Figure 3.127.** CDV for various number of VBR traffic sources and channel delay.

and first-come-first-serve (FCFS) strategies. In this way, when the new connection is setup, the scheduler writes the connection parameters into a linked list. This list is ordered according to priorities of the traffic classes of the connections. The scheduler begins on the top of the list and allocates the slots to services with equal priority in a round robin manner as shown in Fig. 3.125.

First, it serves all CBR connections, then rt-VBR ones and if capacity is available also other connections, until all time slots are allocated. In such a way every connection of the same priority gets a minimum capacity.

### 3.5.2.4 Simulation Setup and Results

The simulation setup consists of VBR traffic sources, which were described in the section about Advanced-PRMA, and satellite as access point with scheduler as described in the previous section. The simulation is realised only on cell level, that means that the sources are active all the time. The traffic is multiplexed onto one uplink carrier according the requests for time slots and scheduler allocation of time slots. The reservation information on time slots is broadcast to all traffic sources on the frame basis.

In Fig. 3.126 the results of the simulation for mean cell delay of one traffic source are shown. The delay gradually increases with the number of traffic sources similarly for the different channel delays. Because of the scheduler technique the traffic source which has first established the does not have the problems with the capacity until the number of sources is much higher than 100 per cent load (the number of the traffic sources multiplied by the mean bit rate of the traffic sources is exactly the carrier capacity). That is not true for the sources which established the connection last. The reason is that the scheduler always allocates the first slot in the frame to the first traffic source in the linked list of traffic sources as described in the previous section. The delays for the sources at the end of the linked list begin to increase very fast when the load is about 100 per cent and over.

Figure 3.127 shows the CDV for various number of traffic sources and channel delays. It can be obtained that the CDV increases with the channel delay.

Recently some new facts about the rt-VBR traffic have been obtained. The following points may be important for the future work:

● The rt-VBR was proposed for transmission of MPEG-2 compressed applications. This standard is mainly intended for broadcast services.

● Recovering the source timing, cell delay variation in the network causes problems for clock recovery, and the cell delay variation is expected to be greater for VBR than CBR.

● ATM Forum has been working on Phase 2 AMS over ATM (Phase 1: CBR MPEG-2 over ATM with AAL-5), a transmission of rt-VBR (also pre-stored Constant Quality VBR MPEG).

● Since February 1998 no new work has been done. In addition, the baseline text is very incomplete. The last living list is concentrated on pre-encoded sequences.

● The ITU-T values for end-to-end CTD and CDV are 400 ms and 3 ms. The WATM Forum Group envisons 10 per cent of these values from the terminal to base station (MAC layer).

● The HIPERLAN/2 frame length will be 2 ms and the capacity of the TDD carrier 25 Mbit/s.

It seems that rt-VBR traffic category is quite uninteresting for wireless systems and particularly for satellite systems.

### 3.5.3 Conclusions

This research activity has dealt with MAC protocols to access satellite resources. An attractive solution, initially proposed for terrestrial microcellular systems, is the PRMA protocol, which exploits silent phases of voice sources to multiplex more conversations on the same channel. In this activity we have also investigated a modification of the classical PRMA scheme in order to improve its performance, that is the PRMA-HS scheme. The focus here is on LEO-MSSs, since they will provide (partly or totally) the satellite coverage of next generation mobile communication systems.

The quality of service has been measured by the packet dropping probability, $P_{drop}$, for VTs and the mean message transmission delay, $T_{msg}$, for DTs. For voice sources we have considered a VBR-RT service and for data sources we have envisaged an ABR service.

We have shown that the PRMA scheme maintains a satisfactory performance in terms of voice multiplexing capabilities while supporting also a data traffic. Moreover, we have proved that also a higher efficiency is possible by means of the PRMA-HS scheme where a terminal does not stop contending while it waits for receiving the outcome of a previous attempt. Therefore, we can conclude that PRMA-HS is a good solution as a unified MAC protocol for the terrestrial and satellite components of future mobile communication systems.

In addition, the A-PRMA was evaluated for rt-VBR traffic. The traffic source models the video-conference VBR traffic. The results were evaluated according CLR and they show that this technique does not support the requests for transmission of multimedia applications. Actually all PRMA variants can support only voice traffic with voice activity detection (quasi VBR source) and data traffic which does not request any time limits (as for example TCP/IP traffic because of TCP protocol).

The section continues with the aspects of scheduling for uplink MAC protocol. The scheduling can be combined with the MF-TDMA technique, since it allows implicit and explicit reservations of needed capacity. The implemented scheduler technique guarantees the CLR, but the CDV still cannot be guaranteed and therefore new approaches are needed. However, the presented scheduling technique with some modifications, such as introduction of weights, is very suitable for non-real time service categories.

The presented work has shown how problematic is the rt-VBR service category for transmission over radio or satellite networks. Future work should focus on further elaboration and definition of interesting multimedia services for satellite multimedia broadband networks. In addition the existing protocols like TCP/IP have to be taken into account since it seems that the evolution is preferred to revolution in networking.

## 3.6 Conclusions

This section has reported the work carried out by the Networks sub-group of the COST 252 project. Three main research areas were identified, of which two led to studies for the approach of providing network functions for a future broadband satellite system. Those two research areas were: network functions based on S-UMTS and S-PCN techniques and functions based on the provision of ATM over satellite. Furthermore, a study on medium access control (MAC) protocols was performed.

The first section presented studies on radio resource management and mobility management. In the first study, a dynamic channel assignment (DCA) approach based on handover queuing was shown to be efficient for LEO satellite systems, compared with fixed channel allocation (FCA) schemes. A mobility model was developed and the performance of both schemes was compared. A later study was presented where a DCA scheme based on maximal packing with optimisation by genetic algorithms was proposed. This scheme has been optimised for use in satellite systems in response to the unacceptable complexity of terrestrial schemes. Novel models have been presented which enable an analytical approach to the evaluation of DCA scheme, which has so far proved difficult. Finally, a mobility management scheme based upon the determination of a guaranteed coverage area around a fixed earth station (FES) was devised. Inter-FES handovers by satellites during calls are not desirable due to the associated routing task. Areas around the FES are identified within which a user is guaranteed to be able to access the FES for a known proportion of time, the resulting effect on the availability of satellites of a call admission policy based on this area is that user terminals are forced to utilise lower elevation satellites when they are far from the FES site, resulting in a higher inter-satellite handover occurrence.

Studies into the operation of ATM networks over satellite have concentrated on routing and call admission which result from such a policy. The requirement for ATM networks to provide pre-determined QoS levels presents new challenges for non-GEO satellite systems and those studied here have included the reservation of channels and dynamic routing for LEO systems which specify inter-satellite links. A protocol architecture and packet header adaptation based on ATM networks has also been optimised for use over satellite systems.

The problems of routing in the ISL segment of non-geostationary satellite systems, characterised by dynamic network topology and fluctuating traffic load experienced by a particular satellite have been addressed, the realisation of a simulation model of a Celestri-like LEO satellite communication system with intersatellite links using a centralised version of the Dijkstra shortest path algorithm for calculation of routing table updates.

Finally, the packet reservation multiple access (PRMA) scheme with adaptations was examined for use over non-GEO satellite, and was shown to produce satisfactory performance, although the provision of VBR services was highlighted as a problem area in which future research must be carried out, particularly in the light of the presumed increase in IP traffic over mobile terrestrial and satellite systems.

# References

(1) Sammut A, Mertzanis I, Tafazolli R, Evans BG (1998) Networking for a future MEO multimedia satellite systems: COST 252 TD(98)01. Fifth Management Committee Meeting, Florence , Italy, 12–13 February 1998.
(2) Meenan C, Tafazolli R, Evans BG (1997) Satellite-PCN mobility management, COST 252 TD(97)11. Fourth Management Committee Meeting, Wessling, Germany, 17–18 November 1997.
(3) Berruto E et al (1997) Architectural aspects for the evolution of mobile communications towards UMTS, IEEE JSAC 15(8): October 1997.
(4) Del Re E, Fantacci R, Giambene G, Walter S (1997) Performance evaluation of an improved PRMA protocol for low earth orbit mobile communications systems, COST 252, TD(97)12, November 1997.
(5) Del Re E, Fantacci R, Giambene G, Cerboni C (1998) Performance evaluation of the PRMA protocol for voice and data transmissions in low earth orbit mobile communications systems, COST 252 TD(98)02. Fifth Management Committee Meeting, Florence, Italy, 12–13 February 1998.
(6) Mertzanis I, Tafazolli R, Evans BG (1997) Protocol architecture scenarios for satellite and B-ISDN network integration. Third Ka-band Utilisation Conference. Sorrento, Italy, September 1997.
(7) Mazzella M (1997) Development of novel satellite mobile applications (1997) The SINUS project, IEEE Colloquium on EU's initiatives in satellite communications-mobile. London, 8 May 1997.
(8) Bostic J, Kandus G, Werner M (1997) MAC for ATM over satellite. COST 252 TD(97)10, Fourth Management Committee Meeting, Wessling, Germany, 17–18 November 1997.
(9) Werner M (1997) ATM concepts for satellite personal communication networks, COST 252 TD(97)02. Second Management Committee Meeting, Brussels, 6–7 February 1997.
(10) Mertzanis I, Tafazolli R, Evans BG (1997) Connection admission control strategy and routing considerations in multimedia (NON-GEO) satellite networks. IEEE VTC'97, Phoenix, USA, May 1997.
(11) Bostic J et al (1998) Multiple access protocols for ATM over low earth orbit satellites. First 252/259 Joint Workshop, University of Bradford, UK, 21–22 April, 1998.
(12) Del Re E (1996) A Coordinated European effort for the definition of a satellite integrated environment for future mobile communications, IEEE Comm Mag 34(2):98–104, February 1996.
(13) Restrepo J and Maral G (1996) Constellation sizing for non-GEO "Earth-Fixed Cell" satellite systems. Proceedings of the AIAA Sixteenth International Communications Satellite Systems Conference and Exhibit, Washington DC, USA, 25–29 February, pp 768–778.
(14) Tekinay S, Jabbari, B (1991) Handover and channel assignments in mobile cellular networks, IEEE Comm Mag 29(11) 42–46, November 1991.
(15) Del Re E, Fantacci R, Giambene G (1994) performance analysis of a dynamic channel allocation technique for satellite mobile cellular networks, Int J Sat Comm 12:25–32, January/February 1994.
(16) Del Re E, Fantacci R, Giambene G (1995) Efficient dynamic channel allocation techniques with handover queuing for mobile satellite networks. IEEE J Selected Areas Comm 13(2):397–405, February 1995.
(17) Del Re E, Fantacci R, Giambene G (1995) An efficient technique for dynamically allocating channels in satellite cellular networks. Proceedings of IEEE GLOBECOM'95, pp 1624–1628, Singapore, 13–17 November 1995.
(18) Hu HF, et al (1998) Satellite-UMTS Traffic Dimensioning and Resource Management Technique Analysis. IEEE Trans Veh Tech 47(4):1329–1341, November 1998.
(19) Del Re E, Fantacci R, Giambene G (1999) handover queuing strategies with dynamic and fixed channel allocation techniques in low earth orbit mobile satellite systems. IEEE Trans Comm 47(1):89–102, January 1999.
(20) Del Re E, Fantacci R, Giambene G (1999) Different queuing policies for handover requests in low earth orbit mobile satellite systems. IEEE Trans Veh Tech 48(2):448–458, March 1999.
(21) Mouly M and Pautet MB: The GSM System for mobile communications, ISBN 2-9507190-0-7.
(22) Race II SAINT Project 2117, Deliverable No. 15.
(23) Lutz, E (1991) The land mobile satellite communication channel – recording, statistics, and channel model. IEEE Trans Veh Tech 40(2).
(24) Parks, M (1993) High elevation angle propagation results, applied to a statistical model and an enhanced empirical model. IEE Electronics Let 29(19).
(25) Sammut, A (1994) Mobility management related signalling for a MAGSS-14-based satellite personal communications network (S-PCN), COST227D, NTUA, Athens Greece.
(26) Meenan C (1997) Intelligent paging schemes for non-GEO satellite personal communication networks, VTC'97, Phoenix, Arizona.

(27) Del Re E, Fantacci R, Giambene G et al (1994) Review of resource management strategies, R2117 satellite integration in future mobile networks. A3300 Resource Management, October 1994.

(28) Jahn A (1998) Performance Evaluation of Resource Management Schemes for Non GSO Satellite Communications, EMPS98, Venice, Italy, pp 16–44, 4–5 November 1998.

(29) Everit D, Manfield D (1989) Performance analysis of cellular mobile communication system with dynamic channel assignment, IEEE J Select Areas Comm 7(8):1172–1179, October 1989.

(30) Santos V et al (1999) Simplified maximum packing a new dynamic channel allocation technique, ECSC. Fifth CDROM proceedings, Toulouse, France, 3–5 November 1999.

(31) Del Re E, Fantacci R and Giambene G (1995) Handover and dynamic channel allocation techniques in mobile cellular networks, IEEE Trans. Vehicular. Tech 44(2): 229–236, May 1995.

(32) Cimini LJ et al (1994) Call blocking performance for dynamic channel allocation in microcells. IEEE Trans Comm 42(8):2600–2607, August 1994.

(33) Aarts E, Korst J (1990) Simulated annealing and Boltzmann machines. John Wiley.

(34) Kirkpatrick S, Gelatt CD Jr Vecchi MP (1983) Optimization by simulated annealing. Science 220 (4598) 671–680.

(35) http://www.ingber.com/

(36) Goldberg D (1989) Genetic algorithms. Addison-Wesley, New York, chs 1–4.

(37) Holland J (1992) Genetic algorithms. Scientific American, July 1992, pp 66–72.

(38) http://www.staff.uiuc.edu/~carroll/ga.html

(39) Mohorcic M et al (1995) Call blocking performance for channel allocation strategies in integrated satellite/terrestrial mobile system, mobile and personal communications. Elsevier Science, 1995, pp 163–171.

(40) Jordan S, Varaiya PP (1991) Throughput in multiple service multiple resource communication networks. IEEE Trans Comm 39(8):1216–1222, August 1991.

(41) http://www.infowin.org/ACTS/

(42) http://www.ee.surrey.ac.uk/CCSR/ACTS/

(43) Elizondo E et al (1996) ASTROLINK system overview. Second Ka- band Utilization Conference and International Workshop on SCGI, Florence, Italy, 24–26 September 1996.

(44) Leamon RG et al (1997) CYBERSTAR. Third Ka-band Utilisation Conference, Sorrento, Italy, 15–18 September 1997.

(45) http://www.teledesic.com

(46) Otsu T et al (1997) Ka-band satellite communication systems operating through NTT's communication satellite N-STAR. Third Ka-band Utilisation Conference, Sorrento, Italy, 15–18 September.

(47) Le Stradic B, et al (1997) The WEST project: exploiting the Ka-band spectrum to develop the global information infrastructure. Third Ka-band Utilisation Conference, Sorrento, Italy, 15–18 September 1997.

(48) Losquadro G (1998) The EUROSKYWAY system for interactive multimedia operating with feed-back-aided traffic management. Seventeenth AIAA International Communications Satellite Systems Conference and Exhibition, Yokohama, Japan, 23–27 February 1998.

(49) Mertzanis I et al (1999) Protocol Architectures for Satellite-ATM Broadband Networks. IEEE Comm Mag 37(3):46–54, March 1999.

(50) Mertzanis I et al (1998) SECOMS interworking scenario and interconnection with B-ISDN, Conference Proceedings. Third ACTS Mobile Communication Summit, Vol. 2, 8–11 June 1998, Rhodes, Greece.

(51) Valadon C et al Code Division Multiple Access for Provision of Mobile Multimedia Services with a Geostationary Regenerative Payload, submitted to the IEEE JSAC.

(52) Akyildiz IF, Jeong S Ho: Satellite ATM Networks: A survey. IEEE Comm Mag, July.

(53) Romanow A and Floyd S: Dynamics of TCP traffic over ATM Networks, IEEE JSAC, Vol.13, No.4, May 1995.

(54) ITU-T Recommendation Q.2763: B-ISDN- Signalling System No. 7, B-ISDN User Part (B-ISUP) – Formats and Codes, February 1995.

(55) ETS 300 374–1: Intelligent Network (IN), Capability Set 1 (CS-1), Core Intelligent Network Application Part (INAP), Part 1: protocol Specification, September 1994.

(56) Guda DK, Schilling DL, Saadawi TN: Dynamic reservation multiple access technique for data transmission via satellites, IEEE INFOCOM'82, pp 53–61.

(57) Pavey CF, Price R Jr, Cummins EJ (1986) A performance evaluation of the PDAMA satellite access protocol. INFOCOM'86, April 1986, pp 580–589.

(58) Kwak KS, Lim KJ (1995) A modified PDAMA protocol for mobile satellite communications systems, IEEE JSAC, VOL.13, NO.2, February 1995.

(59) Bohm S et al (1994) Analysis of a movable Boundary access technique for a multiservice multibeam satellite system, International Journal Of Satellite Communications, VOL.12, 299–312 ,1994.

(60) Le Ngoc T, Krishnamurthy SV (1996) Performance of combined free/demand assignment multiple-access in satellite communications, Int J Sat Comm VOL.14, 11–21 ,1996.

(61) Ha TT (1990) Digital satellite communications, 2nd edn, McGraw Hill. Ch 7.

(62) Ors T, Sun Z and Evans BG (1998) An adaptive random-reservation MAC protocol to guarantee QoS for ATM over satellite, Broadband Communications: the future of telecommunications (IFIP TC6/WG6.2. Fourth International Conference on Broadband Communications, Stuttgart-Germany, pp 107–119, 1–3 April 1998.

(63) Ors T (1998) Traffic and congestion control for ATM over satellite to provide QoS. PhD thesis, University of Surrey, December 1998.

(64) Peyravi H (1999) Medium access control protocols performance in satellite communications. IEEE Comm Mag, March 1999.

(65) Mertzanis I et al (1998) Satellite-ATM networking and call performance evaluation for multimedia broadband services. Fourth Ka-band Utilization Conference proceedings, 2–4 November 1998, Venice, Italy.

(66) Connors DP, Ryu B, Dao S (1999) Modelling and simulation of broadband satellite networks, Part I: Medium access control for QoS provisioning. IEEE Comm Mag, March 1999.

(67) Gelenbe E, Mang X, Onvural R (1997) Bandwidth allocation and call admission control in high-speed networks. IEEE Comm Mag, May 1997.

(68) Saito H (1997) Dynamic resource allocation in ATM networks. IEEE Comm Mag, May 1997.

(69) Bolla R, Davoli F, Marchese M (1997) Bandwidth allocation and admission control in ATM networks with service separation. IEEE Comm Mag, May 1997.

(70) Liu K et al (1997) Design and analysis of a bandwidth management framework for ATM -based broadband ISDN. IEEE Comm Mag, May 1997.

(71) Gibbens RJ, Kelly FP, Key PB (1995) A decision-theoretic approach to call admission control in ATM networks. IEEE, JSAC 13(6), August 1995.

(72) Berger AW, Whitt W (1998) Extending the effective bandwidth concept to networks with priority classes. IEEE Comm Mag, August 1998.

(73) Enomoto O, Miyamoto H (1973) An analysis of mixtures of multiple bandwidth traffic on time division in switching networks. Seventh International Teletraffic Congress Proceedings, pp 635.1–8.

(74) Aein JM (1978) A multi-user-class, blocked-calls-cleared demand access model. IEEE Trans Comm 26(3):378–385.

(75) Roberts JW (1981) A service system with heterogeneous user requirements – application to multi-service telecommunication systems. Pujolle G (ed.) Performance of data communication systems and their applications. North Holland-Elsevier Science Publishers, pp 423–431.

(76) af-tm-0056.000, ATM Forum: traffic management specification Version 4.0, April 1996.

(77) Mertzanis I et al (1999) Satellite-ATM networking and call performance evaluation for multimedia broadband services (extended version), accepted for publication in Int J Sat Comm.

(78) Mertzanis I (1999) QoS Provisioning for broadband satellite-ATM multimedia networks. PhD thesis, University of Surrey, July 1999.

(79) Mertzanis I et al (1999) Satellite-ATM network dimensioning and ABR capacity estimation in the presence of self-similar traffic. Fifth Ka-band Utilisation Conference proceedings, Taormina, Italy 18–20 October 1999.

(80) Anick D, Mitra D, Sondhi MM (1974) Stochastic theory of a data handling system with multiple sources, Bell Sys Tech J 61(8):10–18.

(81) Garrett MW (1996) A service architecture for ATM: from applications to scheduling. IEEE Network, May/June 1996.

(82) Maglaris B et al (1986) Performance models of statistical multiplexing in packet video communications. IEEE Trans comm 36(7), July 1986.

(83) Kleinrock L (1976) Queueing systems. Vols I and II, John Wiley, ISBN 0–471–49110–1 and 0- 471–1976.

(84) Argyropoulos Y et al (1998) GPRS delay and capacity analysis for web browsing application. ICT'98, Vol. II, Porto Carras, Greece, 21–25 June 1998.

(85) Mertzanis I et al (1999) Multimedia service support for WISDOM: the satellite component of an end-to-end ATM network. Fourth ACTS Mobile Summit, 8–11 June 1999, Sorrento, Italy.

(86) Sammut A et al (1997) GIPSE: A global integrated personal satellite multimedia environment. Fourth European Conference on Satellite Communications (ECSC-4) Rome, 18–20 November 1997.

(87) Vatalaro F, Corazza G, Caini C (1995) Analysis of LEO, MEO and GEO global mobile satellite systems in the presence of interference and fading. IEEE JSAC 13(2), February 1995.

(88) Ananasso F, Carosi M (1994) Architecture and networking issues in satellite systems for personal communications, Int J Sat Comm 12:33–44.

(89) Akyol, Cox D (1996) Rerouting for handoff in a wireless ATM network. IEEE Pers Comm, 3(5), October 1996.

(90) Levine D, Akyildiz I, Naghishineh M (1997) A resource estimation and call admission Algorithm for wireless multimedia networks using the shadow cluster concept, IEEE/ACM Trans Networking 5(1), February 1997.

(91) Mertzanis I, Tafazolli R, Evans BG (1997) Performance issues and modelling of mobile executed handoffs in multispot beam dynamic satellite networks using ATM technology. IEE Colloquium in ATM traffic in the personal mobile communications environment, February 1997, London, UK.

(92) Werner M (1997) A dynamic routing concept for ATM based satellite personal communication networks. IEEE, JSAC 15(8) October 1997.

(93) Zhao W, Tafazolli,R, Evans BG (1995) A UT positioning approach for dynamic satellite constellations, IMSC'95, Ottawa.

(94) Dosiere F et al (1993) A model for the handover traffic in low earth-orbiting (LEO) satellite networks for personal communications, Int J Sat Comm 11:145- 149.

(95) Ruiz G, Doumi TL, Gardiner JG (1996) Teletraffic analysis and simulation of mobile satellite systems. IEEE VTC'96 conference proceedings.

(96) Ruiz G, Doumi TL, Gardiner JG (1998) Teletraffic analysis and simulation of mobile satellite systems. IEEE Trans Veh Tech 47(1), February 1998.

(97) Hong D, Rappaport S (1986) Traffic model and performance analysis for cellular mobile radio telephone systems with prioritized and non-prioritized handoff procedures. IEEE Trans Veh tech VT-35 (3), August 1986.

(98) Mertzanis I (1999) QoS provisioning for broadband satellite-ATM multimedia networks. PhD thesis, University of Surrey, July 1999.

(99) Mertzanis I, Tafazolli R Evans BG (1998) A new approach for radio resource management in multimedia dynamic satellite networks. Seventeenth AIAA International Communication Satellite Systems Conference and Exhibition, Yokohama, Japan, 23–27 February 1998.

(100) Roberts JW (1983) Teletraffic models for the telecom 1 integrated services network, In Tenth Int Teletraffic Congress Proceedings, Section 1.1.2: pp 522–525.

(101) Jonson SA (1985) A performance analysis of integrated communications systems, British Telecom Tech J 3(4) pp 514, 517, 525, October 1985.

(102) Tran-Gia P, Hubner F (1993) An analysis of trunk reservation and grade of service balancing mechanisms in multiservice broadband networks. IFIP Workshop TC6: Modelling and performance evaluation of ATM technology, Martinique, pp 517, 524, 525, 533.

(103) Virtamo JT (1988) Reciprocity of blocking probabilities in multiservice loss systems, IEEE Trans Comm, 36(10):1174–1175, pp 516, 517.

(104) M. Werner (1996) ATM Concepts for Satellite Personal Communication Networks. Proc. European Conference on Networks and Optical Communications (NOC'96), pp 247–254, June 1996, Heidelberg, Germany.

(105) Werner M, Delucchi C, Burchard K (1997) ATM networking for future ISL-based LEO satellite constellations. Proc. Fifth Int. Mobile Satellite Conference (IMSC'97), pp 295–300, Pasadena, California, June 1997.

(106) Wauquiez F, Werner M (1998) Capacity dimensioning of intersatellite link networks in broadband LEO satellite systems. COST 252 TD(98)24, September 1998.

(107) Kennedy MD, Malet PL (1997) Application for authority to construct, launch and operate the Celestri multimedia LEO system, Filing to FCC, Washington DC, June 1997.

(108) Werner M (1997) A dynamic routing concept for ATM-Based satellite personal communication networks, IEEE JSAC 15(8):1636–1648, October 1997.

(109) Papapetrou E, Gragopoulos I, Pavlidou FN (1999) Performance evaluation of LEO satellite constellations with inter-satellite links under self-similar and Poisson traffic. Int J Satel Comm 17: pp 51–64.

(110) Chang HS et al (1996) Performance comparison of static routing and dynamic routing in low earth orbit satellite networks in Proc VTC'96.

(111) Ballard AH (1980) Rosette constellations of earth satellites. IEEE Transactions on Aerospace and Electronic Systems, AES-16(5):656–673, September 1980.

(112) Werner M, Maral G (1997) Traffic flows and dynamic routing in LEO intersatellite link networks, in Proc. of IMSC '97, pp 283–288, Pasadena, California, USA, June 1997.

(113) IEEE Pers Comm Mag – Special issue on IMT-2000. Vol. 4(4), August 1997.

(114) Abrishamkar, Siveski Z (1996) PCS global mobile satellites. IEEE Comm Mag 34(9):132–136, September 1996.

(115) Restrepo J, Maral G (1995) Coverage concepts for satellite constellations providing communications services to fixed and mobile users. Space Comm 13(2):145–157.

(116) Leung, Alnuweiri H, Nasiopoulos P: Interworking broadband satellite networks with terrestrial networks subsystems, University of British Columbia – Report published on WWW at the address http://www.ece.concordia.ca/~hamid/majorp4.html

(117) Goodman J et al (1989) Packet reservation multiple access for local wireless communications. IEEE Trans Comm 37:885–890, August 1989.

(118) Goodman J (1991) Trends in cellular and cordless communications. IEEE Comm Mag pp 31–40, June 1991.

(119) Frullone: On the Performance of Packet Reservation Multiple Access with Fixed and Dynamic Channel Allocation, IEEE Trans Veh Tech 42(1):78–86, February 1993.

(120) Wai-Choong Wong Dynamic Allocation of Packet Reservation Multiple access Carriers, IEEE Trans Veh Tech 42(4)385–392, November 1993.

(121) Yalun Li, Steinar Andersen Boning Feng (1995) On the performance analysis of EPRMA protocol with Markov chain model. Proc of GLOBECOM '95, 13 November Singapore, pp 1502–1506.

(122) Koh, Liu MT (1996) A wireless multiple access control protocol for voice-data integration. Proc Int Conf Parallel and Dist Sys, June 1996, Tokyo, Japan, pp 206–213.

(123) Nanda D, Goodman J, Timor U (1991) Performance of PRMA: a packet voice protocol for cellular systems. IEEE Trans Veh Tech 40(3):584–598, August 1991.

(124) Del Re E, et al (1997) Performance Evaluation of an Improved PRMA Protocol for Low Earth Orbit Mobile Communication Systems, Int J Sat Comm 15: pp 281–291.

(125) Del Re E, et al (1999) Performance analysis of an improved PRMA protocol for low earth orbit mobile satellite systems. IEEE Trans Veh Tech 48(3):985–1001, May 1999.

(126) Corovesis, Venieri D: New User and Service Requirements, INSURED Project (AC229).

(127) Del Re E, et al (1998) Performance evaluation of the PRMA protocol for voice and data transmissions in low earth orbit mobile communication systems, 252TD(98)17, COST 252/259 Joint Workshop, University of Bradford, 21–22 April 1998.

(128) Sen, G. Karlsson, B. Maglaris, D. Anastassiou and J. D. Robbins: Packet Models of Statistical Multiplexing in Packet Video Communications, IEEE Trans. on Comm., Vol. 36, pp 834–843, August 1988.

(129) Giroux, Ganti S (1999) Quality of Service in ATM Networks, Prentice Hall, ch. 5, pp 85–119.

(130) Zhang: Service Disciplines for Guaranteed Performance Service in Packet-Switching Networks, Proc. IEE, Vol. 83, pp 1374–1396, October 1995.

(131) Saito: (1994) Teletraffic technologies in ATM networks. Artech House, ch 3, pp 71–98.

(132) Hung, Montpetit MJ, Kesidis G (1998) ATM via satellite: a framework and implementation. ACM-Baltzer Wireless Networks, 4(2):141–153.

# Air Interface Aspects

**F. Cercas** Instituto Superior Técnico, Lisbon (francisco.cercas@lx.it.pt)
**W. Krewel** ENST Paris (krewel@enst.fr)

## 4.1 Introduction

In line with the efforts of ITU to provide global recommendations for IMT-2000, WG3 activities addressed in particular to the W-CDMA approach. The choice of CDMA for the third generation wireless communications is attractive because of its potential capacities to support universal frequency reuse, variable rate heterogeneous traffic and the possibility to use classical time diversity techniques together with Multiuser Detection (MUD) to contrast multipath fading effects.

Additional, further aspects were considered by WG3:

- Synchronous CDMA (S-CDMA) with Trellis Coded Modulations (TCM) and Multilevel Trellis Coded Modulations (MTCM) to overcome the constraint of capacity limitation.
- Direct-sequence spread spectrum techniques (DS-CDMA) and Blind-Adaptive Multiuser Detection.
- Quasi-synchronous condition with Multicarrier CDMA (MC-CDMA).

WG3 research activities aimed to highlight these issues with obtaining performance evaluation in LEO and MEO satellite environments.

WG3 worked out an overview of candidate CDMA environments and provided a comparative performance evaluation, regarding:

- Trellis Coded Modulation (TCM vs. Multilevel TCM; single-carrier vs. multi carrier).
- TCH codes vs. standard FEC codes with simplified receiver design.
- New Direct Spread-CDMA receiver concepts :
  - Multiuser detection based on generalizing the Sliding Window Algorithm (SWA), and coupled with adaptive antenna arrays.
  - Different Schemes of Blind Adaptive Multiuser Detectors.

**Table 4.1.** WG3 working areas and tasks

| WG3100 channel characteristic, propagation, measurements, modelling (TOR, SRU, TZD, NTUA, ENST) | Channel characteristics for non-Geo satellites |
| | Propagation |
| | Measurements |
| | |
| | Modelling |
| WG3200 multiple access techniques (UFI, NTUA, CNET, CSELT, TOR, IST) | Multiple access and inter-system interference |
| | Comparison of TDMA CDMA |
| | PRMA via satellite |
| | Multimedia applications |
| | Modulation |
| | Coding |
| WG3300 receivers (IJS, AUT, UCL, CSELT, SRU, IST, TZD, UFI) | Equalization with variable data rates |
| | Smart antennas for space diversity |

– OFDM receiver in a LEO satellite link with different service classes on the same radio channel.
– TCH codes vs. standard Multiuser CDMA codes.

### 4.1.1 WG3 Working Areas and Tasks

WG3 tasks were grouped into three working areas as shown in Table 4.1, which also shows the corresponding tasks and the institutions involved, according to their expressed will.

## 4.2 Propagation and Diversity Characteristics of L- and EHF-Band Systems

A survey of land mobile satellite channel characteristics is presented in this section. The channel characteristics comprise L-band data for Low Earth Orbiting (LEO) personal communications satellite systems as well as EHF-band data for future high bit rate multimedia services. The characteristics are given for narrowband and wideband applications in different scenarios and environments.

### 4.2.1 Introduction

Several system alternatives are being considered for provision of mobile and personal satellite services in the near future. Many proposals adopt non-geostationary satellite constellations, thus the channel characteristics are not stationary. Furthermore, multiple access techniques under consideration range from narrowband to wideband (e.g. CDMA) solutions. Finally, due to the requirement of being virtually global, a satellite system should provide service in a wide range of environmental conditions. The interesting frequency range covers L/S-band, K/Ka-band and the EHF-band.

Propagation measurements at L-band have previously been made by several organisations, for LEO, MEO, HEO and GEO systems. Aircraft and helicopters have been used to simulate satellites in a wide range of elevation angles and environments (1, 2, 3, 4). Some wideband channel measurements have been performed (5, 6) but results are available to a limited extent. DLR has performed a measurement campaign aiming at the channel characteristics of LMS systems (7).

Today, there exist a lot of different approaches for modelling the LMS channel. They mostly differ in effects and scenarios that can be simulated. A comparison of models is described in (8). In this paper we use only the Lutz model as narrowband model (9) and the DLR wideband model (10, 11) as described in (12).

At higher frequencies, the availability of channel data is scarce. While parameters of the K-band channel have been published, the EHF-band is completely unknown. Furthermore, no wideband data are known at higher frequencies. In the framework of SECOMS (Satellite EHF Communications for Mobile Multimedia Services, an European ACTS project) a satellite propagation experiment at 40.1 GHz has been performed.

Since satellites are not yet available with sufficient link margins and transponder bandwidth to test the wideband and narrowband characteristics, a suitable aircraft carrier (CESSNA 207) was used to simulate the satellite. This approach allows also to test the channel properties for a wide range of environments in full azimuth and elevation ranges. The advantages and limitations of airborne test platforms are discussed in more detail in (13).

The purpose of the measurement campaign was (i) to perform narrowband measurements in order to collect a channel database with a wide range of environments, elevation angles and azimuth angles; (ii) to investigate the wideband characteristics of the land mobile satellite channel; and (iii) to investigate antenna steering algorithms at higher frequencies in presence of shadowing and fading, using high gain antennas.

## 4.2.2 Channel Characteristics at L-Band

### 4.2.2.1 Measurement Set-Up

An RHCP drooping dipole antenna for the handheld and a car-roof mounted RHCP antenna have been used. The receiver has a dynamic range of 40 dB and a filter bandwidth of 1 kHz. For the wideband measurements, a spread spectrum signal using a pseudo noise bit sequence with a bandwidth of 30 MHz was transmitted. This corresponds to a spatial resolution of 10 m. The sampling rate corresponds to 15.6 impulse responses per second. The measurement set-up has been described previously in more detail in (14). There, the environments and the operational scenarios are listed, too.

### 4.2.2.2 Narrowband Results

The narrowband LMS channel behaviour shall be demonstrated using Figs 4.1 and 4.2 as two examples for the urban LMS channel. In Fig. 4.1 the situation is given for quasi-fixed users, i.e. a handheld

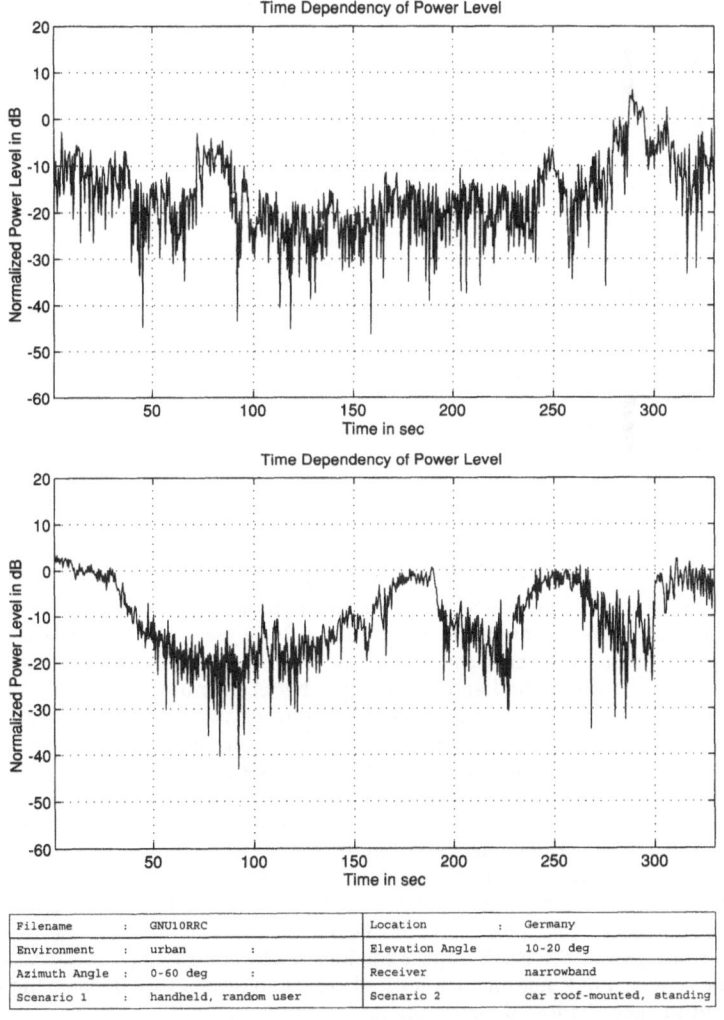

| Filename | : | GNU10RRC | | Location | : | Germany |
|---|---|---|---|---|---|---|
| Environment | : | urban | : | Elevation Angle | | 10-20 deg |
| Azimuth Angle | : | 0-60 deg | : | Receiver | | narrowband |
| Scenario 1 | : | handheld, random user | | Scenario 2 | | car roof-mounted, standing |

**Figure 4.1.** Narrowband power series. Upper graph: handheld, lower graph: car-roof mounted, standing, L-band.

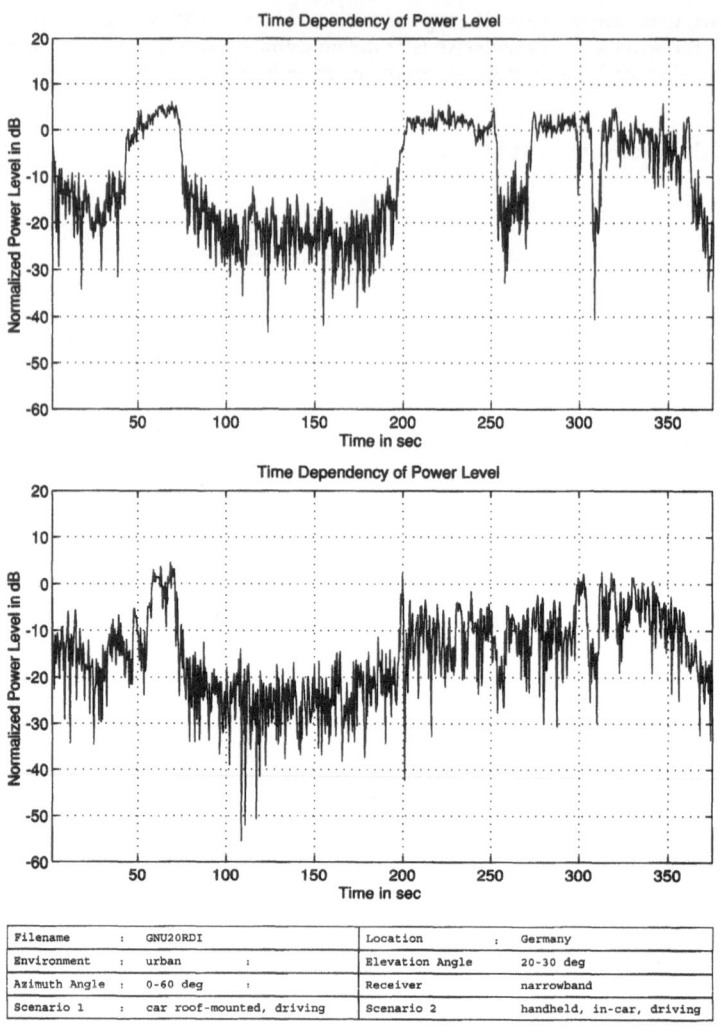

| Filename | : | GNU20RDI | | Location | : | Germany |
|---|---|---|---|---|---|---|
| Environment | : | urban | : | Elevation Angle | | 20-30 deg |
| Azimuth Angle | : | 0-60 deg | : | Receiver | | narrowband |
| Scenario 1 | : | car roof-mounted, driving | | Scenario 2 | | handheld, in-car, driving |

**Figure 4.2.** Narrowband power series. Upper graph: car-roof-mounted, driving, Lower graph: handheld in-car, driving, L-band.

terminal (upper graph) and a standing van (lower graph) whereas in Fig. 4.2 the user is driving. Here, one can compare the performance between the in-car handheld antenna and a car-roof mounted antenna. For the low elevation angles between 10 and 30 degrees, shadowing is the major effect. In shadowed conditions the channel attenuation is 20 ... 30 dB. In both operational scenarios, standing and driving, the car-roof mounted antenna gets the better channel. The handheld terminal suffers from head shadowing and worse line-of-sight (LOS) conditions (since the antenna on the roof of our van has a higher height of approx. 2.3 m). Two-path fading caused by specular reflections is affecting the handheld, but not the car-roof mounted antenna. Fades from specular reflections have a depth ranging from 3...9 dB. Comparing the signal during the driving run, the handheld in-car antenna seems to have an attenuation of 3...10 dB with respect to the roof antenna in LOS conditions.

Cumulative distributions can be derived from the power values. From the CDFs one can easily take link margins for a required grade of service. In Fig. 4.3, values for the required link margin are given for two grades of service, 95 per cent and 98 per cent, in urban and suburban environment. In these environments we expect a large percentage of shadowing. This will increases the average link margin. Moderate link margins of 6...10 dB appear for higher elevation angles above 50 degrees. For low elevations, the required link margins will exceed reasonable values by far. Satellite diversity will be of great benefit in such shadowed environments.

Furthermore, the parameters of channel models can be derived from the measured data. This has

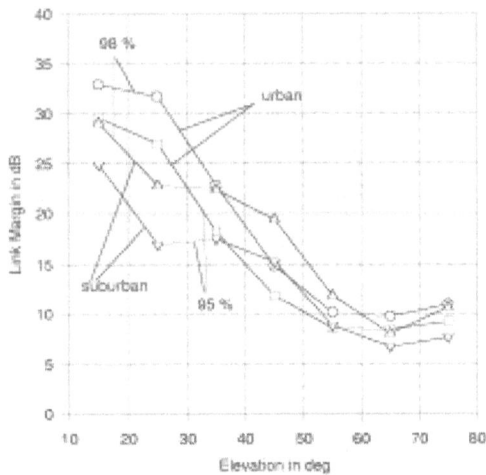

**Figure 4.3** Link margin versus elevation angle for urban and suburban environment. RHCP handheld.

been done for various environments and elevation angles. The parameters of the Lutz-model (15) can be found in (16).

### 4.2.2.3 Wideband Results

The wideband results are presented as channel impulse responses $h(t, \tau)$ (sometimes also referred to as the time-delay system function). The delay of an echo is denoted by $\tau$ with respect to the propagation delay of the direct path. Figure 4.4 shows the impulse responses of the wideband LMS channel for different environments and elevation angles. Common to all environments is that echoes appear in general with small delays, usually smaller than 600 nsec (corresponding to a detour of 200 m). The echoes are attenuated by 10...30 dB. The environment is mainly characterised by shadowing of the direct path and the number and attenuation of echoes. It is also obvious that the power of echoes with small delays decreases exponentially versus delay (note the logarithmical scale).

Figure 4.4a and 4.4b show a handheld in a rural environment, standing on a provincial road. At 15 deg elevation the user is shadowed by trees. At 55 deg elevation, the user has line-of-sight (LOS) conditions most of the time. About 3–5 echoes can be detected with delays up to 300 nsec and an echo attenuation of 15–25 dB. In the mountainous environment, (Fig. 4.4c and d), we can find echoes with longer detours and a higher echo power. In the standing scenario, the echo delay is stationary, the number of echoes depend only on the azimuth angle of the satellite whereas in driving conditions the echo delay is changing monotonously (cf. Fig. 4.4d and e) since the receiver approaches to, or departs from the reflector. The urban environment is again characterised by fewer echoes. On a highway (Fig. 4.4f) the traffic in front of and behind the receiver yields good reflectors. Thus, many echoes can be detected. These echoes have usually short delays.

Although the measurement equipment was able to measure delays of 15 μsec, the figures show power delay profiles for delays up to 2 μsec since most of the echoes appear in the close vicinity of the receiver. The different power statistics of echoes with long and short delays can be seen easily. The power of near echoes is decreasing exponentially. This is not valid for echoes with long delays.

In Fig. 4.5 the delay spread is presented of a handheld phone with RHCP polarisation versus elevation. For higher elevation angles the spread tends to decrease (except for the urban environment). Values range mainly from 500 nsec to 2 μsec.

### 4.2.2.4 Frequency-Selectivity

From the time-delay system function $h(t, \tau)$ with the echo delay $\tau$ and the time $t$, the equivalent Channel Transfer Function CTF $(t, f) = \mathrm{FFT}_{\tau} (h(t, \tau))$ can be derived. It shows the channel properties in the frequency domain. With the CTF, the frequency selectivity of the channel is directly visible. Figure 4.6 gives the CTF for various environments at one moment of time.

All figures are shown for handhelds with RHCP-antenna and a satellite elevation of 25 degrees. In the open environment (a) the CTF is flat over the total system bandwidth of 30 MHz. The channel attenuation is 1 (0 dB). This is an example for an ideal channel behaviour. In more realistic environments the CTF mainly depends on the shadowing of the direct signal. With line-of-sight (LOS) conditions ((c) and (d)), the channel attenuation varies a little around the 0 dB since echoes do not contribute very much to the signal power. In the suburban environment (c) we still find a frequency-

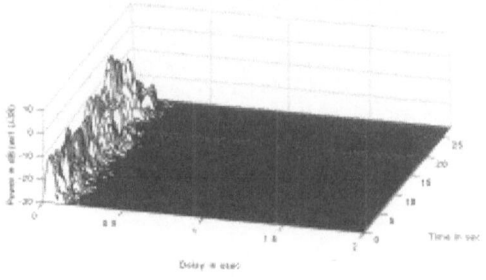

**(a)** provincial road, handheld, 15 degree elevation

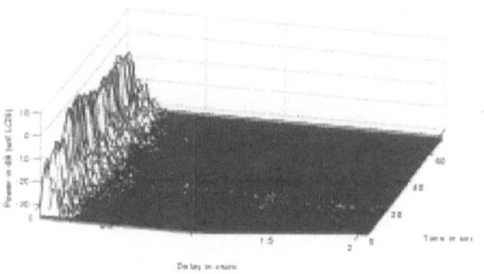

**(b)** provincial road, handheld, 55 degree elevation

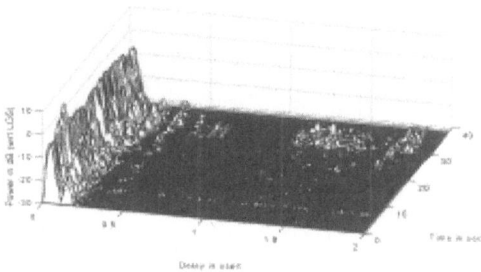

**(c)** mountainous, handheld, 35 degree elevation

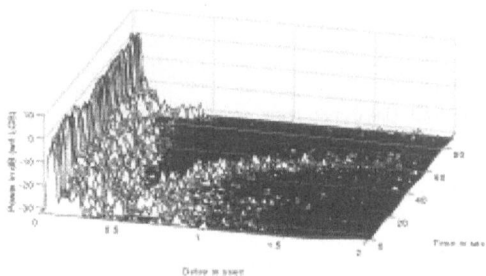

**(d)** mountainous, car driving, 55 degree elevation

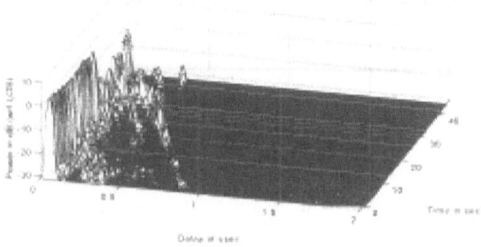

**(e)** urban, car driving, 15 degree elevation

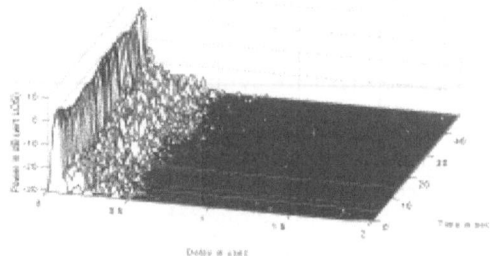

**(f)** highway, car driving, 65 degree elevation

**Figure 4.4.** Power delay profiles for various environments.

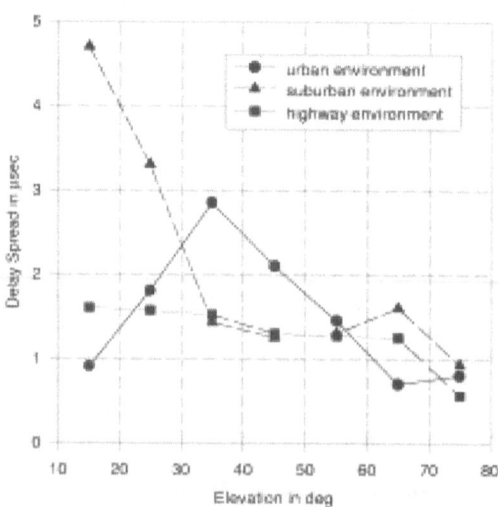

**Figure 4.5.** Delay spread versus elevation angle.

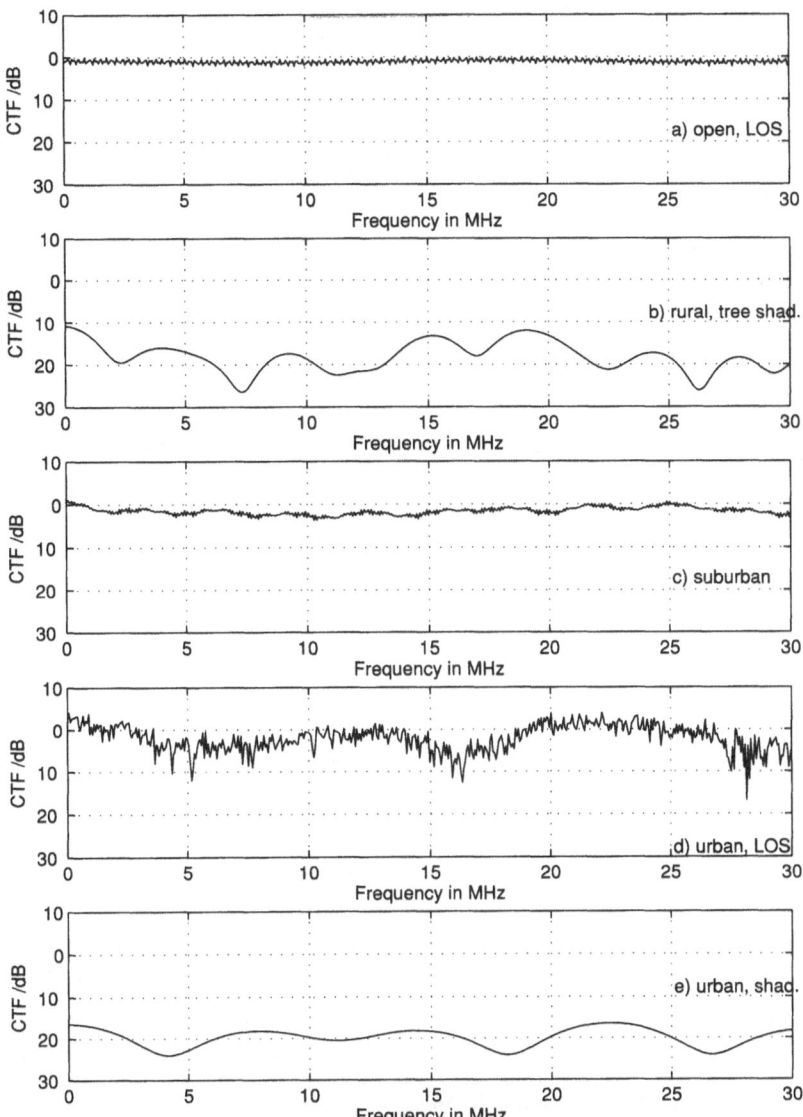

**Figure 4.6.** Channel Transfer Function for different environments.

flat channel for Line-of-Sight (LOS) conditions. Only a small undulation of about 1.5 dB indicates the appearance of some small echoes. The urban environment (d) shows a stronger frequency-selectivity due to the better echo reflections from traffic and buildings. The spectral correlation of 5 MHz indicates that the main echoes are delayed by 200 nsec corresponding to a detour of 60 m. In shadowed conditions in the rural (b) and urban (e) environment, the CTFs are attenuated by 10 ... 20 dB, and the channel becomes frequency-selective just above 3 ... 5 MHz.

Using CDMA with a RAKE architecture, the receiver can benefit from echoes. The maximum theoretical gain corresponds to the difference between the minimum and the maximum of the CTF (about 5...10 dB for worst case environments) when all echoes are detected and combined. Then the system bandwidth should correspond to the frequency bandwidth between the minimum and the maximum value at least. Narrowband TDMA can gain the same when applying slow frequency hopping in shadowed environments when the hopping distance is 3 MHz or higher. This is particularly interesting for call alert or paging services.

**Table 4.2.** Model parameters for the direct path

| Parameter:<br><br>Environment elevation | Time share of<br>shadowing<br>$A$ | $a_l$, LOS<br>Rice<br>$c$ (dB) | $a_l$, shadowed<br>Rayleigh/LN | |
|---|---|---|---|---|
| | | | $\mu$ (dB) | $\sigma$ (dB) |
| Open, 15, degrees | 0 | 6.0 | — | — |
| 25, degrees | 0 | 10.3 | — | — |
| 35, degrees | 0 | 12.0 | — | — |
| 45, degrees | 0 | 10.4 | — | — |
| 55, degrees | 0 | 9.0 | — | — |
| Rural [1], 15, degrees | 0.99 | 7.0 | −11.4 | 1.1 |
| 25, degrees | 0.96 | 10.8 | −9.9 | 3.3 |
| 35, degrees | 0.83 | 4.8 | −6.2 | 3.9 |
| 45, degrees | 0.79 | 4.7 | −5.4 | 2.3 |
| 55, degrees | 0.93 | 4.8 | −7.2 | 3.9 |
| Suburban, 15, degrees | 0.77 | — | −12.6 | 4.8 |
| 25, degrees | 0.59 | 4.7 | −6.0 | 3.5 |
| 35, degrees | 0.54 | 10.7 | −7.6 | 3.2 |
| 45, degrees | 0.43 | 4.0 | −7.2 | 3.2 |
| 55, degrees | 0.35 | 11.8 | −7.7 | 2.6 |
| Urban, 15, degrees | 0.97 | — | −15.2 | 5.2 |
| 25, degrees | 0.79 | 3.2 | −12.1 | 6.3 |
| 35, degrees | 0.6 | 4.8 | −4.4 | 5.1 |
| 45, degrees | 0.56 | 8.5 | −3.0 | 2.7 |
| 55, degrees | 0.3 | 6.0 | — | — |
| Highway, 15, degrees | 0.24 | 9.5 | −9.3 | 5.6 |
| 25, degrees | 0.19 | 8.4 | −5.8 | 1.7 |
| 35, degrees | 0.01 | 8.5 | −5.0 | 3.3 |
| 45, degrees | 0 7.8 | −1.8 | 1.2 | |
| 55, degrees | 0 | 9.0 | — | — |

### 4.2.2.5 Wideband Channel Model

From the measurements a wideband channel model was derived which is explained in (17) in more detail. The model uses three submodels depending on the echo delay: direct signal, near echoes and far echoes. The complex impulse response of the satellite wideband channel can then be superimposed by a sum of all echoes.

The model parameters have been determined using least square fits with measured data. The Gauss–Newton algorithm was adopted for computation. The fits have been performed for a lot of environments and several elevation ranges. A set of about 3000 files and 50 GByte of data have been used to select typical results representing the 50 per cent-quantile of the delay spread values. Numeric values of the parameters of the direct path are given in Table 4.2, the parameters for the near and far echoes are given in Tables 4.3 and 4.4, respectively.

All parameters are given for an handheld user, except for the highway environment where a car-roof mounted antenna was used. The parameter sets are distinguished by the environments. Note that the background environment determines the far echoes and the foreground environment the near echoes, respectively. This model has been also submitted to ETSI for standardisation (18, 19).

A simpler model with fixed number of taps and fixed tap delays has been submitted to ITU (20). Its parameters are given in Tables 4.5 to 4.7. The environments city, rural and suburban correspond to the 90 per cent, 50 per cent and 10 per cent-percentiles of the delay spread values of all measurements.

## 4.2.3 Channel Characteristics at EHF-Band

### 4.2.3.1 Measurement Set-Up

The receiver were roof-top mounted on a van Mercedes 209D. The environments of interest were: open, rural with tree-shadowing (one or two sides), suburban and urban. The measurement set-up basically consists of a test transmitter in an aeroplane and a receiver in a vehicle. The transmitter radiates a test signal, either wideband or narrowband, and the receiver is probing the channel for the transmitted test signal. The carrier frequency is 40.175 GHz. The receiver consists of two-stage demodulators

**Table 4.3.** Model parameters for near echoes

| Parameter: Environment, elevation | $N(n)$ : Poisson $\lambda$ | Max. delay $\tau_e$ (nsec) | Delay $\Delta\tau^{(n)}$ exp. $b$ ($\mu$sec) | $S(\tau)$ | |
|---|---|---|---|---|---|
| | | | | $S_0$ (dB) | $d$ (dB) |
| Open, 15, degrees | 1.6 | 400 | 0.033 | −28.5 | 3.0 |
| 25, degrees | 1.2 | 400 | 0.03 | −28.6 | 1.0 |
| 35, degrees | 1.2 | 400 | 0.027 | −25.7 | 9.5 |
| 45, degrees | 0.5 | 400 | 0.027 | −29.0 | 1.1 |
| 55, degrees | — | 400 | — | — | — |
| Rural, 15, degrees | 1.8 | 400 | 0.061 | −25.9 | 10.7 |
| 25, degrees | 1.5 | 400 | 0.055 | −24.9 | 19.2 |
| 35, degrees | 1.6 | 400 | 0.043 | −25.3 | 14.1 |
| 45, degrees | 1.8 | 400 | 0.051 | −24.5 | 13.4 |
| 55, degrees | 1.6 | 400 | 0.047 | −21.7 | 36.8 |
| Suburban, 15, degrees | 1.2 | 400 | 0.037 | −22.6 | −21.9 |
| 25, degrees | 1.4 | 400 | 0.038 | −23.8 | 23.7 |
| 35, degrees | 1.2 | 400 | 0.039 | −24.9 | 19.4 |
| 45, degrees | 1.5 | 400 | 0.027 | −24.4 | 23.0 |
| 55, degrees | 1.6 | 400 | 0.033 | −24.7 | 18.7 |
| Urban, 15, degrees | 1.2 | 600 | 0.118 | −16.5 | 11.0 |
| 25, degrees | 4.0 | 600 | 0.063 | −17.0 | 26.2 |
| 35, degrees | 3.5 | 600 | 0.069 | −23.6 | 6.5 |
| 45, degrees | 3.6 | 600 | 0.081 | −23.5 | 8.5 |
| 55, degrees | 3.8 | 600 | 0.079 | −26.1 | 6.3 |
| Highway, 15, degrees | 1.2 | 600 | 0.072 | −27.0 | 6.4 |
| 25, degrees | 2.2 | 600 | 0.077 | −25.8 | 7.3 |
| 35, degrees | 2.8 | 600 | 0.091 | −26.8 | 30.6 |
| 45, degrees | 1.8 | 600 | 0.043 | −27.1 | 29.5 |
| 55, degrees | — | — | — | — | — |

**Table 4.4.** Model parameters for far echoes

| Parameter: Environment, elevation | $N(f)$ Poisson $\lambda$ | $a_k(f)$ Rayleigh $2\sigma^2$ (dB) | Max. delay $\tau_{max}$ ($\mu$sec) |
|---|---|---|---|
| Flat terrain, 15, degrees | | | 15 |
| 25, degrees | 0.3 | −26.4 | 15 |
| 35, degrees | | | 15 |
| 45, degrees | | | 15 |
| 55, degrees | | | 15 |
| Rural, 15, degrees | | | 5 |
| 25, degrees | 0.8 | −28.2 | 5 |
| 35, degrees | | | 5 |
| 45, degrees | | | 5 |
| 55, degrees | | | 5 |
| Hilly, 15, degrees | — | — | 10 |
| 25, degrees | 1.2 | −29.0 | 10 |
| 35, degrees | — | — | 10 |
| 45, degrees | — | — | 10 |
| 55, degrees | — | — | 10 |
| Mountaineous, 15, degrees | 0.9 | −29.0 | 15 |
| 25, degrees | 1.8 | −28.5 | 15 |
| 35, degrees | 4.4 | −23.5 | 15 |
| 45, degrees | 4.0 | −21.7 | 15 |
| 55, degrees | — | — | 15 |

**Table 4.5.** Parameters of the ITU model for city environment (90 per cent percentile of delay spread)

| # of echoes | Delay ns | Amplitude distribution | Parameter of distribution | Value dB |
|---|---|---|---|---|
| 1 | 0 | LOS: Rice | $c$ | 5.2 |
|   |   | nLOS: Rayleigh | $2\sigma^2$ | −12.1 |
| 2 | 60 | Rayleigh | $2\sigma^2$ | −17.0 |
| 3 | 100 | Rayleigh | $2\sigma^2$ | −18.3 |
| 4 | 130 | Rayleigh | $2\sigma^2$ | −19.1 |
| 5 | 250 | Rayleigh | $2\sigma^2$ | −22.1 |

**Table 4.6.** Parameters of the ITU model for rural environment (50 per cent percentile of delay spread)

| # of echoes | Delay ns | Amplitude distribution | Parameter of distribution | Value dB |
|---|---|---|---|---|
| 1 | 0 | LOS: Rice | $c$ | 6.3 |
|   |   | nLOS: Rayleigh | $2\sigma^2$ | −9.5 |
| 2 | 100 | Rayleigh | $2\sigma^2$ | −24.1 |
| 3 | 250 | Rayleigh | $2\sigma^2$ | −25.2 |

**Table 4.7.** Parameters of the ITU model for suburban environment (10 per cent percentile of delay spread)

| # of echoes | Delay ns | Amplitude distribution | Parameter of distribution | Value dB |
|---|---|---|---|---|
| 1 | 0 | LOS: Rice | $c$ | 9.7 |
|   |   | nLOS: Rayleigh | $2\sigma^2$ | −7.3 |
| 2 | 100 | Rayleigh | $2\sigma^2$ | −23.6 |
| 3 | 180 | Rayleigh | $2\sigma^2$ | −28.1 |

**Figure 4.7.** Picture of the steered antenna platform with antenna and 40-GHz demodulator (r.) and video camera (l.)

transferring the 40 GHz EHF-signal to a 1.8 GHz intermediate frequency band. Commercially available channel sounders working in this band can now be used to measure the received signal. The wideband equipment uses a signal bandwidth up to 30 MHz and allows the measurement of 15 impulse responses per second. The dynamic range of the measurement system is better than 30 dB.

A high bit rate user terminal at EHF-band will likely need a high-gain self-steering antenna with a beamwidth of 5 ... 7 degrees. For the channel measurements, the steering of the user antenna should be implemented in the most optimal way, in order to eliminate the effect of antenna steering. Therefore, a GPS-based implementation of the steering was selected. The aircraft sends its position through a telemetry link to the measurement vehicle. The receiver is equipped with a steering processor calculating the pointing angle to the aircraft emulating the satellite. A mechanical positioner is used to orient the user antenna, cf. Fig. 4.7. In order to achieve a high pointing accuracy, the assessment of on- line differential GPS is required. A GPS reference station is therefore operated at DLR. This leads to a position accuracy of 1 ... 2 m. A precise positioning reference system in the vehicle, providing the orientation with respect to a reference plane, is mandatory, too. The resulting accuracy of the steering has been proved by a video camera mounted on the antenna platform. The pointing error was less than 0.7 degrees. An advantage of this antenna steering approach is that self-steering algorithms can be implemented in user terminals, e.g. using a signal strength criterion. The

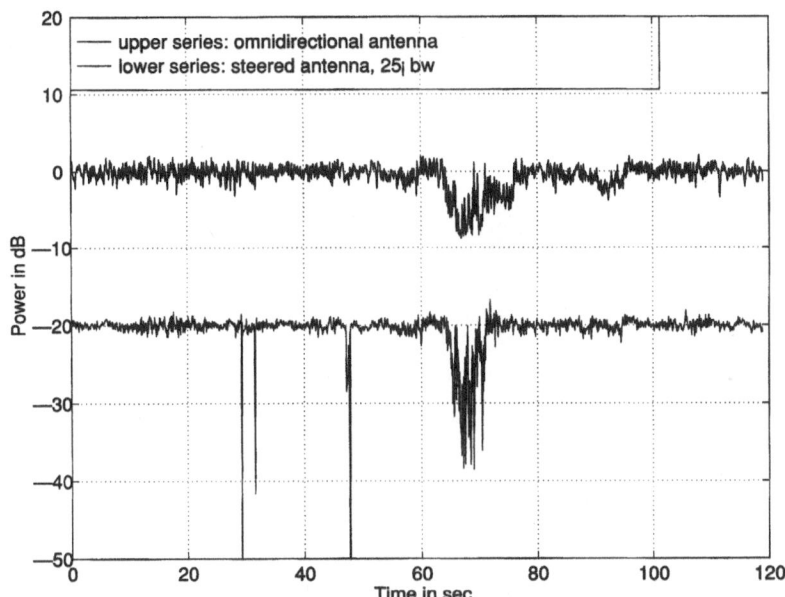

**Figure 4.8.** Narrowband channel series; rural road; 25-degree elevation; upper curve: hemispherical antenna, lower curve: steered antenna.

performance of different steering algorithms can then be compared with the ideal channel behaviour measured with the GPS- based approach.

Two independent receivers allow the simultaneous reception with different antennas or different receiver types. Thus, it is possible to measure narrowband and wideband signals, or narrowband signals with an omni-directional antenna and a high-gain antenna, for instance. High-gain antennas require steering which is only possible on one platform. The non-steerable platform is equipped with an omni-directional antenna. The different combinations of narrowband and wideband receivers with antennas and platforms allow (i) the comparison of different antennas types; (ii) the comparison of narrowband and wideband measurements; and (iii) the investigation of the antenna steering algorithms by comparison with an omni-directional reference antenna.

### 4.2.3.2 Narrowband Results

All measurements have been normalised with respect to the free space propagation loss. Narrowband power plots versus time in Fig. 4.8 compare the performance of a high-gain antenna (25 degree beamwidth) with an omni-directional one in a provincial road environment at an elevation angle of 25 degrees. Little multipath fading can be observed beside short shadowing events. It is obvious that the omni-directional antenna picks up more multipath due to the lower discrimination of echoes, but for the same reason it is less sensible to short blockage. Note that for the purpose of better presentation, the series of the high-gain antenna was offset by −20 dB. The Rice-factor in non-shadowed conditions for the steered high-gain antenna is 2.15 dB, whereas the omni-directional antenna yields to an Rice-factor of 17 dB.

In Figs 4.9 and 4.10, narrowband measurements of the steered high-gain antenna in an urban and city environment are given (the reader should note the different time scaling). The power series is characterised by severe shadowing and blockage. In shadowed conditions, the channel attenuation is 10 ... 20 dB.

This relatively strong multipath power (also visible in the fading during LOS) indicates that the 25 degree antenna can still pick up reflected and diffuse components.

Figure 4.11 shows the cumulative distribution function of the channels given in Figs 4.9 and 4.10. In situations with direct sight to the satellite, a moderate link margin (less than 4 dB) is sufficient for a service availability of more than 95 per cent. However, in shadowed environment the link margin would be as high as 30 dB to get a good availability. This high margins seem unrealistic for satellites from a cost point-of-view.

System simulations often use a channel model to evaluate the performance of a transmission

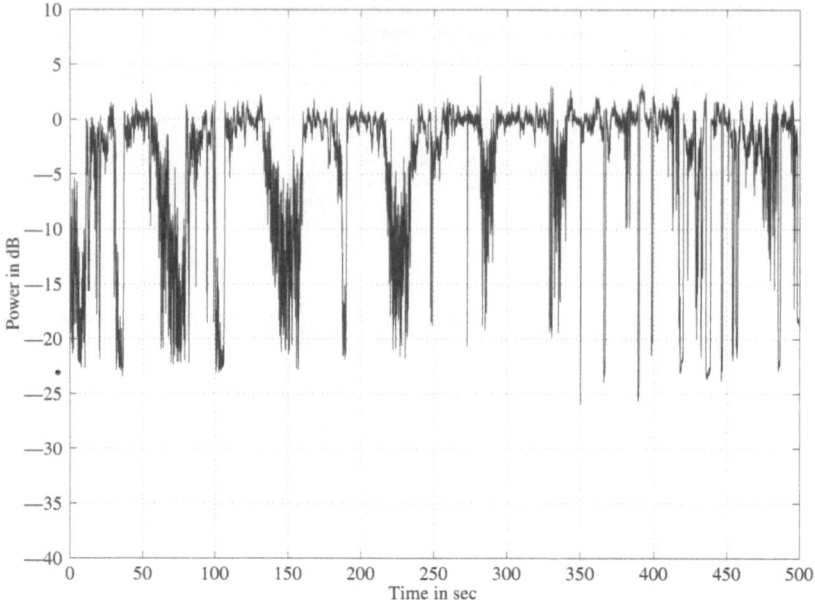

**Figure 4.9.** Narrowband channel measurement in an urban area; 25-degree elevation; steered antenna.

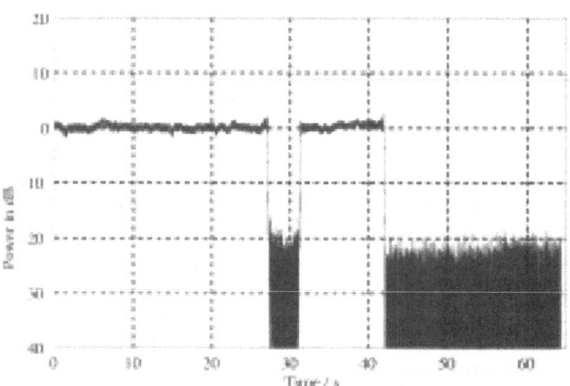

**Figure 4.10.** Narrowband channel measurement in city area; 35° elevation; steered antenna.

system. For LMS channels the Lutz-model is widely used (21). The channel behaviour is characterised by two states, a *good* and a *bad* channel state, corresponding to the non-shadowed and shadowed channel. The bad states occur with probability $A$, named as shadowing factor. In the good state, the channel power $y^2$ is described by a Ricean distribution

$$f_{\text{Rice}}(y^2) = c \exp\left[-c\left(1 + y^2\right)\right] I_0(2c\sqrt{y^2})$$  (4.1)

with $I_0$ being the Bessel-function of first kind and zero'th order, normalised to a carrier power (direct signal) of 1. The parameter $c$ is called Rice-factor and determines the ratio of direct to multipath power. In the bad state, a Rayleigh-distribution is appropriate

$$f_{\text{Rayl}}(y^2) = \frac{1}{2\sigma_y^2} \exp\left(-\frac{y^2}{2\sigma_y^2}\right)$$  (4.2)

with the mean power $P_0 = 2\sigma_y^2$. The mean power follows a lognormal distribution

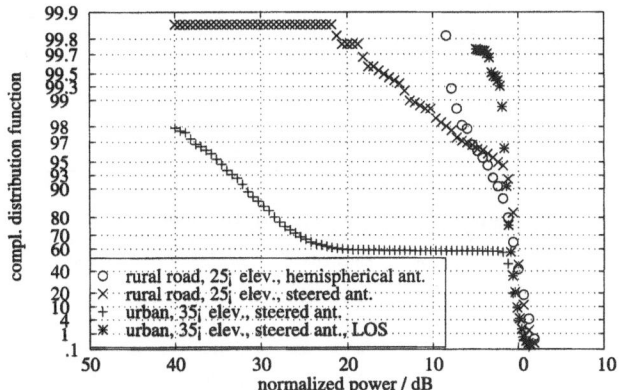

**Figure 4.11.** Cumulative distribution functions of LMS channels.

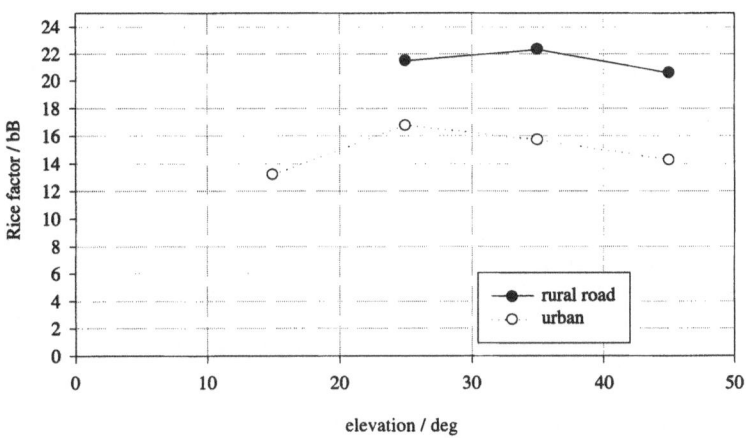

**Figure 4.12.** Rice factor of the LMS channel.

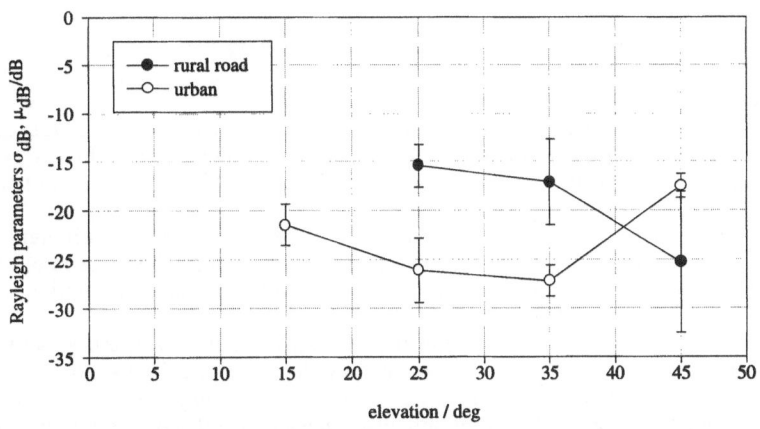

**Figure 4.13.** Rayleigh/Lognormal parameters of the LMS channel.

**Table 4.8.** EHF-band channel parameters, rural road, standing car, steered antenna (note that the increase of the shadowing with elevation is caused by the location of the receiver under trees with foliage)

| Elevation, degrees | Shadowing A | Rice f.c/dB | $\mu_{dB}$/dB | $\sigma_{dB}$/dB |
|---|---|---|---|---|
| 25 | 0.21 | 21.5 | −15.4 | 2.2 |
| 35 | 0.40 | 22.3 | −17.1 | 4.4 |
| 45 | .81 | 20.6 | −25.3 | 7.2 |

**Table 4.9.** EHF-band channel parameters, urban, standing car, steered antenna

| Elevation, degrees | Shadowing A | Rice f.c/dB | $\mu_{dB}$/dB | $\sigma_{dB}$/dB |
|---|---|---|---|---|
| 15 | 0.71 | 13.2 | −21.5 | 2.1 |
| 25 | 0.17 | 16.8 | −26.1 | 3.3 |
| 35 | 0.24 | 15.7 | −27.2 | 1.6 |
| 45 | .08 | 14.3 | −17.5 | 1.2 |

$$f_{LN}(P_0) = \frac{20}{\sqrt{2\pi}\sigma_{dB}\ln 10} \cdot \frac{1}{P_0} \cdot \exp\left(-\frac{(10\log P_0 - \mu_{dB})^2}{2\sigma_{dB}}\right) \tag{4.3}$$

with the parameters $\mu_{dB}$ and $\sigma_{dB}$, often referred to as dB-mean and dB-spread. Thus, the LMS channel in a given environment can be characterised by four parameters. These parameters have been determined for the EHF-band channel by least-square fits to the measured data. Figures 4.12 and 4.13 show a graphical representation of the Rice-factor and the parameters of the lognormal distribution, respectively. Tables 4.8 and 4.9 give the numerical values.

### 4.2.3.3 Wideband Results

In Fig. 4.14 an example of the wideband results is shown. The measurement was taken in an tree-shadowed rural environment at 25-degree elevation using the omni-directional antenna. The 0-dB line corresponds to an undisturbed signal under line-of-sight condition. The signal on the direct path is subject to shadowing. Echoes appear very rarely, if at all (note that signals below −29 dB are caused by the noise floor of the wideband receiver). Compared to results at L-band obtained at the same location, the number of echoes is much smaller and the echo attenuation is higher at EHF-band (22). Fig. 4.15 shows the situation in the urban environment.

The shadowing process for the direct component of the signal is now stronger. The attenuation reaches 15 ... 25 dB if shadowed. Furthermore, the number of echoes is higher. There are few (2–4) echoes with short delays (30...100 nsec), and a couple of echoes with delays of 350, 600 and 850 nsec. The echo attenuation is in the range of 20 ... 30 dB.

## 4.2.4 Shadowing Correlation For Multi-Satellite Diversity

### 4.2.4.1 Introduction

Land mobile satellite (LMS) system models together with numerous propagation measurements carried out using planes or other platforms to simulate the mobile-to-satellite link geometry indicate that signal shadowing is the dominant feature influencing LMS system availability and performance.

While multipath fading can be overcome for a given fade margin by using different transmission techniques, blockage effects can hardly be mitigated resulting in high bit error rates and in temporary unavailability. Given the power limitation, especially in the up-link, the solution to reduce such shadowing effects is path or satellite diversity.

Although most available LMS propagation models can reproduce with fairly good accuracy the various channel effects (fading, time dispersion, Doppler, etc.) and be used in the evaluation of different channel coding, modulations or equalisation techniques, only those models based on a physical approach can be directly used to simultaneously analyse various satellite-mobile links, and thus, evaluate the benefits of satellite diversity.

Physical models try to quantify the magnitudes of the different contributions due to the various propagation mechanisms involved in the LMS channel. Basically, shadowing is due to diffraction

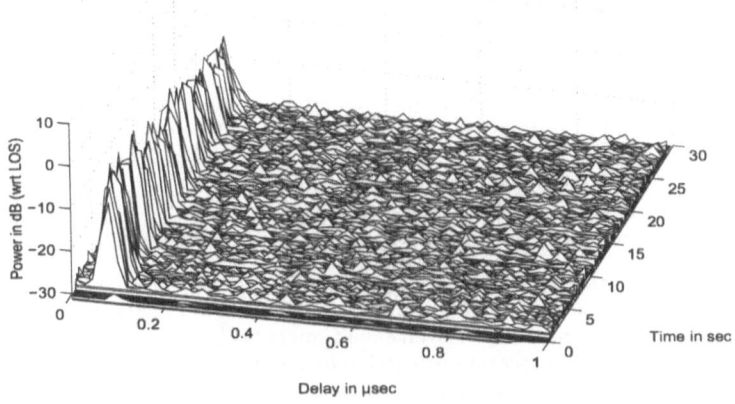

**Figure 4.14.** Wideband impulse response measurements of the LMS channel; open area.

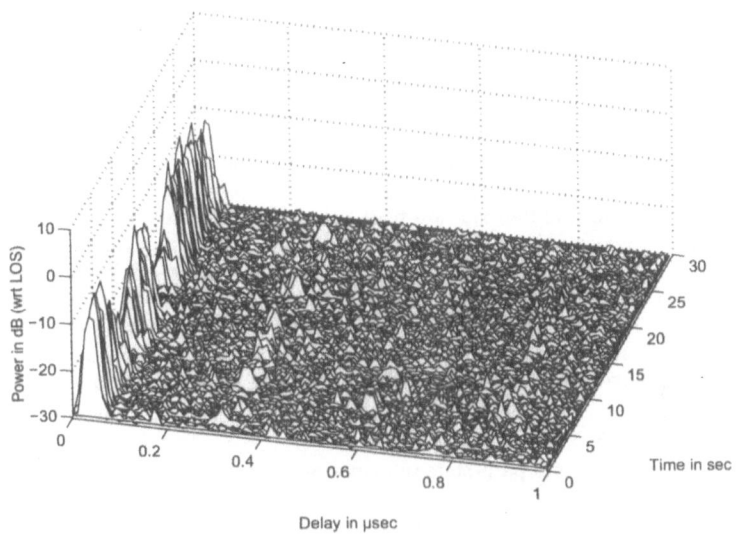

**Figure 4.15.** Wideband impulse response measurements of the LMS channel; urban area; 25-degree elevation.

effects, especially in urban areas, although other mechanisms like absorption losses through vegetation are also a major player in a number of environments.

Concentrating on urban area studies, diffraction based models can be used if detailed urban layout information is available. If this is the case, the simultaneous study of one or several mobile-satellite links is feasible and thus, possible shadowing correlation effects can be accounted for when evaluating macro-diversity gains.

On the other hand, detailed urban layout information can be supplied by means of urban data bases when *deterministic models* are used. Other approach that makes use of synthetic environments is the use of *physical statistical models*. In this case, the input environment is generated from statistical distributions describing building heights and general urban layout.

In both cases, deterministic and physical-statistical models, diversity effects can be directly assessed by extending the single link study to a multi-link study performed for the input scenario. This is not the case in other modelling approaches like *purely* or *empirical models*. For the case of statistical models the quantification of the correlation coefficient for different link angle separations is needed to be "forced" into the models to accurately reproduce the effects of correlated shadowing effects, which are more likely to occur when the angle separation between links is small. A method proposed in (23) is an example of how a single-satellite statistical model can be extended to study the simultaneous behaviour of two partially correlated links by introducing in the modelling the shadowing correlation coefficient, $\rho$.

A number of specific methods are found in the literature that can be used to analyse and quantify this effect. This section describes and discusses these techniques stressing the various conceptual approaches followed by different authors.

It is clear that gains in service availability through satellite diversity can only be achieved if the various links behave in an uncorrelated manner, or even better, if they present negative correlation values.

### 4.2.4.2 Overview of Other Satellite Diversity Studies

In this section a brief summary of a number of relevant studies on satellite diversity is presented:

(a) In (24) an extension of the two-state model proposed in (25) for a single satellite-mobile link to two angle-spaced links was proposed. The approach followed to evaluate the correlation coefficient was to use circular scans within a given environment to obtain numerical landscape pictures in which a "0" or "1" would represent obstruction or visibility respectively. A campaign in rural, suburban and urban environments using a video-camera to record the landscape was carried out. The outcome of this study was the formulation of an empirical model for the correlation coefficient for a number of environments.

(b) The two-state model proposed in (26) was extended by the same author to model two correlated links (27). Lutz proposed a four-state Markov model (Fig. 4.16) to describe the possible combinations of good and bad states in two different links. Equilibrium state and transition probabilities for a four-state model were computed for the correlated and the uncorrelated cases in terms of the individual two-state model probabilities and of the correlation coefficient, $\rho$.

It is worth noting that the range of values of the correlation coefficient, $\rho$, that can be used in this four-state model is limited (depending on the individual two-state model transition probability values).

This model allows the simultaneous study of two satellite links with a given constant correlation behaviour. In his paper Lutz does not provide numerical values for the correlation coefficient. Further studies (28) have provided correlation coefficient values extracted from experimental data. As it is also pointed out later in this section, high correlation coefficient for 90 degrees and 180 degrees were observed in regular grid urban environments.

(c) A so called photogrammetry method for the analysis of path diversity for LEO Satellite-PCS networks in urban environments was presented in (29, 30). The method consists of the following steps:

1. Taking fisheye photos at potential user locations.
2. Extracting from the images path-state information (clear/shadowed/blocked) as a function of look angles.

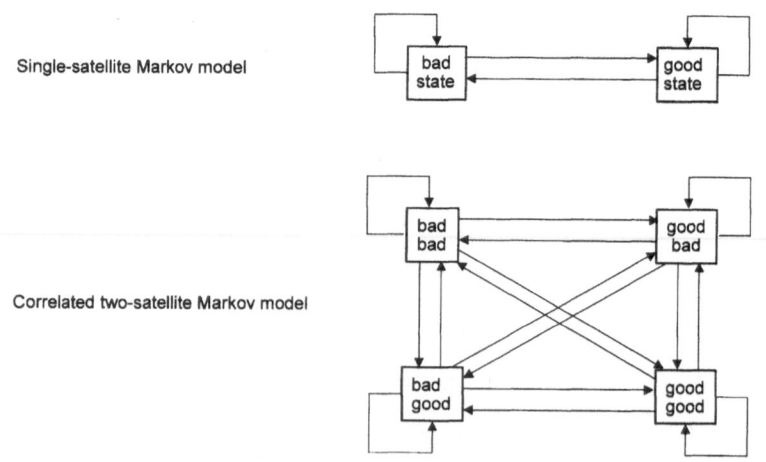

**Figure 4.16.** Two-state and four-state Markov models with transition probabilities.

3. Combining each path-state for single or multiple satellites in a specific constellation using appropriate statistical fade models. Each of the three possible path states of the mobile satellite link is associated with a given fade distribution. The clear state is described using a Rice distribution, the shadowed and the blocked states are modelled using the Loo model. The derived urban three-state model was:

$$f_r(r, \alpha) = C(\alpha)f_{Rice}(r) + S(\alpha)f_{Loo}(r) + B(\alpha)f_{Loo}(r) \tag{4.4}$$

where $\alpha$ is the elevation angle. The Rice and Loo distribution parameters were extracted from measurements. The values (per cent) of C, S and B were estimated from fisheye pictures (31).

Also, the authors extracted from the images of rural, suburban and urban Texas locations, information about the elevation angles at which the sky becomes visible, i.e. what the authors define as the skyline. In the three environments, the ninetieth percentile of the skyline was found to be at 11, 17 and near 55 degrees, respectively.

(d) Another approach for the study of satellite diversity can be found in (32). In this study, a switching diversity scheme was studied. The clear state is described by a Rice distribution, the shadowed state is modelled by means of the Loo distribution and the blocked state is modelled with a Rayleigh distribution.

A three-state Markov model is assumed to account for the large dynamic range of the received signal. The use of uncorrelated Markov models, one per satellite link was proposed (Fig. 4.19) and it is assumed that state occurences for one satellite link are not correlated with those of other satellite links. Results of this study showed a significant improvement in service availability thanks to the diversity effect.

(e) In (33), the availability of the ICO and Globalstar systems was analysed using the correlated four-state model developed by Lutz. From fisheye pictures taken in Guildford, Southampton, London and Los Angeles, blockage and correlation statistics were extracted. In this study it was observed that for azimuth separations smaller than 30 degrees, satellite channels tend to be correlated.

## 4.2.4.3 Azimuth Correlation of Shadowing

Exploiting multiple satellite visibility on earth, the service availability may be improved substantially. Of course, gain in service availability can only be achieved if the considered satellite channels behave *differently*. For the investigation of satellite diversity, the correlation of two channels are therefore important. The correlation can be used to model two statistically dependent satellite channels, cf. (34).

For the definition of the correlation coefficient we consider the amplitude $h_i(t)$ of channel $i = 1$, 2 as a stochastic process which is 0 for the shadowed channel state and 1 for line-of-sight condition:

$$h_i(t) = \begin{cases} 0 & \text{bad channel state} \\ 1 & \text{good channel state} \end{cases} \tag{4.5}$$

The mean value and variance of the channel amplitude are:

$$E\{h_i(t)\} = \overline{h_i} = A = \frac{D_g}{D_g + D_b} \tag{4.6}$$

$$E\{(h_i(t) - \overline{h_i})^2\} = \overline{h_i} = \sigma_i^2 = \frac{D_g D_b}{(D_g + D_b)^2} \tag{4.7}$$

with $D_b$ and $D_g$ denoting the duration of bad and good channel states, respectively. With (2), the correlation coefficient can be defined as time average

$$\rho = \frac{E\{(h_1(t) - \overline{h_1})(h_2(t) - \overline{h_2})\}}{\sigma_1 \sigma_2} \tag{4.8}$$

According to (3), the correlation coefficient can be evaluated from pairs of (time-synchronised) channel measurements with regard to the same mobile terminal. The correlation depends on the user environment, as well as the elevation and azimuth angles of the channels. As shown in (35), the dependency on the azimuth angles may approximately be described as a function of their difference $\Delta \varphi$. With this simplifying limitation, $\rho$ can also be estimated from circular measurements at constant elevation angles or from "fish-eye" photograph (36) for a single fixed user position, according to

$$\rho = \frac{E\{(h_1(\varphi) - \overline{h_1})(h_2(\varphi + \Delta\varphi) - \overline{h_2})\}}{\sigma_1 \sigma_2} \tag{4.9}$$

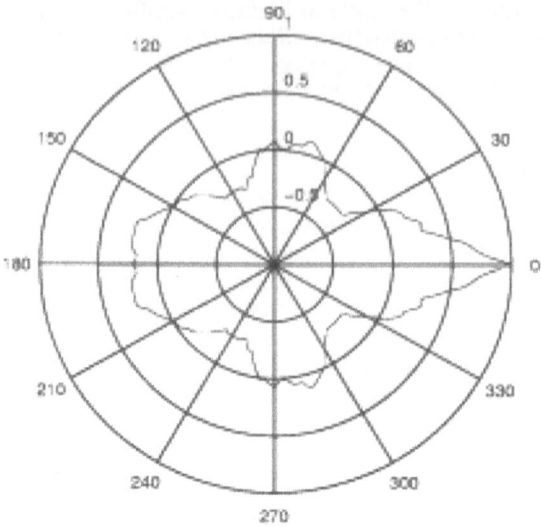

**Figure 4.17.** Azimuth correlation of shadowing in urban environment.

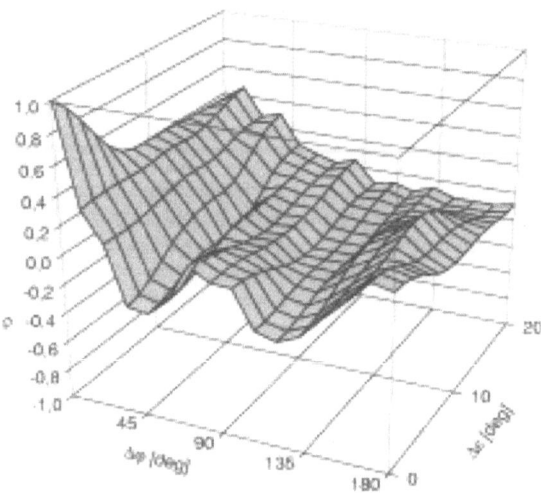

**Figure 4.18.** Elevation and azimuth correlation of shadowing in urban environment.

Here, $\rho$ depends on the user environment, the chosen pair of elevation angles, and the azimuth separation $\Delta\varphi$.

An example for the azimuth correlation of shadowing in an urban area is given in Fig. 4.17 for two channels at 35-degree elevation. It can be easily seen that satellite channels with an azimuth separation of 90 deg and 180 deg are likely to be correlated (due to the geometric structure of streets) whereas at 45 deg and 135 deg the channels are opposite. In Fig. 4.18 the correlation is presented for a 25 deg elevation of one satellite over higher elevations and full azimuth. The correlation decreases with increasing azimuth separation and is smaller if the satellites have different elevation angles.

## 4.2.4.4 COST 252 Studies on Satellite Diversity

Two complementary aspects related to shadowing correlation need attention, they are the characterisation/quantification of the reflection coefficient and its introduction in propagation models, thus extending single satellite models to the more general, multi-satellite case. As pointed out in the introduction to this section, physical models do not require that this information be available as an input

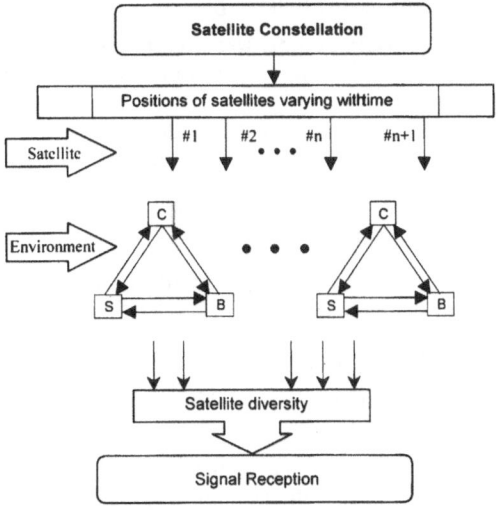

**Figure 4.19.** Evaluation of satellite diversity gain.

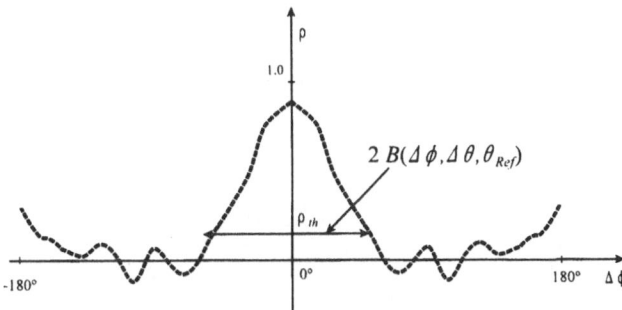

**Figure 4.20.** Definition of the average azimuth correlation sector parameter, $\bar{B}$ ($\Delta\phi$, $\Delta\theta$, $\theta_{ref}$).

since it is inherently present in the geometry of models. For statistical models, however, correlation information has to be known and models modified, as in (37), so as to account for correlation between shadowing events.

Two different approaches to shadowing correlation characterisation/quantification are discussed here. One is based on the use of a physical-statistical model and the other is based on a purely deterministic model. These two approaches produce similar results and can be used to furnish correlation coefficient data to statistical models.

The studies presented here aim at the quantification of the correlation coefficient for different types of environments, and are based on simulations using synthetic scenarios. The synthetic environments to be used should be easily described by a set of degree of urbanisation related parameters that permit a straightforward extrapolation of the obtained results. Very similar results to those obtained using measurements, as in (38), have been reached. The advantage of the two proposed approaches over the use of measurements and other techniques (e.g. photogrammetry method) is that they can be run on a computer at a much lower cost.

In the two proposed approaches, simulations are carried out using physical models rather than experimental data to gain insight into the correlation behaviour. Simulations are limited to urban scenarios given its relevance and ease of simulation. In the two cases, either an explicit mathematical expression or tabulated data as a function of the simple and easy to observe urbanisation parameters are sought.

The two approaches proposed here are the following:

1. In (39) a physical-statistical propagation model was used providing explicit expressions relating the average correlation coefficient, $\bar{\rho}$, for a given environment characterised by the height distribution of buildings on both sides of the road.
2. In (40) the average *azimuth correlation sector*, $\bar{B}$ ($\Delta\phi$, $\Delta\theta$, $\theta_{ref}$), is defined to describe correlation effects. This parameter indicates what is the average angular separation for two satellites to become uncorrelated for a given environment. A deterministic model employing simple ray-tracing (LOS/non-LOS) techniques applied to synthetic environments (urban databases) was used to compute the $\bar{B}$ parameter (Fig. 4.20).

Each approach starts off with the same definition of the stochastic process under study (Fig. 4.21) although each follows a different mathematical procedure, leading to equal results as it is shown below.

A stochastic process $S_i(t)$ for satellite-$i$ is defined as follows:

$$S_i(t) = \begin{cases} 1 & \text{if line-of-sight conditions} \\ 0 & \text{if shadowed conditions} \end{cases}$$

The line-of-sight and shadowed states are defined by the existence/non-existence of the direct ray. The stochastic process is considered to be in the shadowed state, with value 0, when the direct ray is blocked and 1 when there is direct visibility to the satellite.

Random variables result from the collection of a number of received signal values (converted to ones or zeroes) simultaneously observed at a specific time $i$. These values correspond to the state of the direct ray for a sufficiently large number of directions (azimuths and elevations) around the receiver (circular scans) located within the test environment.

The objective is, thus, to find the correlation as a function of the azimuth increment, $\Delta\phi$, among different circular scans corresponding to satellites at the same ($\Delta\theta = 0°$) or different elevations ($\Delta\theta \neq 0°$).

The average behaviour is, in principle, the information needed. Averaging has been performed here according to two different procedures as illustrated in Fig. 4.21.

Approach 1. By solving the mathematical expectation found in the definition of the correlation coefficient (Eq. 4.2) (normalised second-order joint moment). The option is only feasible if simplifying assumptions are made of the geometry of the urban environment. The approach followed in (41) is based on a physical-statistical model (42) (Fig. 4.22) where, basically, physical propagation models, including diffraction, are performed on synthetic environments (made up of plane screens representing buildings) with heights following distributions obtained by direct observation of the urban environment.

The physical-statistical model used proposes a canonical geometry for the environment traversed by a mobile receiver, typically a street canyon (Fig. 4.21a and Fig. 4.22), composed of buildings on both sides of the street. For this approach, the average correlation coefficient between the two random variables is defined for a single observation point and follows the expression

$$\bar{\rho} = \frac{COV(S_1, S_2)}{\sigma_2 \sigma_2} = \frac{E[(S_1 - \bar{S}_1)(S_2 - \bar{S}_2)]}{\sqrt{E[(S_1 - \bar{S}_2)^2]}\sqrt{E[(S_2 - \bar{S}_2)^2]}} \tag{4.10}$$

where the expectation (averaging process) is taken over all possible satellite-to-mobile azimuths assuming an uniform distribution in the interval $(0, 2\pi)$ and a given distribution for the building heights. The result is an analytic formula that depends only on the azimuth angle separation between the two satellites. The mathematical procedure (43) used is described in detail in the Appendix on page 000).

Approach 2. A second alternative is to obtain different realisations (azimuth scan series) of the random process by means of a simple ray-tracing algorithm performed on synthetic environments in the form of building data bases created according to controlled parameters/distributions.

In (44) and (45) this methodology is described. Only the blockage of the direct ray is computed and only street canyons are considered. In this case, spatial averaging is performed, i.e. the averaging is performed over different circular scans for a large number of evenly spaced points along the test street. This approach is another way of performing the averaging process described in the first approach taking into consideration the street building height distribution and a uniform azimuth distribution (Fig. 4.21).

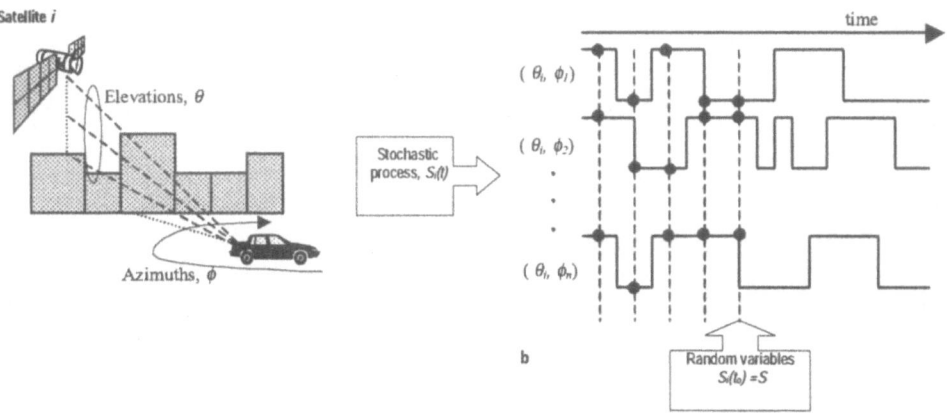

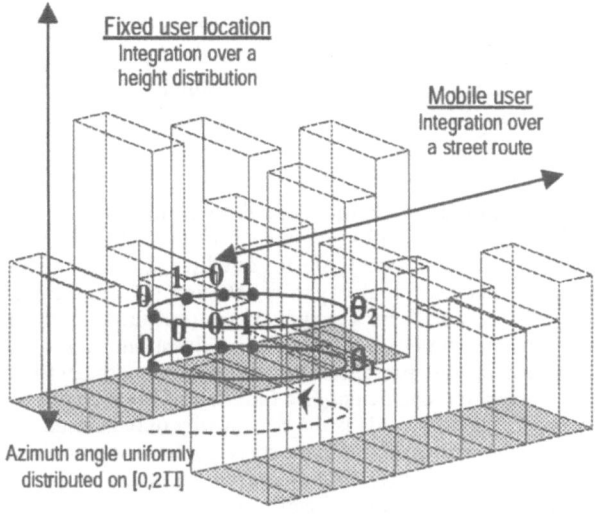

**Figure 4.21.** Shadowing stochastic process, showing the two approaches followed.

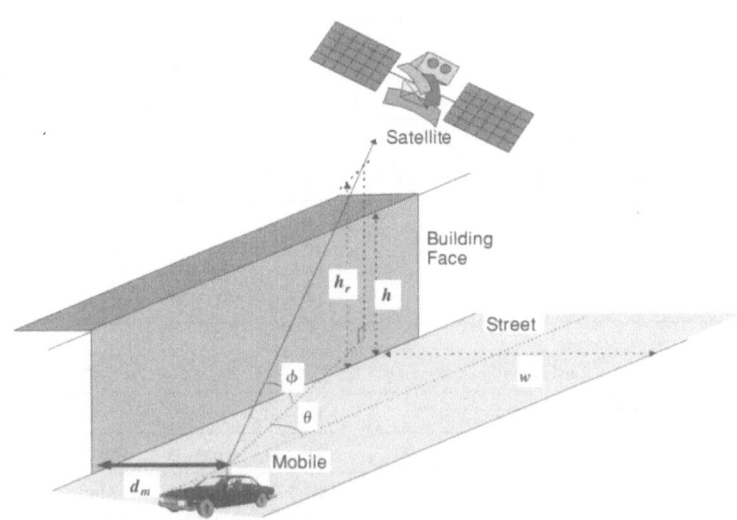

**Figure 4.22.** Mobile-satellite canonical street geometry.

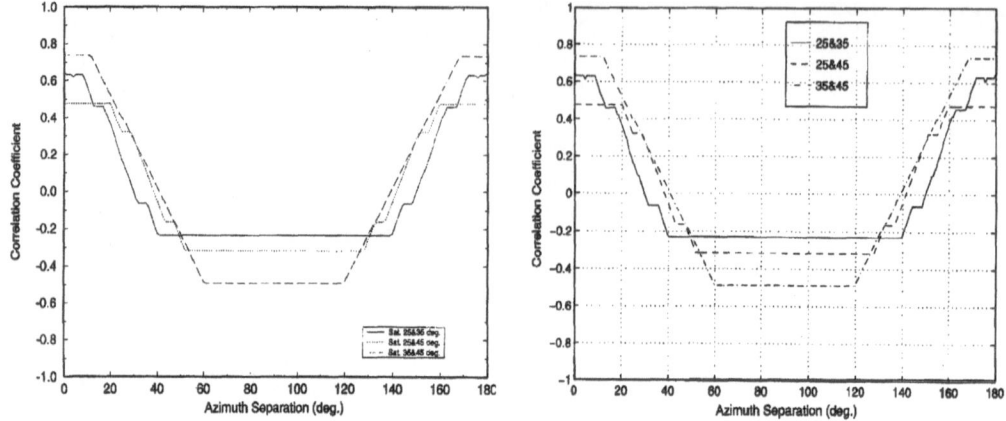

**Figure 4.23.** Correlation coefficient for a test canyon, computed by using Approach 1a and Approach 2b.

This second approach can be expressed mathematically by means of the equation below:

$$\bar{\rho} = \frac{COV(S_1, S_2)}{\sigma_2 \sigma_2} = \frac{\langle (S_1 - \bar{S_1})(S_2 - \bar{S_2}) \rangle}{\sqrt{\langle (S_1 - \bar{S_2})^2 \rangle} \sqrt{\langle (S_2 - \bar{S_2})^2 \rangle}} \qquad (4.11)$$

Both methods have been used for a number of test environments and exactly the same results were obtained. Fig. 4.23 shows an example where $\bar{\rho}$ was computed with both methods for a given environment, as a function of the azimuth offset, $\Delta\phi$, for satellites at different elevations, $\theta_1$ and $\theta_2$.

Note that for azimuth separations in the vicinity of 0° and 180° high correlation values are observed in the street canyon geometry, both for the auto-correlation function (same satellite elevations $\theta_1 = \theta_2$) and cross-correlation functions (different satellite elevations $\theta_1 \neq \theta_2$). For street intersections, a further peak at 90° would also be present in the correlation coefficient plots.

Also note that the correlation coefficient as a function of the azimuth increment, $\Delta\phi$, has circular symmetry about $\Delta\phi = 0$ (Fig. 4.20), this is the reason why only azimuth increments between 0° and 180° are plotted. The shape and values of these correlation curves are in very good agreement with results presented by other researchers using measured data (46).

As regards the distributions of building heights used in the simulations, studies (47) have shown that the log-normal distribution is in good agreement with actual building height distributions. For the correlation studies presented below, the log-normal parameters characterising street canyons in London and Guildford (UK) were a mean of 17.6 and a standard deviation 0.31 for London and a mean of 7.2 and standard deviation of 0.26 for Guildford.

**Table 4.10.** Values of $\bar{\rho}_{Max}$ and $\bar{B}$ in the vicinity of $\Delta\phi = 0°$ for $\Delta\theta = 0°$ and different street widths, Ws(m)

| $\theta_{ref}$ | LN(7.2, 0.3) | | | LN(17.6, 0.3) | | |
|---|---|---|---|---|---|---|
| | Ws = 10 | Ws = 15 | Ws = 20 | Ws = 10 | Ws = 15 | Ws = 20 |
| 20° | 20 | 24 | 30 | 6 | 10 | 14 |
| 30° | 30 | 30 | 22 | 10 | 15 | 20 |
| 40° | 30 | 30 | LOS | 20 | 22 | 25 |
| 50° | LOS | LOS | LOS | 24 | 30 | LOS |

**Table 4.11.** Values of $\bar{B}$ and $\bar{\rho}_{Max}$ in the vicinity of $\Delta\phi = 180°$ for $\Delta\theta = 0°$ and different street widths, Ws(m)

| | LN(7.2,0.3) | | | LN(17.6,0.3) | | |
|---|---|---|---|---|---|---|
| | Ws = 10 | Ws = 15 | Ws = 20 | Ws = 10 | Ws = 15 | Ws = 20 |
| 20° | .8/20 | 0.7/25 | 0.5/30 | 0.7/5 | 0.7/7 | 0.8/10 |
| 30° | .6/30 | 0.5/20 | — | 0.7/10 | 0.7/15 | 0.8/5 |
| 40° | .5/20 | — | — | 0.8/18 | 0.8/20 | 0.8/22 |

**Table 4.12.** $\bar{B}$ and $\bar{\rho}_{Max}$ for Ws = 15 m

| | $\Delta\theta$ | Around $\Delta\phi = 0°$ | | | | Around $\Delta\phi = 180°$ | | | |
|---|---|---|---|---|---|---|---|---|---|
| | | LN(7.2, 0.3) | | LN(17.6, 0.3) | | LN(7.2, 0.3) | | LN(17.6, 0.3) | |
| | | 10° | 20° | 10° | 20° | 10° | 20° | 10° | 20° |
| Ref. elevn | 10° | 0.5522 | 0.35/22 | 0.65/10 | 0.5/18 | 0.55/20 | 0.35/22 | 0.6/10 | 0.5/13 |
| | 20° | 0.6/25 | — | 0.73/15 | 0.55/18 | 0.5/22 | — | 0.7/14 | 0.55/17 |
| | 30° | 0.5/30 | — | 0.8/20 | 0.55/22 | 0.5/30 | — | 0.7/20 | 0.55/22 |
| | 40° | 0.8/20 | n/a | 0.7/23 | n/a | 0.7/20 | n/a | 0.6/22 | n/a |

n/a: not available.

**Table 4.13.** Summary of geometrical conditions for which correlation can be expected in street canyon scenarios

| Reference elevation | Elevation increment | Azimuth increment | Existence of correlation |
|---|---|---|---|
| $10° \leqslant \theta_{ref} < 50°$ | $\Delta\theta = 0°$ | $\|\Delta\phi\| < 30°$ $170° < \|\Delta\phi\| < 180°$ | Yes |
| $10° \leqslant \theta_{ref} < 50°$ | $\Delta\theta = 0°$ | $30° < \|\Delta\phi\| < 170°$ | No |
| $10° \leqslant \theta_{ref} < 50°$ | $0° < \Delta\theta \leqslant 20°$ | $\|\Delta\phi\| < 15°$ $170° < \|\Delta\phi\| < 180°$ | Yes |
| $10° \leqslant \theta_{ref} < 50°$ | $\Delta\theta > 20°$ | Any | No |
| $\theta_{ref} \geqslant 50°$ | Any | Any | LOS |

The results obtained for simulations using the second approach are summarised in Tables 4.10, 4.11, 4.12, and 4.13. The (−) symbol in Table 4.11 and Table 4.12 means that the values of $\bar{\rho}$ were below the threshold considered, in this case, $\rho th = 0.3$. Table 4.10, Table 4.11, Tables 4.12 and 4.13, show the average azimuth correlation sector values, $\bar{B}$ ($\Delta\phi$, $\Delta\theta$, $\theta_{ref}$), obtained from simulations using the second approach. For satellite-to-mobile links having angle separations greater than $\bar{B}$, loss of correlation is to be expected and, thus, there will be the possibility of achieving some diversity gain.

The dependence of the correlation parameter with mean street height and width is apparent from the tables. It can be observed that, for satellites at the same elevation, loss of correlation is reached around $\Delta\phi = 30°$. It can also be observed how correlation is lost for two satellite links with elevation offsets greater than 20°. Relevant results of this study (Table 4.10, Table 4.11, Table 4.12) are summarised in Table 4.13.

### 4.2.4.5 Variability of the Correlation Coefficient

In the preceding section a study has been presented of the average correlation coefficient and the average azimuth correlation sector. However, from the study of correlation coefficient plots obtained from individual azimuth scans performed at different points along the mobile route significant differences were apparent.

Figure 4.26 shows a 3D representation of various correlation plots corresponding to different mobile route sampling points along a given street canyon, also in the figure the averaged correlation coefficient for the street under study is shown in polar form. Furthermore, in the same figure the distribution of the correlation coefficient variations for the same street are presented.

In order to further illustrate the correlation coefficient variability, Figs 4.24 and 4.25 show histograms computed for various azimuth increments, $\Delta\phi$. This is further illustrated in a schematic form in Fig. 4.26. From Figs 4.25 and 4.26 it is clear that, for simulation purposes, the average correlation coefficient is not representative of the conditions undergone by two closely spaced mobile-to-satellite links. The use of correlation coefficients in statistical models of the Markov type have to be made with care taking into account the variability of the correlation coefficient.

Also from Fig. 4.25, a very promising feature can be observed in the correlation coefficient plots which is the fact that negative values of this parameter occur with significant probabilities. It is also clear form the observation of these figures that the issue of shadowing correlation characterisation is in no way closed and that new in depth studies are still required.

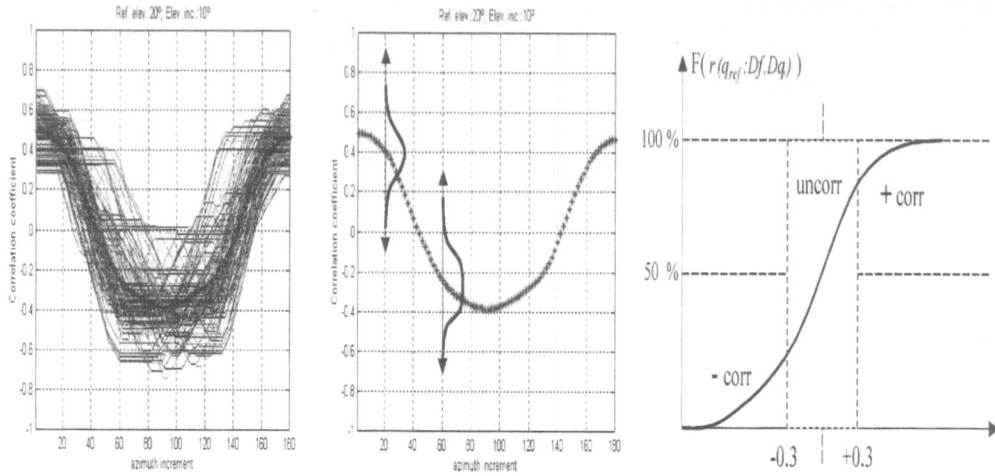

**Figure 4.24.** Single route point and route averaged correlation coefficient plots in a street canyon.

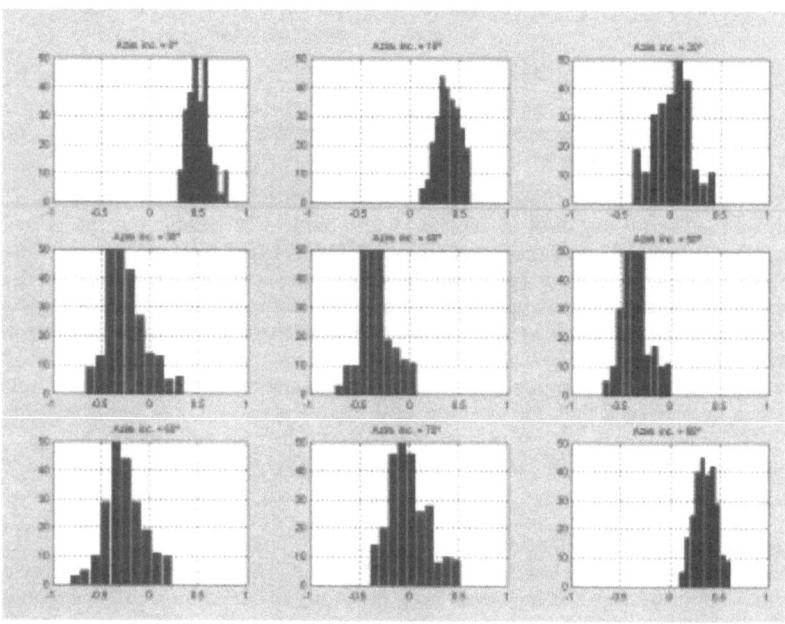

**Figure 4.25.** Histograms showing the variability of $\rho$ for different values of $\Delta\phi$ for the case illustrated in Fig. 4.24.

## 4.2.4.6 Conclusions

In this section a study of shadowing effects and its correlation for links with a given angle separation has been addressed. First, a review of the most relevant studies dealing with these issues has been made after two new approaches have been proposed yielding both of them similar conclusions, one using a physical-statistical model while the other is based on a purely deterministic approach.

The obtained results can be used in the assessment of different proposed constellations for third generation personal communications systems. Other application of the obtained results is for them to be used as additional inputs in statistical models to evaluate system availability levels.

It has been shown that correlation coefficients tend to vary about their mean value presenting, for a significant number of cases, small or even negative values. The characterisation of correlation coefficient variability and other related issues requires further studies to be carried out in the framework of ensuing COST actions.

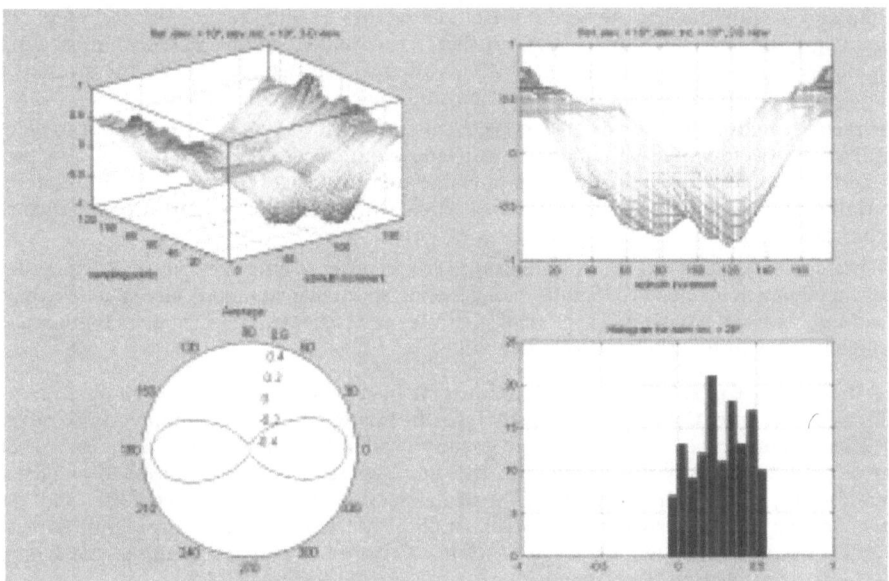

**Figure 4.26.** Schematic representation of the variability of around ρ its average value: probability density function and cumulative distribution function.

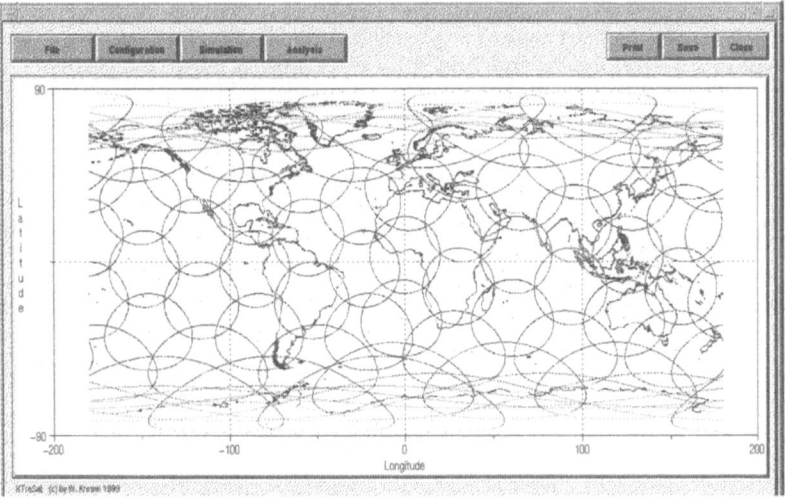

**Figure 4.27.** Instantaneous coverage of Iridium ($E_{min} = 8.2°$).

## 4.2.5 Satellite Diversity and Capacity

### 4.2.5.1 Introduction

Satellite diversity is the general case where at least two satellites cover a given location on earth. It results in an overlap of satellite footprints that delineate the area where the satellites are seen above a specified elevation angle, $E_{min}$ (Fig. 4.27).

Although satellite diversity is a common feature to all non-GEO constellations providing continuous coverage of the service area, the design of non-GEO constellations may emphasize to more or less satellite diversity at different earth locations depending on the objective aimed at. Examples for such objectives are:

1. Minimising satellite diversity to avoid waste of coverage so as to reduce the cost of the constellation (e.g. number of satellites). ICO, for instance, provides single coverage with minimum elevation angles $E_{min} \geq 18°$ up to latitude 62° (48)) with only 10 satellites, but satellite diversity occurs with a probability of less then 80 per cent beyond the equator (Fig. 4.29).

2. Maximising satellite diversity as an intrinsic feature to increase service availability. Indeed, satellite diversity was identified as an appropriate fading mitigation technique (49, 50). Satellite diversity furthermore decreases the duration of congestion that arises from capacity limitations, since the traffic can be routed over alternative satellites whenever all channels of one visible satellite are blocked.

3. Restricting a satellite's footprint to a minimum size in order to enhance communications capacity: Active antenna elements, for example, can generate spot beams dynamically so as to concentrate power into areas whose sizes are so small that they just cover the service area continuously. This ultimately leads to no satellite diversity at all.

Here, the influence of satellite diversity on the system capacity will be evaluated quantitatively for the indicated strategies. Capacity will be expressed in Erlang per square kilometer (allowable traffic load density). Numerical examples will be given for typical LEO constellations. Section 4.2.5.2 outlines aspects of satellite diversity illustrated with two first-generation satellite constellations for mobile communications, namely Iridium and Globalstar. Section 4.2.5.3 discusses fixed and dynamic coverage. Section 4.2.5.4 derives the allowable traffic load density under consideration of satellite diversity for fixed and dynamic coverage. Section 4.2.5.5 gives numerical examples for Iridium and the Globalstar constellation. The conclusions are outlined in Section 4.2.5.6.

## 4.2.5.2 Satellite Diversity

### 4.2.5.2.1 Diversity Performance of Commercial Systems

Various non-GEO satellite constellations have been designed for voice and data communications. The chosen constellation parameters (number of satellites, orbit height, inclination, etc.) provide continuous coverage of the service area, and temporary or permanent satellite diversity at a given earth location. Figs 4.27 and 4.28 show the instantaneous coverage of the near polar (inclination $i = 86.4°$) Iridium constellation (Fig. 4.27) and the non-polar ($i = 52°$) Globalstar constellation (Fig. 4.28).

The footprint boundaries are displayed for the considered constellation's intrinsic minimum elevation angle, i.e. the elevation angle that provides continuous coverage all over the service area. For Iridium ($E_{min} = 8.2°$), the figures indicate higher satellite diversity above the intrinsic minimum elevation angle $E_{min}$ at the poles, whereas satellite diversity is maximal at european/north american latitudes for Globalstar ($E_{min} = 10°$).

The time ratio of coverage by at least two satellites measures the corresponding probability, $P_{v2}$. Results of a quantitative analysis for both constellations are shown in Fig. 4.29. This figure shows the

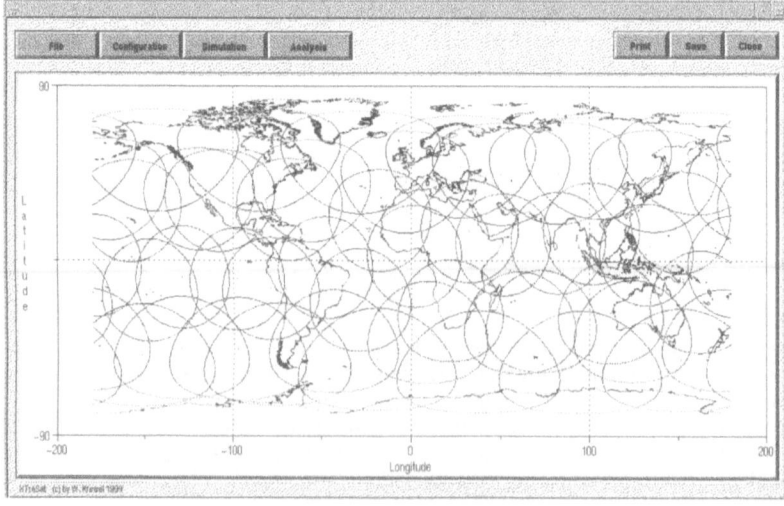

**Figure 4.28.** Instantaneous coverage of Globalstar ($E_{min} = 10°$).

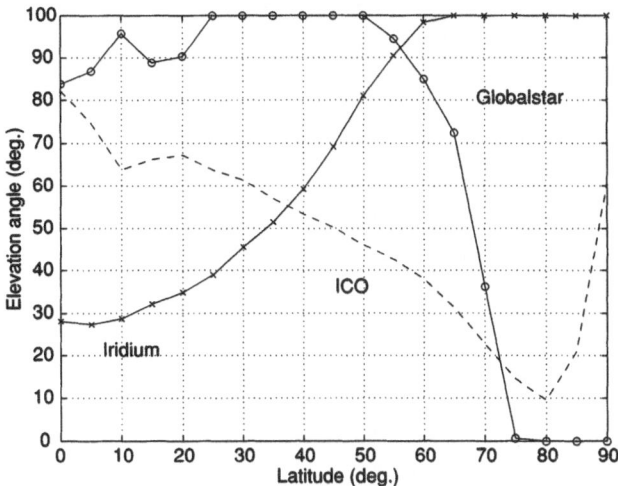

**Figure 4.29.** Probability of at least dual coverage for ICO ($E_{min}$ = 20°), Iridium ($E_{min}$ = 8.2°), and Globalstar ($E_{min}$ = 10°).

probability of at least dual coverage, $P_{v2}$, above $E_{min}$ as a function of the latitude $l$ (the influence of the longitude is negligible, and statistics are alike for northern and southern hemisphere).

For Iridium, permanent satellite diversity is performed in regions above $l = 65°$. Globalstar provides permanent satellite diversity within the latitude band of $25° < l < 50°$.

### 4.2.5.2.2 Diversity-based Mitigation Techniques

The satellite diversity concept as a propagation impairment mitigation technique has been proposed to mitigate fading caused by shadowing or blocking, and to improve path availability (51, 52). Satellite diversity allows for either combining signals transmitted via several satellites (*combining* diversity) or selecting the signal from a non-blocked satellite path among all paths from the covering satellites (*selection* diversity).

*Combining Diversity.* Combining diversity as a fade mitigation technique imposes the implementation of special means to result in a combined signal superior to any single signal. Satellite systems using those facilities will hereinafter be called *diversity based systems*. Diversity based systems (such as Globalstar, for instance) are characterised by high multiple visibility statistics (dual visibility about 100 per cent of time), whereas non-diversity based systems (e.g. Iridium) are characterised by high fade margins (about 10 dB higher than in diversity based systems). Combining diversity is of importance in shadowed environments (sub-urban and tree shadowed), where fading mostly does not result in complete blockage. In this case, already dual satellite diversity yields system cost reduction and higher user acceptance (smaller size of terminals, for instance) (53). Due to the smaller link margin, however, the availability of *two* non-blocked satellites are required. The margin reduction resulting from diversity (diversity gain) has to be paid by a decrease of the allowable traffic load density, since simultanous coverage by two non-blocked/non-saturated satellites is required for communications.

*Selection Diversity.* Communications in rural and urban environments are mainly effectuated under LOS-conditions. Subscriber terminals of all candidate systems (diversity based or not) will then operate with a single satellite in view. Satellite diversity may therefore be available but signal combining is not used, either because fading is negligible or so large that it cannot be overcome by diversity combining. Diversity then confines in selecting one of all the covering satellites, according to some selection diversity schemes. In the simplest scheme (scheme a) the terminal selects the nearest satellite providing the strongest signal and sends an access request. If no channel is available, the access is denied otherwise the call can proceed (Fig. 4.30).

A more elaborated scheme (Scheme b) involves the second nearest satellite in case all channels of the nearest satellite are busy.

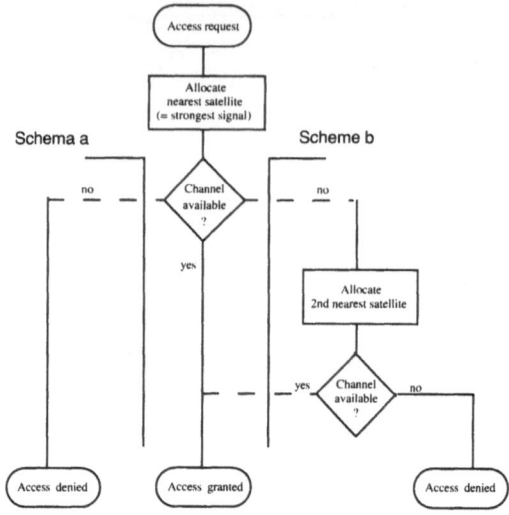

**Figure 4.30.** Selection diversity schemes.

## 4.2.5.3 Beam Forming

### 4.2.5.3.1 Coverage Area

The size of the area that is covered by a satellite beam is determinated by the lower limit of the elevation angle, $E_{min}$, (54):

$$F(E_{min}) = 2\pi R_e^2 \left[ 1 - \cos\left( \frac{S(E_{min})}{R_e} \right) \right] \qquad (4.12)$$

where the surface distance, $S$, is given by

$$S(E_{min}) = R_e \left[ \cos^{-1}\left\{ \frac{R_e \cos(E_{min})}{R_e + h} \right\} - E_{min} \right] \qquad (4.13)$$

where $R_e$ is the earth radius ($R_e = 6378$ km), and $h$ the orbit height.

Satellite antennas can be configured either to cover the service area with fixed beams or with beams dynamically varying in shape and size using active antenna arrays. In both cases, continuous coverage of the service area is considered to be the most important requirement.

### 4.2.5.3.2 Predefined Fixed Coverage

If $(E_{min})_{glob}$ is the minimum elevation angle for continuous coverage all over the service area, then the fixed size of the coverage area is $F = F((E_{min})_{glob})$. This possibly is too much demanding as a higher elevation angle, $(E_{min})_{loc}$, would locally suffice to provide the required coverage. For example, with Iridium the lowest minimum elevation angle providing global coverage is 8.2° (55). A higher minimum elevation angle is obtained for non-equatorial locations, so that $(E_{min})_{loc} \geqslant (E_{min})_{glob}$, and $F((E_{min})_{glob}) \geqslant F((E_{min})_{loc})$.

However, predefining a fixed coverage with $(E_{min})_{glob}$ as the minimum elevation angle results in an overlap of several coverage areas, as shown in Fig. 4.31. This provides an opportunity for satellite diversity (see Section 4.2.5.2).

### 4.2.5.3.3 Dynamic Coverage

Active array antennas allow the generation of beams with variable shape and size in time (56, 57, 58, 59). Figure 4.32 shows that the size of the coverage area reduces to the minimum area that provides continuous coverage at a specific terminal location.

The corresponding coverage area is then $F = F((E_{min})_{loc})$, where $(E_{min})_{loc}$ is the minimum elevation angle providing continuous coverage at a given terminal location.

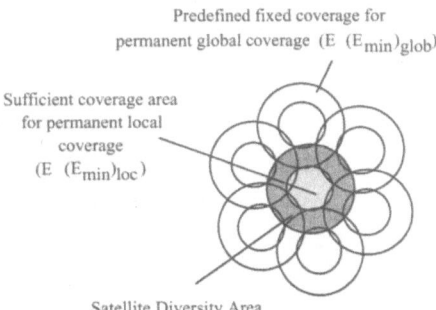

Figure 4.31. Predefined fixed coverage of the service area.

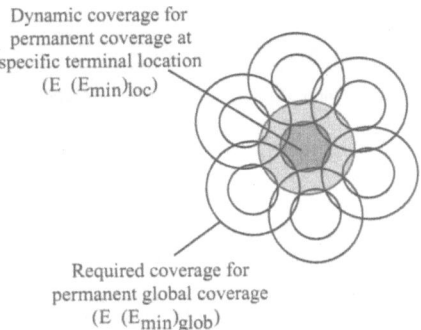

Figure 4.32. Dynamic coverage of the service area.

### 4.2.5.4 Allowable Traffic Load Density

*4.2.5.4.1 Definition*
The allowable traffic load is the traffic load that can be served by the offered number of satellite channels without exceeding a given call blocking probability. The ratio of allowable newly generated traffic load, $A_n(P_b)$, to the size of the satellite's coverage area, $F(E_{min})$, is referred to as the allowable traffic load density, $\rho$ (Erl/km$^2$):

$$\rho = \frac{A_n(P_b)}{F(E_{min})} \tag{4.14}$$

where $E_{min}$ is the minimum elevation angle.

*4.2.5.4.2 Allowable traffic load*
Traffic load is the product of mean medium access requests (mean arrival rate, $\lambda$), and mean channel holding time, $1/\mu$. In mobile cellular radio communications systems, such as terrestrial cellular or non- GEO satellite systems, the mean arrival rate builds up from the newly generated traffic arrival rate, $\lambda n$, and the handed over traffic arrival rate, $\lambda h$. Similarly, the mean channel holding time is determined by the mean call duration $T_{call}$ as well as the mean terminal residing time in a cell, $T_c$. Two basic features, however, differ from terrestrial networks: First, the terminal residing time in a cell is conditioned by the speed of the satellite rather then by the user mobility. Second, the direction of the satellite motion is deterministic. This results in a simplified distribution of the satellite dwell time that was derived by (60) based on the model of the street of coverage (61).

The total allowable traffic load per satellite as a function of the blocking probability for the non-priority case (same blocking probabilities for new call requests and handed over calls) can be derived as

$$A_n(P_b) = A_t(P_b)[(1 - P_{h1}) + G(P_b)(1 - P_{h2}]^{-1} \tag{4.15}$$

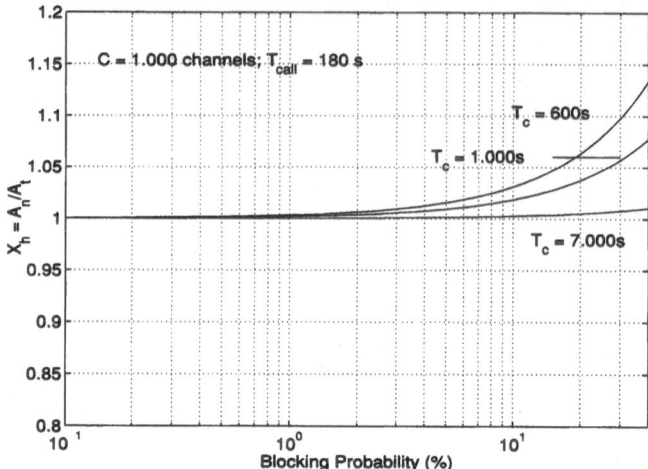

**Figure 4.33.** Newly generated to total carried traffic ($X_h = A_n/A_t$) vs. blocking probability for different residing times.

$G(P_b)$ is calculated according to

$$G = \frac{\lambda_h}{\lambda} = \frac{(1 - P_{b1})P_{b1}}{1 - (1 - P_{b2})P_{h2}} \qquad (4.16)$$

where $P_{h1}$ is the probability that a newly generated call faces a hand-over, and $P_{h2}$ is the probability that a handed-over call faces a further hand-over. These probabilities are respectively given by

$$P_{h1} = \gamma\left[1 - \exp\left(-\frac{1}{\gamma}\right)\right] \qquad (4.17)$$

$$P_{h2} = \exp\left(-\frac{1}{\gamma}\right)$$

where the user mobility $\gamma$ is given by $\gamma = T_{\text{call}}/T_c$.

The factor $X_h = (1 - P_{h1}) + G(1 - Ph_2)^{-1}$ in Eq. 4 determines the ratio of newly generated to total carried traffic, thus $X_h = A_n/A_t$. The variation of $X_h$ as a function of the blocking probability is shown in Fig. 4.33 for residing times in different constellations where $T_c = 600$s for Iridium, $T_c = 1.000$s for Globalstar, and $T_c = 7.000$s for ICO.

It can be seen that $X_h$ varies only slightly with the residing time for small blocking probabilities ($P_b \leq 0.01$), and that $A_n \sim A_t$ up to this boundary. The allowable total traffic load $A_t$ is determined by the blocking probability, $P_b$, and the selected channel allocation scheme. It also can be inferred that the influence of fixed or dynamic coverage on the total allowable traffic load is marginal for $P_b^2$ 0.01.

### 4.2.5.4.3 Allowable Traffic Load Density
Expressions for the allowable traffic load density (Erl/km²) will now be given according to the considered diversity technique (see Section 4.2.5.2), which impacts the call blocking probability.

### 4.2.5.4.3.1 Combining Diversity

(1) Fixed Coverage. The coverage area is of constant size all over the service area, and provides continuous coverage by at least one satellite. This is achieved for $(E_{1\text{min}})_{\text{glob}}$, and the corresponding coverage area is $F = F((E_{1\text{min}})_{\text{glob}})$. Furthermore, two non-blocked satellite channels are required for combining diversity. Call blocking and dual coverage are statistically independent. The allowable traffic load density is therefore:

$$\rho = \frac{A_n\left(P_{b2} = 1 - \sqrt{\frac{1 - P_{bc}}{P_{v2}}}\right)}{F((E_{1\text{min}})_{\text{glob}})} \qquad (4.18)$$

where $Pv_2$ is the probability that at least two satellites are visible. $P_{bc}$ is the overall call blocking probability, and $P_{b2}$ is the blocking probability for each satellite.

The fixed coverage of an area corresponding to $(E_{1min})_{glob}$ is a suitable allocation scheme only for systems performing (at least approximately) continuous dual coverage at $(E_{1min})_{glob}$, since (Eq. 4.18 requires $Pv_2 > (1 - P_{bc})$. Assuming a call blocking probability of $P_{bc} = 1$ per cent, dual visibility of at least $P_{v2} = 99$ per cent at $(E_{1min})_{glob}$ is required. In case of Globalstar, for example, $P_{v2} \geqslant 99$ per cent at $(E_{1min})_{glob} = 10°$ is given for terminals located within a latitude band of $25° < l < 50°$ (see Fig. 4.28). If coverage beyond this band is required, a higher call blocking probability has to be accepted. At the equator, for instance, $P_{bc}$ exceeds $(1 - P_{v2}) = 0.16$ due to a dual visibility of $P_{v2} = 84$ per cent at latitude $l = 0°$ for $(E_{1min})_{glob} = 10°$.

A more reasonable coverage area corresponds to the minimum elevation angle that performs continuous (i.e. $P_{v2} = 1$) coverage from at least two satellites all over the service area. The minimum elevation angle above which at least *two* satellites are permanently visible all over the service area is hereinafter named $(E_{2min})_{glob}$. The allowable traffic load density is then given by:

$$\rho = \frac{A_n(P_{b2} = 1 - \sqrt{1 - P_{bc}})}{F((E_{2min})_{glob})} \tag{4.19}$$

(2) Dynamic Coverage. The disadvantage of the fixed coverage area originates in the fact that a smaller coverage area would be sufficient for continuous coverage, since $(E_{min})_{loc}$ is larger than $(E_{min})_{glob}$ almost all over the service area. In case of dynamic coverage the allowable traffic load density then becomes:

$$\rho = \frac{A_n(P_{b2} = 1 - \sqrt{1 - P_{bc}})}{F((E_{2min})_{glob})} \tag{4.20}$$

where $(E_{2min})_{loc}$ is the minimum elevation angle above which at least two satellites are permanently visible at a given terminal location.

### 4.2.5.4.3.2 Selection Diversity (Scheme a)

In this scheme, the call blocking probability is directly conditioned by the blocking probability of the nearest satellite (and only of this satellite, since only the best can be selected). Thus $P_{b2} = P_{bc}$, regardless of whether the service area is covered by one or several satellites.

(1) Fixed Coverage. In case of a fixed coverage $(F = F((E_{min})_{glob}))$ the allowable traffic load density is given by

$$\rho = \frac{A_n(P_{b2} = P_{bc})}{F((E_{min})_{glob})} \tag{4.21}$$

(2) Dynamic Coverage. Dynamic coverage takes into account the local value of $E_{min}$, so that

$$\rho = \frac{A_n(P_{b2} = P_{bc})}{F((E_{min})_{glob})} \tag{4.22}$$

### 4.2.5.4.3.3 Selection Diversity (Scheme b)

(1) Fixed Coverage. Similarly to the combining diversity scheme, the size of the fixed coverage area is determined by $(E_{1min})_{glob}$ or $(E_{2min})_{glob}$ with consequences on the blocking probability. We will first consider the case of a fixed single coverage. The probability that a call is blocked is then the sum of the probability to be covered by only one satellite that is blocked, on the one hand, and the probability to be covered by two satellites, both being blocked, on the other hand. The allowable traffic load density is therefore given by:

$$\rho = \frac{A_n\left(P_{b2} = \sqrt{\dfrac{P_{bc}}{P_{v2}} + \left[\dfrac{1 - P_{v2}}{2P_{v2}}\right]^2} - \dfrac{1 - P_{v2}}{2P_{v2}}\right)}{F((E_{1min})_{glob})} \tag{4.23}$$

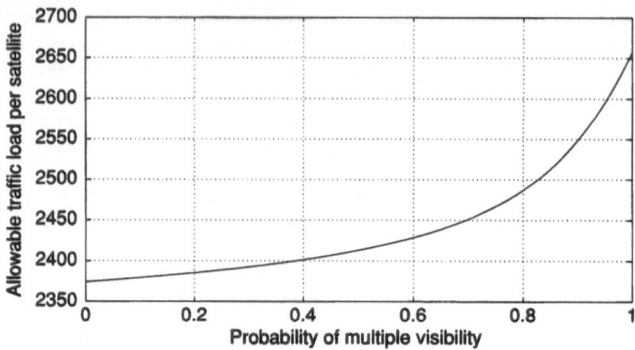

**Figure 4.34.** Allowable traffic load per satellite for the Erlang-loss system ($C = 2400$ channels per satellite, $P_{bc} = 1$ per cent).

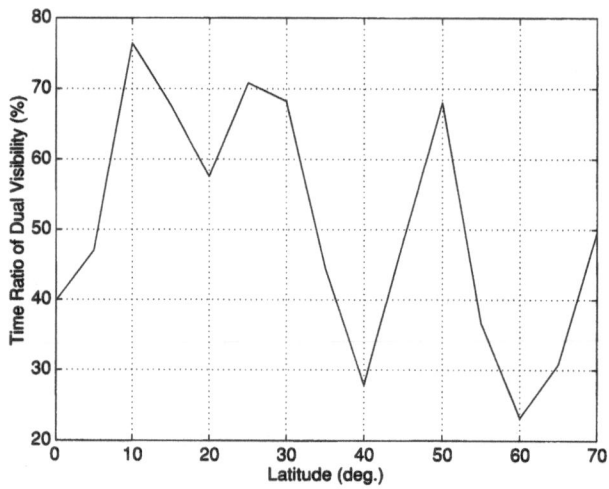

**Figure 4.35.** Probability that two satellites exceed the local minimum elevation angle (Globalstar).

Figure 4.34 visualises the allowable traffic load per satellite as a function of the multiple visibility probability, $P_{v2}$, for a call blocking probability $P_{bc} = 1$ per cent ($C = 2400$ channels). In this chart, the minimum value ($A(P_{v2}) = 0$)) indicates the allowable traffic load for single visibility, whereas the maximum (($A(P_{v2} = 1)$) is the allowable traffic load for continuous coverage of at least two satellites.

We now consider the coverage area corresponding to $(E_{2min})_{glob}$, i.e. continuous coverage from two satellites. The call blocking probability is then the probability that both satellites are blocked. The allowable traffic load density is given by:

$$\rho = \frac{A_n(P_{b2} = \sqrt{P_{bc}})}{F((E_{2min})_{glob})} \tag{4.24}$$

(2) Dynamic Coverage. In the case of dynamic coverage area allocation to $F = F((E_{1min})_{loc})$, the allowable traffic load increases to (since $F((E_{1min})_{loc})^2\, F((E_{1min})_{glob})$)

$$\rho = \frac{A_n\left(P_{b2} = \sqrt{\dfrac{P_{bc}}{P_{v2}} + \left[\dfrac{1 - P_{v2}}{2P_{v2}}\right]^2} - \dfrac{1 - P_{v2}}{2P_{v2}}\right)}{F((E_{1min})_{loc})} \tag{4.25}$$

The percentage of time, $P_{v2}$, when two satellites are in visibility above the local minimum elevation angle ($E_{1min})_{loc}$ is exemplary plotted in Fig. 4.35 for the Globalstar constellation.

Table 4.14. Parameters of considered systems

| | Iridium | Globalstar |
|---|---|---|
| No of satellites | 66 | 48 |
| Orbit height (km) | 780 | 1414 |
| Channels/satellite | 550 | 2400 |
| $(E_{1min})_{glob}$ (deg) | 8.2° | $10^{o2}$ |
| $F((E_{1min})_{glob})$ (km²) | 1.53E7 | 2.64E7² |
| Min. offered traffic density ($10^{-5}$ Erl/km²) | 3.4 | $5.9^{2,3}/8.8^{2,4}$ |
| Diversity-based | no | yes |

[1] $P_{bc}$ 1 per cent assuming Erlang-B.
[2] In latitude band $-70° \leqslant 1 \leqslant 70°$
[3] If combining.
[4] If no combining.

Table 4.15. Allowable traffic load density in $10^{-5}$ Erl/km² (Globalstar, $1 = 40°$, $P_{bc} = 1$ per cent)

| Coverage | Combining | Selection (a) | Selection (b) | |
|---|---|---|---|---|
| Fixed | $F((E_{1min})_{glob})$ | (8.7) | 8.8 | 9.8 |
| | $F((E_{2min})_{glob})$ | 5.9 | 5.9 | 6.6 |
| Dynamic | $F((E_{1min})_{loc})$ | — | 30.4 | 30.6 |
| | $F((E_{2min})_{loc})$ | 12.4 | 12.5 | 14.0 |

Table 4.16. Allowable traffic load density in $10^{-5}$ Erl/km² (Iridium, $1 = 40°$, $P_{bc} = 1$ per cent)

| Coverage | Combining | Selection (a) | Selection (b) | |
|---|---|---|---|---|
| Fixed | $F((E_{1min})_{glob})$ | — | 3.4 | 3.5 |
| | $F((E_{2min})_{glob})$ | — | — | — |
| Dynamic | $F((E_{1min})_{loc})$ | — | 4.4 | 4.6 |
| | $F((E_{2min})_{loc})$ | — | — | — |

The dynamic allocation to $F = F((E_{2min})_{loc})$, as discussed for the case of combining diversity allocation, results in

$$\rho = \frac{A_n(P_{b2} = \sqrt{P_{bc}})}{F((E_{2min})_{loc})} \tag{4.26}$$

### 4.2.5.5 Examples

Numerical examples for the derived cases will now be given for Iridium and Globalstar. The assumed parameters for these constellations are given in Table 4.14.

In Table 4.15, the numerical values of the allowable traffic load density are given for Globalstar at latitude $1 = 40°$.

Combining diversity and fixed or dynamic dual coverage leads to an allowable traffic load density of $5.9 \cdot 10^{-5}$ Erl/km² and $12.4 \cdot 10^{-5}$ Erl/km² respectively. Non-shadowed users apply either the diversity scheme (a) or (b). Should there be dynamic beam forming, the allowable traffic load density would increase up to about $12.5 \cdot 10^{-5}$ Erl/km² and $14.4 \cdot 10^{-5}$ Erl/km², respectively. These values are lower than the theoretical values from continuous local single coverage ($30.4 \cdot 10^{-5}$ Erl/km² and $30.6 \cdot 10^{-5}$ Erl/km²), but are still higher than the values obtained from fixed coverage. Thus, dynamic coverage increases the allowable traffic load density significantly (by a factor of nearly two), but the diversity gain due to signal combining has to be paid by a decrease of the allowable traffic load density. The slight increase (about 10 per cent) of the allowable traffic load density with Scheme B probably does not justify the increase in the system complexity.

In Table 4.16, numerical values of the allowable traffic load density are given for Iridium at latitude $l = 40°$.

Due to a small difference (about 5°) between the local and global minimum elevation angle up to latitude $1 = 40°$, dynamic coverage leads to smaller "gain" in the allowable traffic load density that, however, is still about 30 per cent more.

### 4.2.5.6 Conclusion

This section has outlined the influence of satellite diversity on the capacity of non-GEO satellite constellations for mobile communications, taking into account both fixed coverage and dynamic coverage, the latter implying reconfigurable satellite antenna patterns. The applied method provides means for comparing the capacity for diversity and non-diversity based systems. Dynamic coverage is shown to bring a significant increase in the capacity for systems with high satellite diversity probability ($P_{v2} > 90$ per cent), and therefore such a feature is an appealing one for future system design.

## 4.3 Novel Modulation Techniques: Variable Rate CPFSK Modulation

### 4.3.1 Introduction

This section presents a new modulation technique called a variable rate M-CPFSK modulation. M- CPFSK consists of 2-CPFSK, 4-CPFSK and 8-CPFSK modulation schemes with modulation indices $h = 1/2, 1/4$ and $1/8$ respectively. The number of modulation levels M varies in accordance with the mobile radio channel impairments experienced by a radio signal.

Additive white Gaussian noise, Rayleigh fading and Doppler frequency shift are considered in the simulation model. An eye pattern is used as a basis for the switching thresholds. The results show that a variable rate M-CPFSK modulation with the average of 2 bits per symbol outperforms the constant rate 4-CPFSK modulation giving the same average transmission rate.

Continuous phase modulation is bandwidth efficient as well as resistant to channel nonlinearity and fading. The use of multilevel modulation can further increase spectral efficiency and accelerate data rate. On the other side, multilevel modulation requires greater signal-to-noise ratio and is more sensitive to co-channel interference.

The performance of variable M-CPFSK modulation scheme, which varies the number of modulation levels in accordance with the interference intensity, is investigated. The modulation index h and the modulation level M are related by the equation $h = 1/M$. When the interference level is low, 8-CPFSK modulator is active and the system transmits 3 bits per symbol interval. When the interference increases 4-CPFSK modulator is activated, what means that 2 bits are transmitted in the symbol interval. In the worst case, when the interference level is very high, the most robust 2-CPFSK modulation scheme is used.

The advantage of the proposed system is that the number of modulation levels and the modulation index values are chosen in such a way that it is possible to generate various level CPFSK signals by the same modulator.

### 4.3.2 Variable Rate M-CPFSK System

CPFSK signal is described by the following equation

$$s(t) = \sqrt{\frac{2E}{T}} \cos\left(2\pi f_c t + \pi \frac{t}{T} h a_i\right) \tag{4.27}$$

where $E$ is the symbol energy, $T$ is the length of a symbol interval, $f_c$ is the carrier frequency, $h$ is the modulation index and $ai$ is the input symbol. The input symbols are chosen from the alphabet $a_i \in \{\pm 1, \pm 3, \ldots, \pm(\dot{M}-1)\}$, where $M$ is the number of modulation levels.

In our investigation, we dealt with three different CPFSK schemes. We considered M-CPFSK signals with $M = 2, 4, 8$ modulation levels and $h = 1/2, 1/4, 1/8$ modulation indices respectively. In our particular case the relation between the modulation index $h$ and modulation level $M$ is determined by the equation $h = 1/M$.

The signal phase shift in one symbol interval is defined by the product between the modulation index $h$ and the input data in that symbol interval. If all possible phase shifts for modulation levels $M = 2,4,8$ and input data $a_i \in \{\pm 1, \pm 3, \pm 5, \pm 7\}$ are recalculated to least common denominator $\pi/8$, it is evident, that the phase shift values for M-CPFSK modulation schemes are uniformly distributed between $+7\pi/8$ and $-7\pi/8$. In this particular case only one CPFSK modulator with modulation index $h = 1/8$ is sufficient to generate all three M-CPFSK ($M = 2, 4, 8$) modulated signals. However, the input data have to be encoded in an appropriate way to obtain the desired modulated signal. If the encoded input data values are chosen from following sets: $a_i \in \{\pm 4\}$ for $M = 2$, $a_i \{\pm 2, \pm 6\}$ for $M = 4$ and $a_i \in \{\pm 1, \pm 3, \pm 5, \pm 7\}$ for $M = 8$, all three modulation schemes can be implemented with the same modulator having a modulation index $h = 1/8$ and input symbol set $a_i \in \{\pm 1, \pm 2, \pm 3, \pm 4, \pm 5, \pm 6, \pm 7\}$.

The modulation parameters are determined by the physical and MAC layer of the system. The

**Table 4.17.** Data encoder mapping functions

| M = 2 | | M = 4 | | M = 8 | | | |
|---|---|---|---|---|---|---|---|
| Input | Output | Input | Output | Input | Output | Input | Output |
| 0 | −4 | 00 | −2 | 000 | −1 | 100 | +1 |
| 1 | +4 | 01 | −6 | 001 | −3 | 101 | +3 |
| | | 10 | +2 | 010 | −5 | 110 | +5 |
| | | 11 | +6 | 011 | −7 | 111 | +7 |

simplest duplex arrangement for real variable rate modem is time division duplex (TDD), where both base station and mobile station share the same frequency channel, but at different times. If the fading time intervals are long enough in the comparison with the symbol interval, then the modem can track the mobile channel fading changes.

The transmitter part of the system consists of the information source, elastic memory, data encoder and CPFSK modulator. The input signal shaping filter can be added to improve the system spectral performance.

Data source usually generates bit stream with constant bit rate. Information bits are buffered in an elastic memory, which provides the required bit rate to the data encoder. The encoder converts the $N$ serial bits into $N$ parallel bits forming one multilevel symbol. $N$ is the number of transmitted bits per symbol interval. Actual number of transmitted bits per symbol depends on channel distortion in that moment. There are several algorithms to determine the $N$ value. Mapping functions of the encoder are shown in Fig. 4.36. The most significant bit determines the sign of the encoder output for all three $M$ = 2, 4, 8 modulation schemes.

Received signal is demodulated and decoded afterwards. The decoded signal is passed to the elastic memory. For the constant rate services, such as telephony, a constant bit rate is required. Elastic memory buffers the bits, which are sent when the transmission rate over the channel is reduced so that the constant average bit rate is provided for the user. A variable rate M-CPFSK communication system is shown in Fig. 4.28.

We chose a time division duplex (TDD) transmission mode for our variable rate M-CPFSK simulation model. If the channel variations are slow in comparison with the length of the symbol interval, the bit error rate of the received signal can be the criterion for the choice of the modulation level.

The base station and the mobile station are informing each other permanently about the current transmission rate. Modulation level value is transmitted usually in the header part of the block and is always binary CPFSK modulated.

Two modulation level switching approaches are known for variable rate modulation techniques: error detector switching approach and received signal strength indicator (RSSI) approach. The first approach gives inferior results because the system reacts not earlier than the errors are detected by

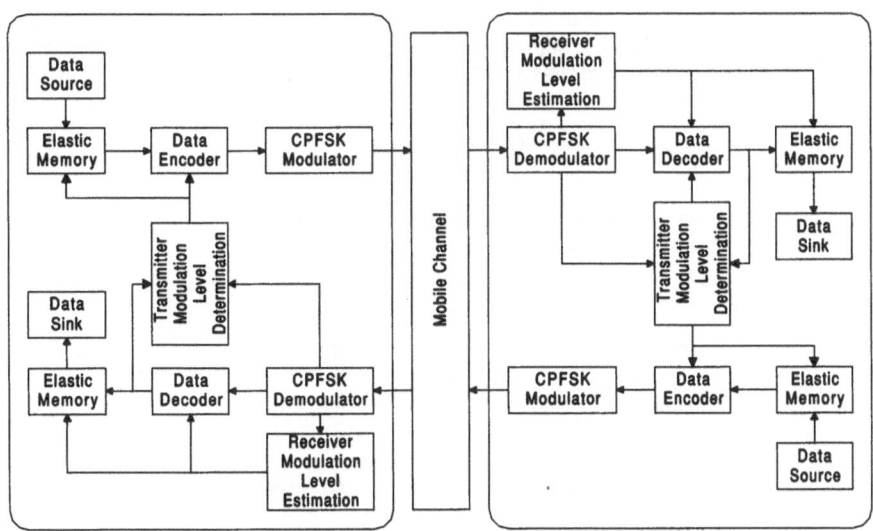

**Figure 4.36.** Block diagram of variable rate CPFSK communication system.

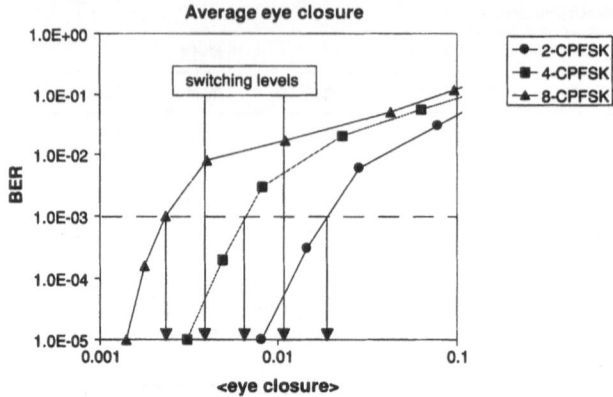

**Figure 4.37.** BER as a function of eye closure.

error detection codes. The second approach is based on the received signal strength and assumes that the majority of distortions are added to the signal at the receiver. The RSSI approach does not performs well in the case of cochannel and adjacent channel interference.

Bearing in mind that all impairments such as white Gaussian noise, fast Raileigh fading, cochannel interference, adjacent channel interference and Doppler frequency shift are reflected on eye pattern, we propose a new approach for modulation level switching thresholds based on eye pattern observation.

It is possible to calculate the bit error rate on the basis of the average eye closure. Average BER value is drawn as a horizontal line in the BER versus eye closure diagram. The middle of the intersection points represent the eye closure values corresponding to the thresholds for the modulation levels. The threshold values for 2, 4 and 8-CPFSK system are shown in Fig. 4.37.

When the eye closure value is lying between the threshold values, 4-CPFSK modulated signal is transmitted. 8-CPFSK signal is activated when the eye is closed less and the binary CPFSK is signal activated when the eye is closed more then at the threshold values.

### 4.3.3 Simulation Results

In order to analyse the performance of a variable rate M-CPFSK modulation scheme the S/N ratio and Doppler frequency shift values should be varied during the simulation.

In Fig. 4.38 the results for constant rate 4-CPFSK and variable rate M-CPFSK ($M$ = 2, 4 and 8) are shown. The 128 bits long sliding block is used to get eye closure for variable rate M-CPFSK modulation scheme. The eye closure switching margins are set to 0.002 to switch from $M$ = 8 to $M$ = 4, and to 0.09 to switch from quaternary $M$ = 4 to binary $M$ = 2 modulation scheme. The probability of error for constant rate 4-CPFSK is approximately ten times higher than for variable rate M-CPFSK. The throughput of the constant rate 4-CPFSK is two bits per symbol. The throughput for variable

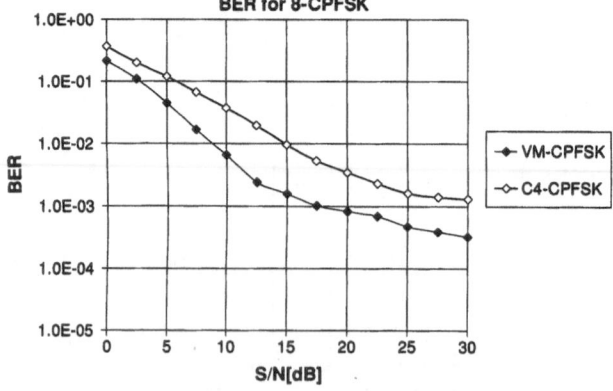

**Figure 4.38.** BER for variable rate M-CPFSK scheme.

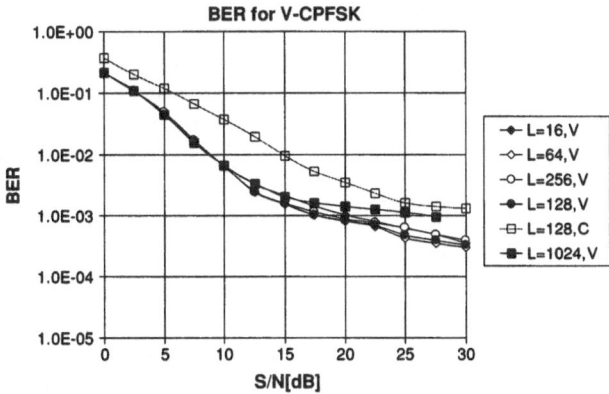

**Figure 4.39.** CPFSK system performance for different block lengths.

rate M-CPFSK varied from 1 bit per symbol interval for low S/N ratios to 3 bits per symbol interval for higher S/N ratios. Average throughput in variable rate M-CPFSK system is 2 bits per symbol interval.

We analysed also the influence of block length onto BER performance of the system. The simulation results are shown in Fig. 4.39. The system performance is getting worse when the block size goes over 256 symbols. The system reaction becomes slow in this case and adaptation to the bad channel state is being late. Too short block length is also inconvenient because it leads to instability in regard with modulation level switching. It seems that the optimum block size is around 128 symbols.

### 4.3.4 Conclusions

The eye pattern has been used to establish the threshold values for the switching among M-CPFSK modulation schemes. It was found out that the variable rate M-CPFSK modulation scheme outperforms constant rate 4-CPFSK modulation scheme in variable S/N ratio environment if a moving of the mobile is not too fast.

The proposed approach is particularly suitable for data transmission where the constant rate and transmission delay can be tolerated to certain extent.

## 4.4 Novel Coding Techniques: TCH Codes

### 4.4.1 Introduction

With the increasing use of multimedia services and personal communications, the design of communication systems is being pushed to the limits of channel capacity with the use of very efficient modulation and coding schemes. Since most of these systems tend to be portable and mobile, these terminals should be as small as possible with low gain antennas. The power of these terminals must also be kept as low as possible not only for portability reasons but also for reducing the interference on neighbour systems. However, these restrictions should not compromise the overall performance in a real environment, that is, when adverse propagation conditions are considered.

For example, we take into account propagation with multi-path and other sources of fading that result in low signal-to-noise ratios, namely a Ka or higher frequency bands. This is even more critical in satellite receivers as they operate under limited power of the satellite.

The new TCH (Tomlinson, Cercas, Hughes) family of codes was devised for such applications, namely for FEC (Forward Error Correction), and it was shown that they can exhibit good performance and undertake maximum likelihood soft-decision decoding with a very simple decoder structure using DSP techniques known as TCH receiver (62).

A further analysis of the correlation properties of TCH sequences, i.e. TCH codewords taken as sequences, revealed that it is possible to identify some sets of sequences that exhibit good auto and cross-correlation. These are very important properties for this family of sequences since, as we know, the performance of CDMA (Code Division Multiple Access) systems depends not only on the cross-correlation properties of the sequences in order to minimise inter-user interference, but also on their autocorrelation because of the synchronisation process. Therefore, TCH codes can also be used for spreading taking advantage of its simplified correlation receiver to detect different users in a CDMA environment.

It is not difficult to design a multiple access system that uses the inherent properties of TCH sequences such as a DS-CDMA (Direct Sequence CDMA) system, for example. The use of TCH sequences in a CDMA system simplifies the structure of the receiver and may accelerate the synchronism process. Besides, some sets of TCH sequences are nearly orthogonal and, if used in a synchronous CDMA system, allows a relatively large number of users to communicate simultaneously.

The above applications of TCH codes are under investigation and this section presents results in which the performance of TCH codes for FEC is considered. Later in the report we will find some results showing the application of TCH codes and sequences in a synchronous and asynchronous CDMA system.

## 4.4.2  TCH Codes and Sequences

These codes were firstly intended for use in low cost satellite receivers (63) since they can provide good coding gains, comparable with those of similar length and rate, and considerable higher gains when concatenated with RS codes while maintaining a very low implementation complexity, i. e. with a very small number of correlators. In addition, the required codes should perform soft-decision maximum likelihood decoding for best performance with no extra complexity of the receiver.

These and other requirements listed in were successfully achieved using an analytical method together with several auxiliary methods namely the shift-and-add method. The resulting codes are denominated as TCH$(n, k)$ where $n$ is the code length and $k$ is the number of information bits in a codeword. A TCH code can be described by a set of $h$ polynomials that mutualy obey a given value of minimum distance:

$$\text{TCH}(n, k) = \begin{bmatrix} p_1(x) \\ p_2(x) \\ \cdot \\ \cdot \\ \cdot \\ p_h(x) \end{bmatrix} \tag{4.28}$$

A TCH sequence can be a polynomial or a codeword from a TCH code set. TCH codes are binary, nonlinear, non-systematic cyclic codes length $n = 2^m$, $m$ being any positive integer. The best TCH codes known are derived from polynomials known as Basic TCH or B-TCH Polynomials.

### 4.4.2.1  Analytical Generation of B-TCH Polynomials

B-TCH polynomials have a high algebraic structure. The problem of finding such polynomials was reduced to the search of a polynomial $p(x)$ length $n = 2^m$ with coefficients $a_i$ i = 0, 1, ..., $n-1$ from Galois Field GF(2):

$$p(x) = \sum_{i=0}^{n-1} a_i \cdot x^{k_i} = a_0 \cdot x^{k_0} + a_1 \cdot x^{k_1} + \ldots + a_{n-1} \cdot x^{k_{n-1}} \tag{4.29}$$

where the non-zero $k_i$ coefficients define the polynomial and need to be determined. If their number equals half the code length we obtain a balanced code.

Finite field theory, as described in (65), tell us that we can associate these polynomials of degree n with elements of a GF($q$), where $q$ must be a prime $p$ or the power of a prime. Since we are not interested in the all-zero codeword, which is the zero element of any GF, that prime number must verify the following Eq.:

$$p = n + 1 = 2^m + 1 \tag{4.30}$$

From number theory, as shown in (66), prime numbers obeying this condition are very restrict and are exactly the Fermat numbers $F_i$:

$$F_i = 2^{2i} + 1 \tag{4.31}$$

where $i$ is a non-negative integer. Unfortunately, most Fermat numbers are not prime as they have been found to be factored in some way. The highest known prime number obeying this equation is the Fermat number $F_4 = 65537$ which is suitable for generation of B-TCH sequences length 65536. The other prime numbers are shown in Table 4.18 and can generate B-TCH sequences length 2, 4, 16 and 256.

**Table 4.18.** Possible values for the length of B-TCH polynomials

| i | $F_i$ | n |
|---|-------|---|
| 0 | 3 | 2 |
| 1 | 5 | 4 |
| 2 | 17 | 16 |
| 3 | 257 | 256 |
| 4 | 65537 | 65536 |

These B-TCH sequences length $n$ can be written as a polynomial of the form:

$$p(x) = \sum_{i=0}^{[(p-1)/2]-1} x^{k_i} = x^{k_0} + x^{k_1} + \ldots + x^{k_{[(p-1)/2]-1}} \qquad (4.32)$$

Since we prefer to generate perfectly balanced sequences, i.e. with the same number of zeros and ones and therefore weight $n/2$, the non-zero coefficients were chosen from a set of quadratic residues. Knowing that for any primitive element $\alpha$ of a GF($p$), $p$ an odd prime, there are exactly $n/2$ elements which are the odd quadratic residues and $n/2$ which are the even quadratic residues the choice of one of these sets served this purpose.

Another important condition imposed to these B-TCH sequences is the SAA (Shift-And-Add) property, i.e. the addition modulo 2 of a sequence with a shifted version of itself should result in a different sequence with low cross-correlation with its generator and, if possible, also balanced and with good auto-correlation. In fact, if the resulting sequence keeps the balance property then the cross- correlation with its generator sequence for that particular shift is zero by definition.

The SAA and balance properties were essential for determining the non-zero coefficients $ki$ in $p(x)$ which can be synthesized in this very simple equation:

$$\alpha^{k_i} = 1 + \alpha^{2 \cdot i + 1} \qquad (4.33)$$

where $\alpha$ is any primitive element in GF($p$).

As shown in (67) only the set of odd quadratic residues could guarantee the elimination of the all-zero sequence. Each in $\alpha$ GF($p$) is able to generate a different B-TCH polynomial $p(x)$ with the mentioned properties. It was verified that for any TCH sequence length n and the mentioned Fermat numbers it is possible to generate $n/2$ different B-TCH sequences with the above properties. So, with $F_3$ we can generate 128 B-TCH sequences and with $F_4$ we can generate 32 768 different sequences length 65 536. In turn, each of these sequences can easily generate other good sequences using the SAA method that, as we saw, is inherent to their structure.

The large number of existing B-TCH sequences length 65 536, whose actual generation was only tried recently due to the high computational requirements involved, is very important for spread spectrum systems not only by their great ability to generate more sequences but also because of their auto-correlation. A general and very important property of B-TCH polynomials is that their auto-correlation is always three-valued with the following non-normalized distribution: $-4$ for $n/4$ even shifts, 0 for all odd shifts and n/4–1 even shifts and n for no shift.

#### 4.4.2.1.1  TCH Sequence Derivation with the SAA Method

If $p_1(x)$ is a B-TCH polynomial then an excellent method to get other TCH sequences with the same length and balance is through the use of a SAA procedure, which consists of cyclically shifting $p_1(x)$ and adding it with the original one, in order to get a second polynomial $p_2(x)$ or TCH sequence. The maximum number of TCH sequences length n that can be generated from a B-TCH polynomial length $n$ is $n/2$ and that is because the auto-correlation is an even function, therefore generating the same sequences after half the number of shifts.

Generally, the nice auto-correlation properties of B-TCH sequences, as shown in Fig. 4.40, are not preserved in the derived sequences. However, these new TCH sequences still have very interesting cross-correlation properties. Each B-TCH sequence can easily generate a large set TCH sequences with mutual low cross-correlation which forms a given TCH code and those sequences can even be selected in order to meet some minimum distance requirements.

The first polynomial in a TCH code is generated by that analytical method and is then extended to increase the code set. That is why this polynomial was designated as *basic TCH polynomial*.

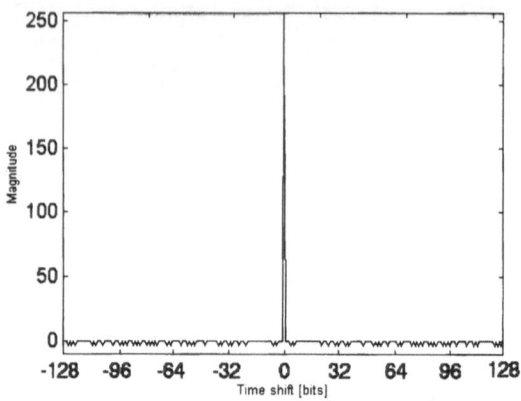

**Figure 4.40.** Typical non-normalised autocorrelation function of a basic TCH polynomial length 256.

### 4.4.2.2 *Efficient Receiver Using FEC with TCH Codes*

The fact that the length of TCH codes is a power of 2 makes it very easy to make correlations in the frequency domain with very fast FFT radix 2 operations. On the other hand, and since any cyclic shift of a TCH codeword or its inverse, which is obtained by simple inversion of the bits and is also a valid codeword in any TCH code, has the same magnitude spectra, we can perform 4n correlations with only one complex FFT, a complex multiplication and a complex IFFT.

In practical terms what this means is that a complete SD MLD (Maximum-Likelihood Decoder) can be implemented with a bank of correlators in the frequency domain with a small number of correlators.

If fact, a TCH($n$, $k$) code can be represented by only $h$ different polynomials and therefore this code whose total number of codewords can be expressed as:

$$2^k = 2nh \tag{4.34}$$

needs a bank with only $h$ different correlators, that is, the total number of codewords divided by twice the code length $n$.

Since a complex FFT can perform two simultaneous real correlations, it is shown in (68) that the scheme shown in Fig. 4.41, which is a basic TCH receiver, needs only $h/2$ correlators to perform SD-MLD.

For example, if we take a TCH(256, 15) code that has 32768 codewords, the TCH receiver needs only a maximum of 32 correlations in the frequency domain to identify the most likely codeword sent, that is, a complexity gain of about 1000!

In order to use this decoding efficiency, the TCH receiver needs to evaluate the FFT magnitude for just h polynomials, select the greatest and from this read the displacement number corresponding to the maximum peak. This value is the address of a ROM table containing the estimated codeword that is sent to the output of the circuit.

Figure 4.42 shows the result of this process for a TCH code length 256 with the outputs of the FFT showing clearly that the estimated codeword is associated with the polynomial correlated by the real part and that position, 150, gives the right shift to get the desired codeword:

The theoretical performance of these codes was first determined via an upper bound for the probability of information bit error $P_e$ or BER (Bit Error Rate) given by:

$$\text{BER} \leq \frac{2^{k-1}}{2^k - 1} \sum_{d=d_{\min}}^{(n/2)-2} N_d \left[ erfc \sqrt{\gamma_b R_c w_d} + erfc \sqrt{\gamma_b R_c (n - w_d)} \right] + N_{\bar{n}/2} \, erfc \sqrt{\gamma_b \frac{k}{2}} \tag{4.35}$$

where:

$d_{\min}$ is the minimum distance of the code,

$N_d$ is the number of coefficients of weight $d$ from the weight distribution of the code,

$\gamma_b = E_b/N_0$ is the signal-to-noise ratio per bit

$R_c = k/n$ is the code rate.

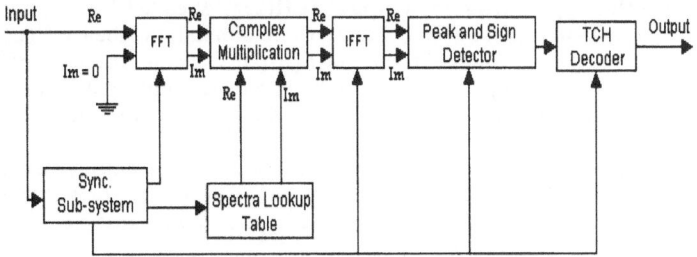

**Figure 4.41.** TCH receiver or TCH maximum likelihood decoder.

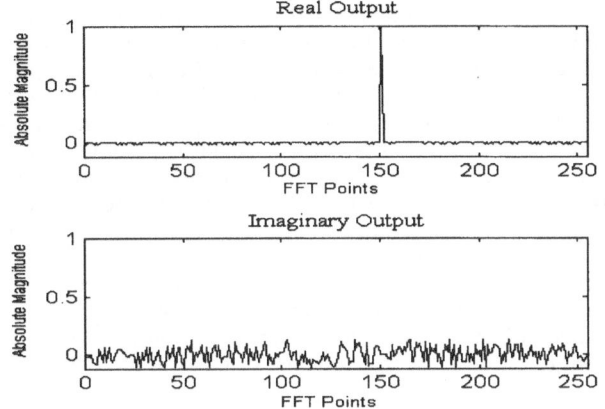

**Figure 4.42.** IFFT outputs of the TCH receiver for a code length 256.

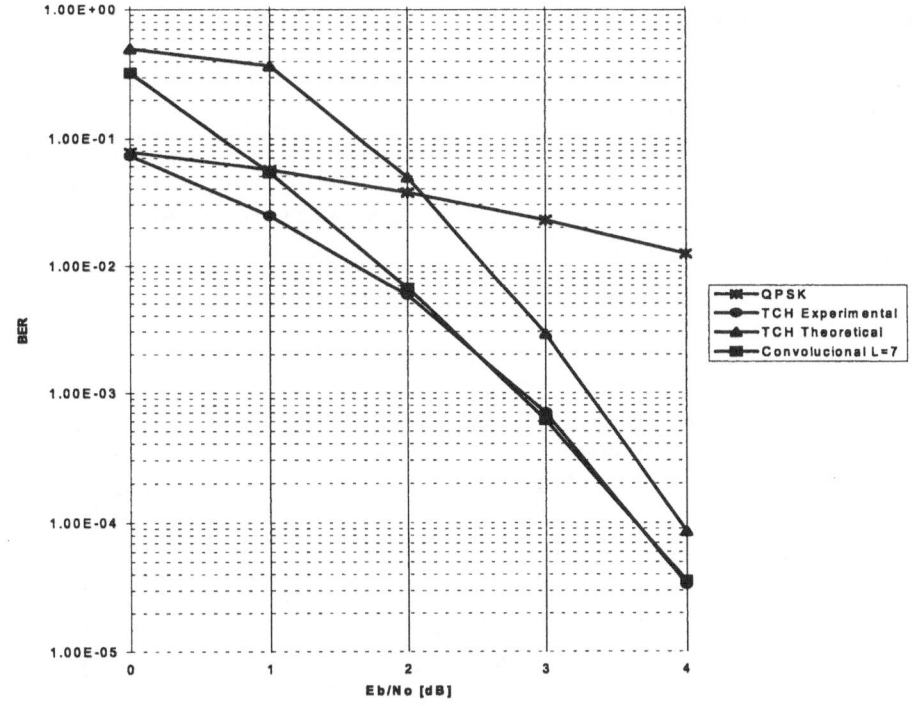

**Figure 4.43.** Comparison of BER for a TCH (256, 14, 55) code.

Figure 4.43 shows this upper bound for a typical TCH(256,14) code labelled as "TCH theoretical". This picture also plots a curve obtained by Monte Carlo simulation labelled "TCH experimental". The obtained curve is compared with the traditional half-rate convolutional code of constraint length 7 and SD Viterbi decoding. Although the rate of this TCH code is about eight times lower, its performance is similar and its implementation in a receiver can be achieved with much less complexity and cost.

When we are interested in higher coding gains we can use a concatenated scheme with a RS (Reed Solomon) code that makes possible to obtain an upper-bound limit for $Pe$ better than $10^{-8}$ with Eb/N0 of only 3dB, which can be very interesting for a VSAT application for instance.

### 4.4.2.3 Performance of TCH Codes with Independent and Burst Errors Using Efficient Simulations

The performance of TCH codes was previously determined using the Union Bound approach for an Additive White Gaussian Noise (AWGN) channel (69). The aim of this section is to present its actual performance with Soft Decision (SD) decoding in channels where errors occur independently and in bursts. The reason for evaluating the performance of TCH codes only for SD and not for hard decision (HD) decoding is twofold: first of all the performance is better and, secondly, for TCH codes SD decoding is done automatically, so the required circuitry is even simpler for SD than for HD.

Although the use of classic simulation techniques could be used, the simulation time would be prohibitive and so a more efficient technique was adopted for this purpose. The chosen technique is the Interval Simulation Technique (IST) that is used with a two states *Markov* model for the bursty channel. This technique is presented and its results are validated. This section finalises with the presentation of some results for the performance of TCH codes, both for independent and bursty channels, and the corresponding trade-offs for the simulation time. This technique is very efficient due to the fact that only the code words that are received with more errors than the error correcting capacity of the code need to be submitted to the whole simulation process. Other code words are only taken into account to estimate the total number of bits and then the *BER*.

#### 4.4.2.3.1 System Model

For the modelled system we have considered QPSK modulation, channel coding with TCH codes and SD decoding. An optimised and very fast maximum likelihood SD decoder for TCH codes is used in which correlation is efficiently computed in the frequency domain (70).

A Discrete Channel Model (DCM) describes the channel without and with memory, as depicted in Fig. 4.44. The first kind of DCM corresponds to the occurrence of independent errors in an AWGN channel. A two states Markov model, known as the Gilbert model (71), describes the DCM with memory, also known as bursty channel.

The performance of TCH codes was then assessed by applying the IST technique to the Gilbert model in two different situations, one for independent errors and another for the occurrence of burst errors, both modelled by the two states Markov model but for different transition probabilities. The two sates Markov model is composed by a "good" state G, without the occurrence of errors and a "bad" state B used to model independent and burst errors.

For the case of the channel with independent errors we used the IST technique as described in (72), Section 4.9.4. For example, the interval length $l-1$ between independent errors is given by:

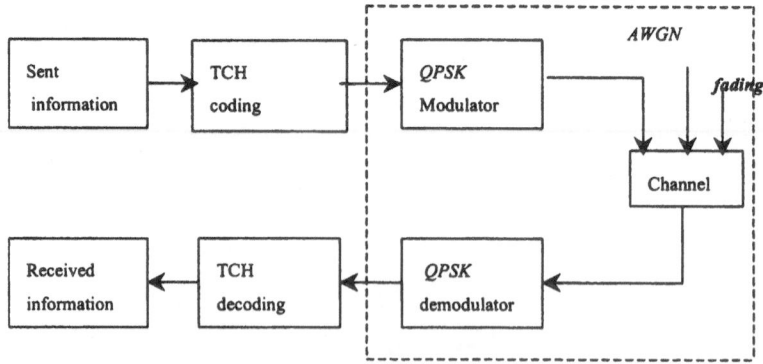

**Figure 4.44.** Discrete channel model.

$$l = \left[ 1 + \frac{\log \xi}{\log (1 - p)} \right] \text{ [bits]} \tag{4.36}$$

where $\xi$ is a random variable uniformly distributed in $(0, 1)$ and $p$ is the error probability of the channel in accordance with the modulation format used (73).

Considering SD decoding in an AWGN channel and taking into account the properties of TCH codes, namely its weight difference distribution, an expression for their BER performance was already derived (74).

### 4.4.2.3.2 Simulation

The simulations were performed following the technique described in (75) and therefore the noise samples were generated according to the modified probability density functions therein referred.

Assuming an AWGN channel, where the signal can have two amplitudes (+1 and −1), the transition probability function in the channel is given by (76):

$$P_{RD} = \frac{1}{\sqrt{2\pi\sigma}} \int_1^\infty e^{-x^2/2\sigma^2} \, dx \tag{4.37}$$

with

$$\sigma = \left( R_c \frac{E_b}{N_o} \right)^{-1/2} \tag{4.38}$$

where $E_b$ is energy of an information bit and $N_o$ the power spectral density of noise. The generation of noise samples that may be responsible for the introduction of errors $(x_e)$ or not $(x_{ne})$, are evaluated according to the following expressions:

$$x_e = \sqrt{2} \, \sigma erf^{-1} (2p_{DB} U - 1) \quad x_e \leqslant -1 \tag{4.39}$$

$$x_{ne} = \sqrt{2} \, \sigma erf^{-1} [2(1 - p_{DB})U - 1] \quad x_{ne} \leqslant -1 \tag{4.40}$$

where $U$ is a random variable uniformly distributed in $(0, 1)$. These equations were obtained from the truncated Gaussian Probability Density Functions (PDF) depicted in Fig. 4.45.

Depending on the code performance, these amplitude noise samples may or may not result in an actual error at the receiver.

In a real channel, as for example a radio channel like the one used in satellite mobile communications, we must consider not only AWGN but also interference, fading from several sources, reflections, multipath, co-channel interference, non-linear distortion and other effects that degrade the received signal. As a result we can have one or more errors in the received code word and, if the signal is severely degraded, a series of burst errors can occur. A good way to model this situation is through the use of a Markov chain (77). Figure 4.46 depicts the mentioned Gilbert model.

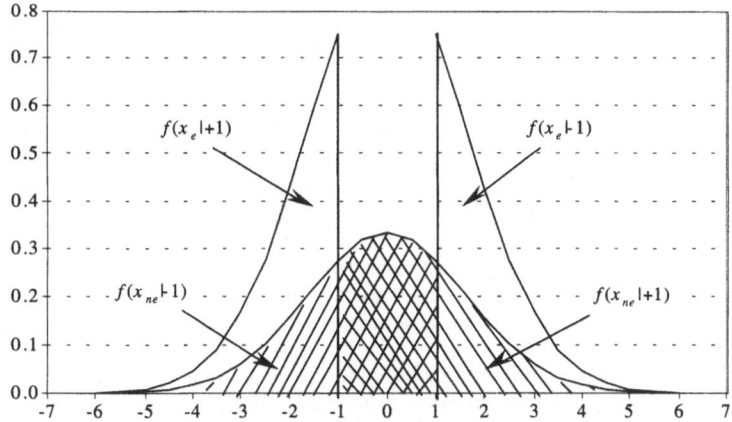

**Figure 4.45.** PDF of the samples $x_e$ and $x_{ne}$.

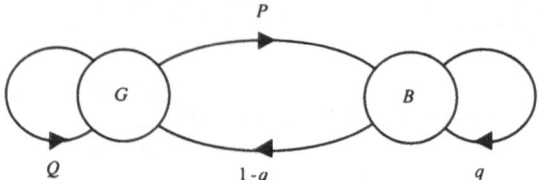

**Figure 4.46.** Gilbert model.

This picture shows the "good" G and "bad" B states and also the corresponding transition probabilities. $Q$ is the probability of remaining in state G, $q$ is the probability of remaining in state B, $P$ is the probability of going to state B given it is in state G and $1-q$ is the probability of changing to state G given it is in state B. If the probability of error in the channel is $p$ and $(1-h)$ is the probability to make a bit error in state B, these variables are related by:

$$p = \Pr(B)(1-h) = \frac{1-Q}{2-Q-q}(1-h) \tag{4.41}$$

The matrix of state transition probabilities $A$ is defined by Gilbert as follows:

$$A = \begin{bmatrix} q_{11} & q_{12} \\ q_{21} & q_{22} \end{bmatrix} = \begin{bmatrix} Q & P \\ 1-q & q \end{bmatrix} = \begin{bmatrix} 0.997 & 0.003 \\ 0.034 & 0.966 \end{bmatrix} \tag{4.42}$$

In the steady state, the elements $a_{ij}$ are given by (78)

$$A^n_{n \to \infty} = \begin{bmatrix} a_{11} & a_{12} \\ a_{21} & a_{22} \end{bmatrix} \cong A^{252} = \begin{bmatrix} 0.9189 & 0.0811 \\ 0.9189 & 0.0811 \end{bmatrix} \tag{4.43}$$

so the probability of being in each state is given by (79)(80):

$$\Pr\{X_n = G\} = a_{21} = 0.9189 \tag{4.44}$$

$$\Pr\{X_n = B\} = a_{22} = 0.0811 \tag{4.45}$$

When starting a simulation, the initial state can be determined by generating a sample of a uniform random variable in the interval (0,1) and checking it with one of those values.

To simulate the Gilbert model the simulation technique described in (81) is followed. Once the signal-to-noise ratio in the channel is known, the probability of error in the channel $p$ is determined for the modulation scheme used. Inverting Eq. (4.41), the probability of bit error for the bad state can then be evaluated.

To check the validity of this technique used to simulate the Gilbert model, the relative frequency of various system parameters was evaluated and compared with the theoretical predictions. The theoretical expressions can be found in (82) except for the probability of a burst interval with dimension $R$ that is given by:

$$\Pr\{R\} = qR(1-h)R^{-2} \tag{4.46}$$

Figures 4.47 and 4.48 show the relative frequencies for state B of the interval without errors and for the interval with adjacent errors, respectively, both for 10 million simulated bits, and the respective theoretical predictions.

Excellent agreement is obtained for all dimension of intervals without errors, as shown in Fig. 4.47 and up to intervals with five adjacent errors for the case of bursts with adjacent errors shown in Fig. 4.48. In this case the probability of the bursts of errors decreases with the burst length, so there is a much higher probability to have bursts with 2 or 3 bits than with 6 or 7, for instance.

In Fig. 4.49 the probability of intervals without errors was determined for both independent and burst errors. It can be noticed that, for the same probability of error in the channel, the probability of occurrence of intervals without errors of length 15 is greater for the two state model.

Figure 4.50 shows the relative error of each element of the estimation obtained for each parameter of the two state model. As can be seen for a simulation of 10 million bits, the relative error is less than 0.5 per cent for all parameters.

Special care was put on the simulation accuracy to obtain good estimates of BER. Three simulations were performed for each signal-to-noise ratio (SNR) and each simulation was stopped when 100 information words in error were decoded. So, three estimates of the BER were obtained and the estimated BER for each SNR is given by the average of those estimates.

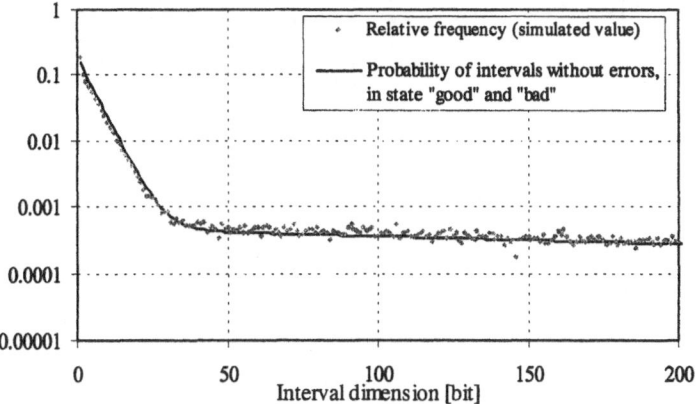

**Figure 4.47.** Relative frequency of the interval without errors ($1 \times 10^7$ bits).

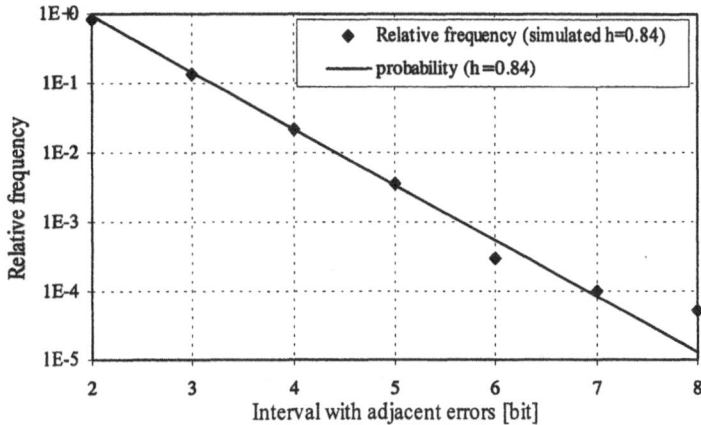

**Figure 4.48.** Relative frequency of the burst intervals with errors ($1 \times 10^7$ bits).

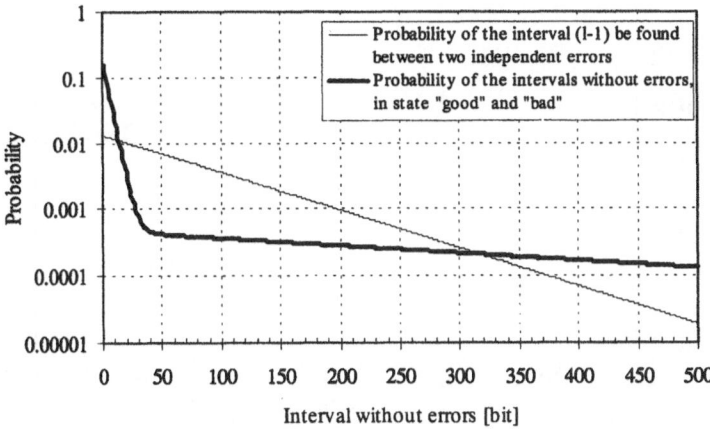

**Figure 4.49.** Probability of the intervals without errors for the independent errors and two state models.

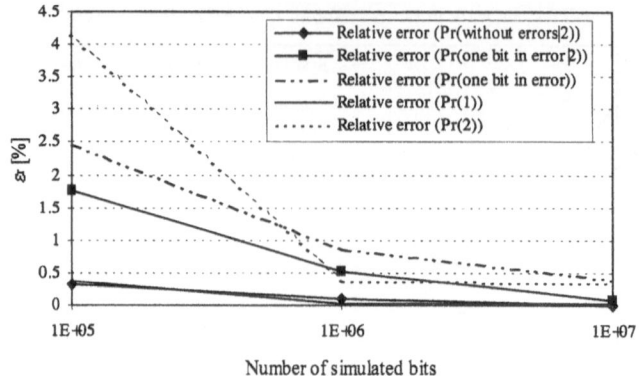

**Figure 4.50.** Relative errors for the two state model.

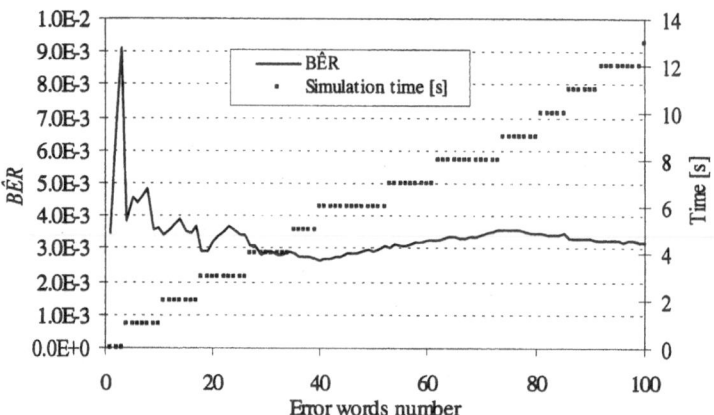

**Figure 4.51.** Estimated BER evolution versus words in error for 2 state model, TCH(16,6,2), $E_b/N_0$ = 7dB.

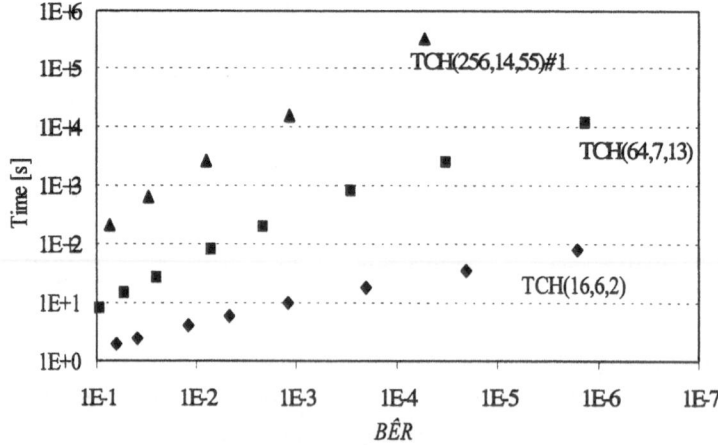

**Figure 4.52.** Time required to simulate 50 information words in error with the independent errors model.

As can also be observed in the example shown in Fig. 4.51, for an $E_b/No = 7$ dB, although after 40 words in error the BER estimate is close to its final value, we decided to stop the simulation only after 100 words in error because this final estimate may vary from simulation to simulation. The final estimate for the BER is the average of three identical but independent simulations.

Figure 4.52 shows the increase of the simulation time as a function of decreasing estimated BER for several TCH codes. The simulations run on a Pentium at 166 MHz with 16 MB of RAM. In spite of using this efficient technique, the time required for estimating the BER of long codes with high SNR is also long.

### 4.4.2.3.3 Numerical Results and Discussion

The BER performance for the AWGN channel is shown in Fig. 4.53. This figure reveals that, for all TCH codes, the simulated results are in quantitative agreement with the theoretical analytical upper bound given by expression (4.35). Furthermore, Fig. 4.53 indicates that these theoretical upper bounds are very tight for SNR greater than about 2 dB, depending on the code.

Figure 4.54. shows the BER of TCH(16, 6, 2) code using SD and the Gilbert model as a function of the SNR. In this case the BER upper-bound is below the simulated values as a consequence of burst errors. Since the occurrence probability of bursts with size greater than the error correcting capability of the code is greater than with independent errors, the BER will increase.

## 4.4.3 Conclusion

New results for the BER of TCH codes have been obtained using a simulation method that gives estimates of the BER with good accuracy in an acceptable time. The system model was carefully

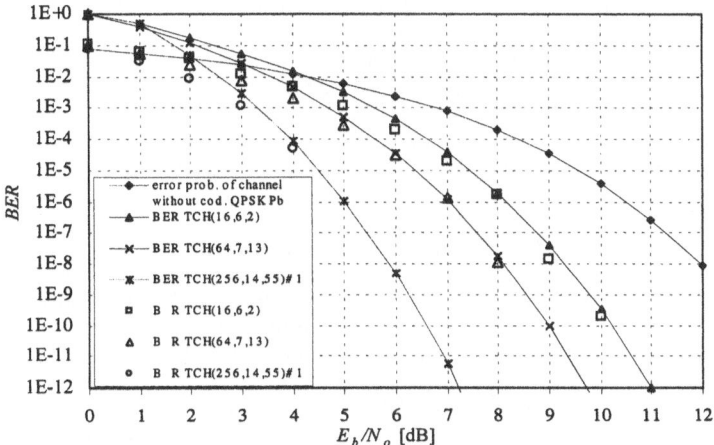

**Figure 4.53.** Estimated performance for TCH (16, 6, 2), TCH (64, 7, 13) and TCH (256, 14, 55) codes using SD and independent errors model.

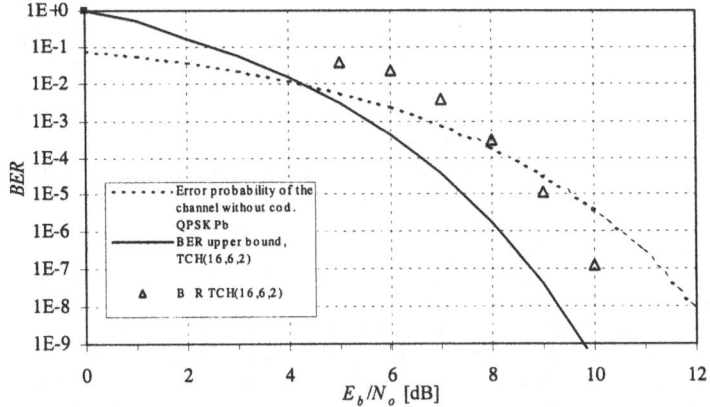

**Figure 4.54.** Estimated performance for the code TCH (16, 6, 2) using SD and two states model.

tested so as to validate the results obtained. These new results for TCH codes have a very good quantitative agreement with the upper bound expressions for an AWGN channel previously determined from the weight difference distribution of these codes(83), which confirm their accuracy. On the other hand it was also verified a good qualitative agreement for the performance of TCH codes in a bursty channel. The simulation technique used in this application can be used for other systems with different types of codes and exhibits a good efficiency in terms of simulation time, which is always desirable despite the constant increase of computing speed from new processors.

## 4.5 CDMA Receiver Techniques for Future MSC Systems

### 4.5.1 Introduction
Evolution of mobile communication systems is strictly related to communication capacity increase: in particular, in multiple access communications suitable transmission techniques and advanced multiuser receivers can take great benefits for performance increase and consequent spectrum exploitation.

The future third-generation wireless communication system IMT-2000, currently in definition within International Telecommunication Union (ITU), is supposed to integrate different services and traffic at variable bit-rate with optimum resources management: in this scenario distinctions between terrestrial and satellite communications segments and between fixed and mobile networks are supposed to vanish towards a global coverage wireless system: so mobile Internet, packet data and file transfer, low rate video/voice services and a considerable re-use of the terminal hardware and software will be allowed. Furthermore IMT-2000 will provide fine granularity in multimedia services by granting data rates from 16 kbit/s to 2 Mbit/s.

ITU recently approved key characteristics for the radio interfaces of third-generation mobile standards: it has been reached the agreement that IMT-2000 will be a single flexible standard with a choice of multiple access methods. However, for what concerns IMT-2000 satellite component several radio interfaces are considered in the 3G standard agreement due to the constraints on satellite system design and deployment and because it does not make sense to harmonise satellite proposals since they are already global.

### 4.5.2 Satellite-Wideband CDMA Multirate Communications
The European Space Agency (ESA) is actively participating into the standardisation of the future 3G satellite mobile systems with an autonomous proposal, defined as Satellite Universal Mobile Telecommunication systems (S-UMTS) (84). In S-UMTS two different access techniques, both based on CDMA, are considered: in particular the so-called Satellite Wideband CDMA (SW-CDMA) option is very similar to one of the two air interface present in Terrestrial UMTS (T-UMTS) proposal.

In SW-CDMA spreading operation is realised by two signature sequences: a *channelisation* code is used to allow users with different bit-rates to reach the same chip-rate, while complex *scrambling* code is cell specific and protects the transmitted signal from interference of other satellites or spotbeams. The channelisation codes are Orthogonal Variable Spreading Factor (OVSF) codes (85) that maintain orthogonality also among codes of different length and for different spreading factors (from 4 up to 256). It is worth stressing that OVSF codes ideal orthogonal properties are unfortunately damaged by multipath fading phenomena.

As it is well known, CDMA techniques most important features are capacities to support universal frequency reuse, variable rate heterogeneous traffic and the possibility to use a time diversity structure joint with Multi-user Detection (MUD) to contrast multipath fading effects. The conventional single- user receiver for CDMA communications is optimum for complexity, but it doesn't show acceptable performance due to the Multiple Access Interference (MAI).

Although optimum multi-user algorithms (86) offer huge potential capacity and significantly improved performance, so mitigating the disadvantages associated with conventional scheme, they get too complex for practical systems; for this fact researches have been spurred on suboptimal solutions like decorrelating detectors or multistage receivers: in particular, a multi-user cancellation detection scheme for SW-CDMA communication systems will be considered (87) in the following.

#### 4.5.2.1 Multi-user Cancellation for SW-CDMA Communications
In literature Interference Cancellation (IC) is accomplished by two different methods: subtracting off the interfering signals successively (88, 89) or in parallel (90, 91) before detecting data.

Successive Interference Cancellation (SIC) approach is a good solution in order to counteract the near- far problem on the condition that the cancellation is quick enough to keep up with the bit-rate

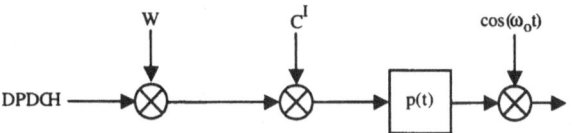

**Figure 4.55.** SW-CDMA Structure.

and not introduce an excessive delay; generally, in several practical applications, the number of cancellation is limited, so increasing Bit Error Rate (BER).

In Parallel Interference Cancellation (PIC) receivers the processing delay constraint is relaxed by collective cancellation of all interfering users from the user of interest, before the final symbol decision. It is easy to understand that, since IC is in parallel for all interfering users, the required delay is ideally very short.

Nevertheless, interference cancellation on user of interest can be demonstrated effective only if interfering signals are powerful enough while weaker interfering users cancellation effects are quite bad: as a consequence, in wireless environment, multipath fading phenomena cause time varying received signal energies so that PIC receivers usually obtain poor performance for modest number of users. Both cancellation schemes have been mainly considered with spread spectrum signals obtained through a single spreading sequence and for single rate communications.

An original PIC receiver, defined as One-shot detector, having a single stage architecture in which the interference cancellation is performed on a selective basis (S-PIC) has been developed (92, 93). On the first step the received user signals are processed by a bank of matched filters, one for each signal replica: matched filter output is compared with a suitable threshold and assumed reliable or not, depending on its value. The MAI effects of reliable signals on the other signals are cancelled at the second step on a one shot basis. Matched filter outputs are used as received signals amplitude estimates in reconstruction and cancellation operation. Therefore, the basic assumption of the proposed detector, named hereafter as One-shot receiver, is to divide matched filter outputs into two different groups according to the received signals power level. The reliable ones are directly detected and cancelled from the whole received signal before making the decision on not reliable signals or replicas, without any further processing delay.

Communication system considered hereinafter is based on SW-CDMA structure, as outlined in Fig. 4.55: main differences are to be found in control channel (DPCCH) suppression, in real (not complex) spreading operation and in the elimination of the optional scrambling code.

All this simplifications aiming to obtain a clearer and deeper evaluation of One-shot positive effects in facing impairments due to MAI and multipath fading phenomena. As described in the following, both synchronous and asynchronous users are considered, so taking into account two different scenarios, namely the mobile to satellite communications and the gateway-based option. The proposed receiver requires knowledge of user delay (in case of asynchronous users), attenuation and phase offset.

Ideal Automatic Frequency Control (AFC) (94) is supposed to be used in order to compensate Doppler frequency shift, inherent to LEO and MEO satellite communications (95).

Performance comparisons with classical rake receiver scheme have been also performed by computer simulations in order to highlight the advantages of the proposed One-shot IC receiver.

### 4.5.2.2 System and Channel Model

Received signal $r(t)$ can be expressed as:

$$r(t) = \sum_{nr=1}^{N_r} \left( \sum_{k(n_r)=1}^{K(n_r)} \sum_{l=1}^{L} \sum_{i=-\infty}^{\infty} \sqrt{P^{k(n_r)}} \, \alpha_{l,k(n_r)} \left[ b_{k(n_r)}(i) \, F_{k(n_r)}(t - iT_b(n_r) - \tau_{kl,k(n_r)}) \right] e^{j\vartheta_{l,k(nr)}} \right) + n(t) \quad (4.47)$$

where $N_r$ is the number of different rates in the systems at the same time, $K(n_r)$ is the total number of active users transmitting with $n_r$th bit rate, assuming $K = \sum n_r K(n_r)$ as the total number of users, $L$ is the signal replicas number, assumed constant without any generality limitation, $\sqrt{Pk(n_r)}$ the transmitted power of $k(n_r)$th user, $T_b(n_r)$ the bit duration of $n_r$ bit rate sources, $\alpha_{l,k(nr)}$ is the $l$th replica $k(n_r)$th transmitted signal attenuation whose statistical distribution is Rice if $l = 1$ (Line-of-Sight, LOS), Rayleigh otherwise (Non-Line-of-Sight, NLOS), $\vartheta_{l,k(nr)}$ is the $l$th replica $k(n_r)$th transmitted signal phase, uniformly distributed in $[0,2\pi)$ and $\tau_{l,k(nr)}$ is the $l$th replica $k(n_r)$th transmitted signal delay, defined according to model recently proposed in COST252 (96) for satellite link. Moreover, $b_{k(nr)}(i) \in \{\pm 1\}$ with equal probability, $k(n_r)$th transmitted signal $i$th informative bit, chip and bit shapes are rectangular, $n(t)$ is the complex AWGN with zero mean and one-sided power spectral density $N_0$. It is worth stressing that, for the sake of simplicity we have avoided to indicate time dependency of parameters $\alpha_{l,k}$, $\vartheta_{l,k}$ and $\tau_{l,k}$. Moreover, $k(n_r)$th user spreading sequence is obtained multiplying OVSF code with a very large Kasami set scrambling code: while first code makes to obtain same chip rate for all the users, the multiplication to the second sequence can be supposed to be performed on a chip-by-chip basis.

Parameters $\alpha_{l,k}$, $\vartheta_{l,k}$ and $\tau_{l,k}$ are assumed known at the receiving end, i.e. tracked accurately. Moreover, knowledge of the spread sequences of all the users is assumed.

Since perfect phase recovery is assumed before despreading operation, output signal of the $\bar{l}$th replica $\bar{k}(\bar{n}_r)$th user matched filter is equal to

$$y_{\bar{l},k(n_r)}(i) = \sqrt{P^{k(n_r)}T_b(n_r)}\, \alpha_{l,k(n_r)}b_{k(n_r)}(i) + \sum_{j=1}^{i+1}\sum_{\substack{l=1 \\ l\neq\bar{l}}}^{L}\sqrt{P^{k(n_r)}T_b(n_r)}\alpha_{l,k(n_r)}b_{k(n_r)}(j)$$

$$\hat{I}_{k(n_r),l,k(n_r),l}\left((i+1)T_b(n_r) + \tau_{l,k(n_r)}, j\right) + \sum_{\substack{k(nr=1) \\ k(nr)\neq k(nr)}}^{K(nr)}\sum_{l=1}^{L}\sum_{j=i-1}^{i+1}\sqrt{P^{k(n_r)}T_b(n_r)}\alpha_{l,k(n_r)}b_{k(n_r)}(j)$$

$$\hat{I}_{k(n_r),l,k(n_r),l}\left([(i+1)T_b(n_r)] + \tau_{l,k(n_r)}, j\right) + \sum_{\substack{k(nr=1) \\ k(nr)\neq k(nr)}}^{K(nr)}\sum_{l=1}^{L}\sum_{j=i-1}^{i+1}\sqrt{P^{k(n_r)}T_b(n_r)}\alpha_{l,k(n_r)}b_{k(n_r)}(j)$$

$$\tilde{I}_{k(n_r),l,k(n_r),l}\left([(i+1)T_b(n_r)] + \tau_{l,k(n_r)}, j\right) + \tilde{n}_{l,k(n_r)}(i) \tag{4.48}$$

where

$$\hat{I}_{k(n_r),l,k(n_r),l}\left((i+1)T_b(n_r) + \tau_{l,k(n_r)}, j\right)$$

expresses the crosscorrelation between same rate signals (97), while

$$\tilde{I}_{k(n_r),l,k(n_r),l}\left([(i+1)T_b(n_r)] + \tau_{l,k(n_r)}, j\right)$$

takes into account cross-correlation terms between different bit rates informative flows: so, it is evident that first term in (2) is due to the transmitted signal of the desired user, the second term is the contribution of self-noise, the third term is MAI produced by same bit-rate interfering users, the fourth term is MAI produced by different rate interfering users and the last is due to the AWGN.

It is worth stressing that time domain of function $\tilde{I}(\cdot,\cdot)$ is strictly based on desired users bit epoch $Tb(\bar{n}_r)$: in particular, higher bit-rate contribution is determined by a number of bit greater than one, while lower bit-rate interference is caused only by a fraction of a single bit. Therefore a two-slot buffer is required to accomplish a correct parallel interference cancellation, as illustrated in Fig. 4.56.

Each single replica contribution can be used to derive $k(\bar{n}_r)$ user Rake receiver output: in particular, a soft detection approach is adopted by using matched filter outputs as signal amplitude rough estimates

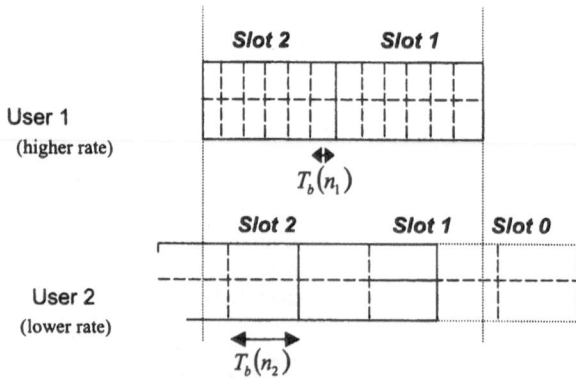

**Figure 4.56.** Asynchronous multi-rate communications

$$Z_{k(n_r)}(i) = \sum_{l=1}^{L} \left| y_{l,k(n_r)}(i) \right| y_{l,k(n_r)}(i) \qquad (4.49)$$

As can be deduced from results shown in the following, conventional rake receiver based on Maximum Ratio Combining algorithm at the end of the up-link channel is vulnerable to MAI impairments. As a consequence, this near-far effect gives rise to an irreducible error floor, even for less than half loaded communication systems.

### 4.5.2.3 One-shot Receiver in Satellite Channel

Matched filter outputs $y_{l,k(nr)}(i)$ obtained in (Eq. 4.29), are compared with a proper threshold in order to define the reliability term reliability term $\tilde{R}_{l,k(nr)}(i)$ as:

$$\tilde{R}_{l,k(nr)}(i) = \begin{cases} -1 & -S_{SD} \geq y_{l,k(nr)}(i); \\ 0 & -S_{SD} < y_{l,k(nr)}(i) < S_{SD}; \\ 1 & y_{l,k(nr)}(i) \geq S_{SD}. \end{cases} \qquad (4.50)$$

If $\tilde{R}_{l,k(nr)}(i) = 1$, a reliable signal replica is under consideration. For all the reliable replicas, MAI, self-noise and AWGN influences are assumed negligible and their effects on all the other signals are cancelled on the second step. Let replica $\bar{l}$ of user $k(n_r)$ be unreliable: for this signal the cancellation term due to the other interfering replicas is equal to:

$$C_{l,k(n_r)}(j) = \sum_{k(n_r)=1}^{K} \sum_{l=1}^{L} R^{-}_{l,k(n_r)}(i) y_{k(n_r),l}(i) F_{k(n_r)}(t - jT - \tau_{k(n_r),l}) e^{j\vartheta_{k(nr),l}(j)} \qquad (4.51)$$

where choice of index $j$ depends on user relative time offset. It is worth emphasising that in (8), by (9), only reliable replicas cause a non-zero contribution to $C\bar{l}$, $\bar{k}(j)$. Therefore cancellation operation gives:

$$\hat{r}(t) = r(t) - \sum_{j=i-1}^{i+1} C_{l,k(n_r)}(j) \qquad (4.52)$$

For reliable replicas cancellation is not performed, i.e. $C_{l,k(nr)}(j)$ is set to zero, so that $\hat{r}(t) = r(t)$.

Cancelled signal $\hat{r}(t)$ is input to correlator, combined with the other replicas, detected through a zero threshold device. The flow chart of SD cancellation process in satellite fading environment is shown in Fig. 4.57.

Due to bandwidth specifications only direct path and one attenuated echo is discriminated at receiving end: besides, reflected echo is nearly negligible. Perfect MAI cancellation cannot be implemented, for this SD structures: in particular, lack of a good received signals parameters estimates leads to approximate operations.

From (6–10) it is evident that the effectiveness of the proposed interference cancellation method is dependent on threshold $SSD$. In considered simulations following conditions have been assumed:

- Satellite up-link channel for 2 GHz band;
- considered users bit-rate equal to 64 kbit/s or 512 kbit/s;
- spreading obtained through a channelisation code called Orthogonal Variable Spreading Factor (OVSF), that is a Hadamard Walsh 256 followed by a scrambling code of O-Gold 256 type; the SF is variable according to the bit rate, i.e. SF = 8 for a bit rate of 512 kbit/s and SF = 64 for 64 kbit/s;
- Doppler spread equal to 100 Hz (ideal AFC is supposed).
- The channel model is derived from (98): this non-stationary fading model is based on continuos variations.

In Figs 4.58 to 4.59 conventional receiver performance is reported in comparison with One- Shot for four equal power users, in synchronous and asynchronous system, respectively: considered bit rate is equal to 64 Kb/s. As it can be seen from reported diagrams, One-shot receiver achieves better BER performance especially in fully asynchronous environment. In Figs 4.60 to 4.61 the two considered detectors are compared for a twelve 64 Kb/s equal power users systems: conventional and One-Shot receiver BER performance is reported in synchronous and asynchronous environment, and the S-PIC approach improvement is remarkable.

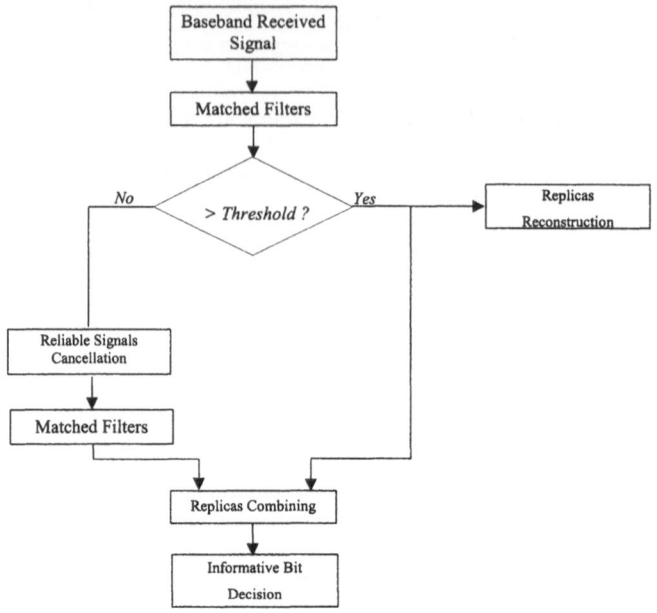

**Figure 4.57.** One shot receiver flow chart.

**4 Asynchoronous Users (64 Kb/s)**

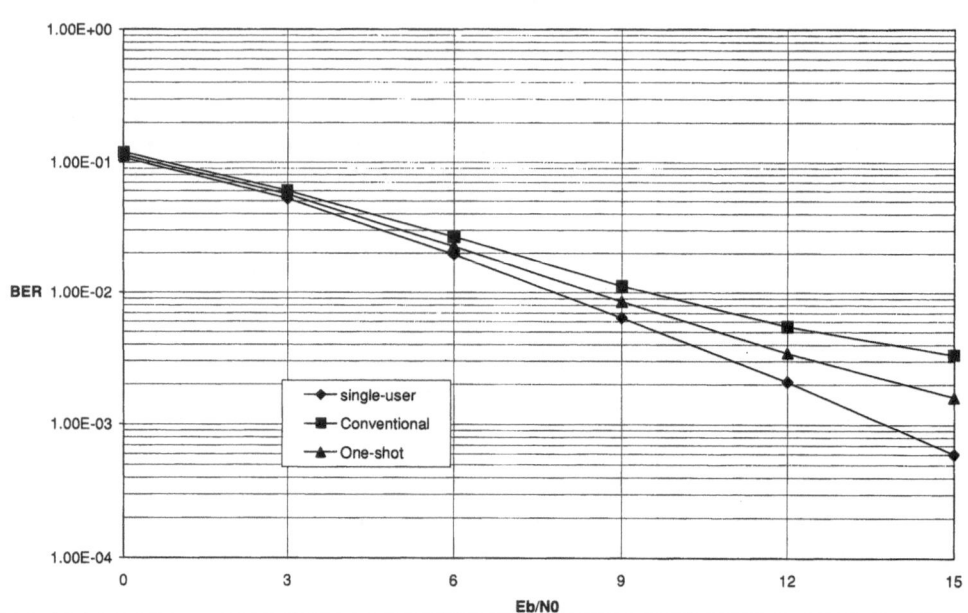

**Figure 4.58.** BER comparison.

In Figs 4.62 to 4.63 the two considered detectors are compared in the case of a system with four users signalling with a bit rate of 64 kbit/s and one at 512 kbit/s: it is worth stressing that, for what concerns band-width occupation, this situation is equivalent to the case of twelve users, all at 64 kbit/s. The performance improvement achieved by one-shot canceller is again evident.

In Figs 4.64 to 4.65 One-shot performance is analysed in the case of a system with four eight users transmitting with a bit rate of 64 kbit/s and two at 512 kbit/s, with a system load equal to 37 per cent of system capacity.

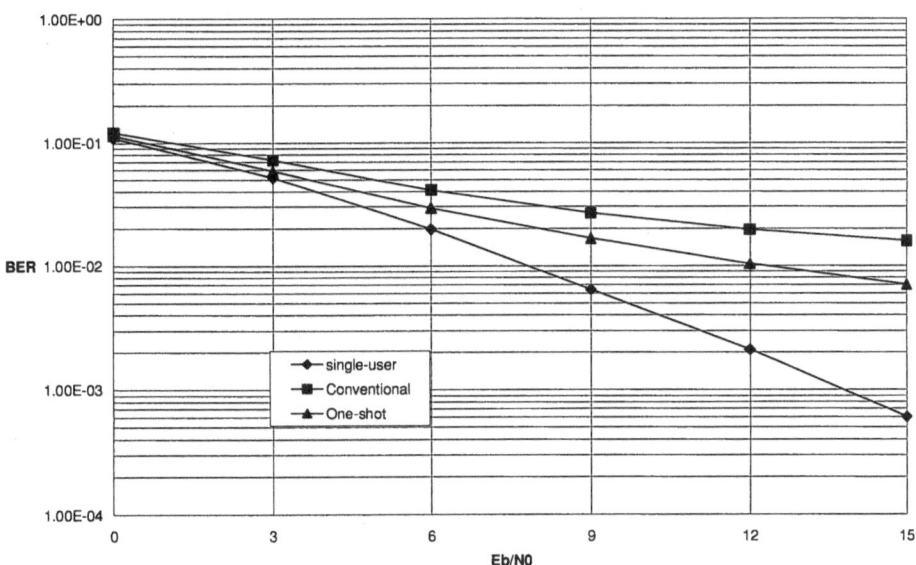

**Figure 4.59.** BER comparison.

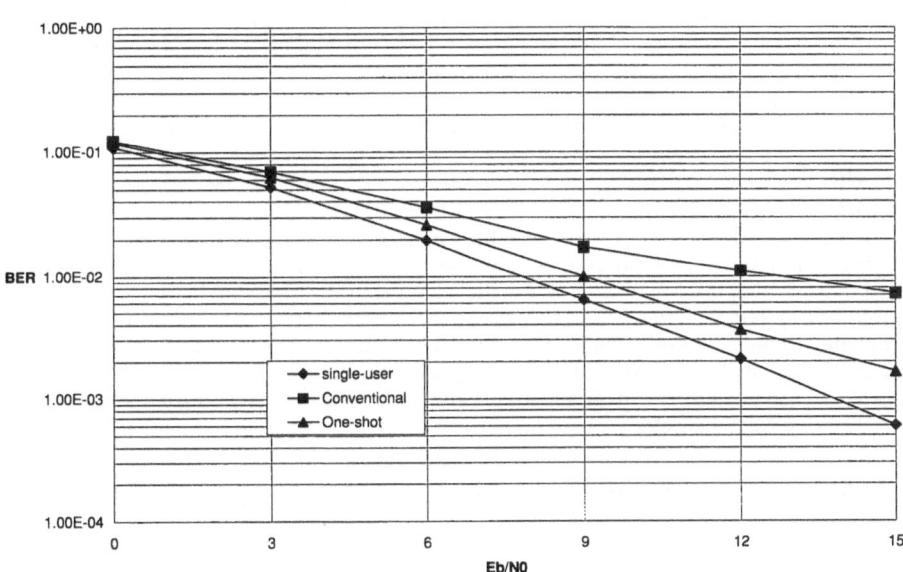

**Figure 4.60.** BER comparison.

Finally, proposed receiver is robust with regards to threshold definition: threshold value is about the same for different system loads, as it can be seen in Fig. 4.66. Moreover, one-shot detector robustness with respect to eventual estimation error in the determination of the optimum threshold value can be deduced.

Proposed interference cancellation approach, therefore, achieves better performance than Rake receiver while keeping low complexity implementation.

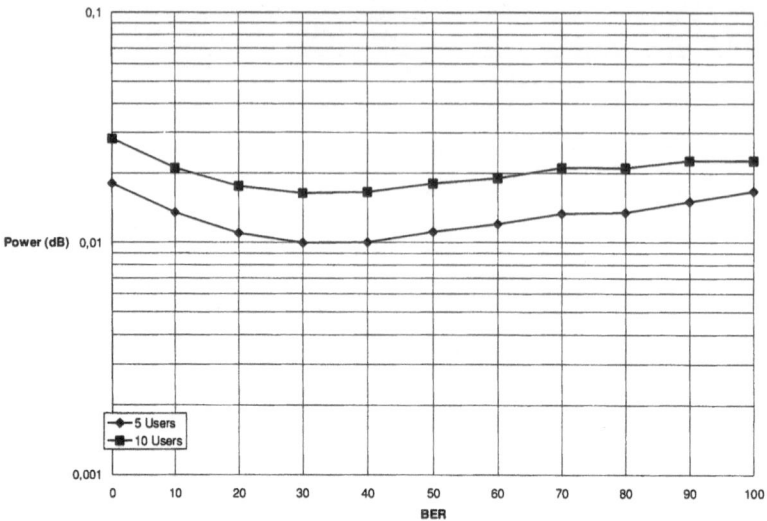

**Figure 4.61.** BER comparison.

### 4.5.3 CDMA Satellite Communication Multiusers Receivers Design

As it is described in previous paragraph, over the years researchers have sought ways to extend the user capacity of CDMA systems either by exploiting optimum (maximum likelihood, ML) multiuser detection, interference cancellation (IC) methods and other techniques such as decorrelating receiver.

The complexity of the classical suboptimal multiuser detectors, lower than optimal multiuser receivers, but still too high for practical implementation, can be lowered by Interference Cancellation (IC) receivers and adaptive blind multiuser techniques introduction (138–145).

#### 4.5.3.1 One-Shot Interference Cancellation for Up-Link DS-CDMA Communications

IC have been considered for DS/CDMA satellite communication systems (99), in which the BPSK modulation technique is used in transmission: as in the previous system, One-shot detector is supposed to be used at the end of up-link satellite channel in the base station, so that the considered communication system is characterised by lack of synchronism between the users in multipath fading satellite channel. One-shot performance is evaluated in BER terms by means of simulations, proving that the proposed satellite receiver exhibits very good behaviour for uniform users power distribution and shows reasonable near-far resistance; channel model is derived according to (100). Received signal $r(t)$ can be expressed as:

$$r(t) = \sum_{k=1}^{K} \sum_{l=1}^{L} \sum_{i=-\infty}^{\infty} \frac{\sqrt{E}}{T} \alpha_{l,k} [b_k(i) F_k(t - iT_b - \tau_{l,k})] e^{j\vartheta_{i,k}} + n(t) \tag{4.53}$$

where $K$ is the total number of active users, $L$ the signal replicas number, assumed constant without any generality limitation, $E$ the energy of every transmitted bit, $T_b$ the bit duration, $\alpha_{l,k}$, $\vartheta_{l,k}$ and $\tau_{l,k}$ $l$th replica $k$th transmitted signal attenuation phase and delay, whose statistical distribution are, respectively, Rayleigh, uniform in $[0, 2\pi)$ and uniform in $[0, T_b)$, $b_k(i) \in \{\pm 1\}$ with equal probability $k$th transmitted signal informative bit, $n(t)$ is the complex AWGN with zero mean and one-sided power spectral density $N_0$. Once again, for the sake of simplicity we have avoided to indicate time dependency of parameters $\alpha_{l,k}$, $\vartheta_{l,k}$ and $\tau_{l,k}$. Moreover, $k$th transmitted signal spreading sequence $F_k$ is Gold codes of length 127, $T_c$ is the chip duration, chip and bit shapes are rectangular.

Parameters $\alpha_{l,k}$, $\vartheta_{l,k}$ and $\tau_{l,k}$ are assumed known at the receiving end, i.e. tracked accurately. Moreover, knowledge of the spread sequences of all the users is assumed.

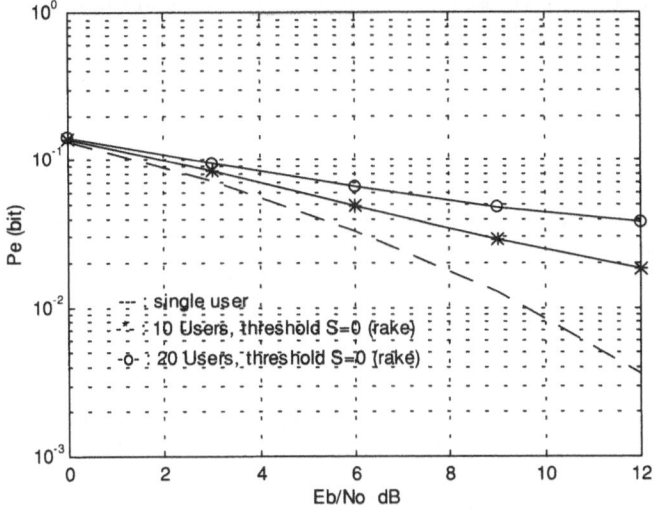

**Figure 4.62.** BER comparison

Since perfect phase recovery is assumed before despreading operation, output signal of the $\bar{l}$th replica $k$th user matched filter is equal to

$$y_{l,k}(i) = \sqrt{E}\,\alpha_{l,k}b_k(i) + \sum_{j=1}^{i+1}\sum_{\substack{l=1\\l\neq l}}^{L}\sqrt{E}\,\alpha_{l,k}b_k(j)I_{k,l,k,l}\big((i+1)T_b + \tau_{l,k},j\big) +$$

$$+ \sum_{\substack{k=1\\k\neq k}}^{K}\sum_{j=1}^{i+1}\sum_{l=1}^{L}\sqrt{E}\,\alpha_{l,k}b_k(j)I_{k,l,k,l}\big((i+1)T_b\big) + \tau_{l,k},j + \tilde{n}_{l,k}(i) \qquad (4.54)$$

where

$$I_{k,l,k,l}\big((i+1)T_b\big) + \tau_{l,k},j$$

takes into account cross-correlation terms, defined as:

$$I_{k,l,k,l}\big([(i+1)T_b] + \tau_{l,k},j\big) = \frac{1}{T}\int_{iT_b+\tau_{l,k}}^{(i+1)T_b+\tau_{l,k}}[t - \tau_{k,l} - iT]F_k[t - \tau_{k,l} - jT]\cos\big(\vartheta_{l,k} - \vartheta_{l,k}\big)\,dt \qquad (4.55)$$

The first term in (2) is due to the transmitted signal of the desired user, the second term is the contribution of the other interfering replicas of the same user $\bar{k}$, the so-called self-noise, the third term is MAI produced by other interfering users and the fourth is due to the AWGN.

Each single replica contribution can be used to derive $\bar{k}$th user Rake receiver output: in particular, according to SD decision approach, matched filter outputs are used as signal amplitude rough estimates

$$Z_k(i) = \sum_{l=1}^{L}\big|y_{i,k}(i)\big|\,y_{l,k}(i) \qquad (4.56)$$

As can be deduced from Fig. 4.67, conventional rake receiver based on Maximum Ratio Combining algorithm at the end of the up-link channel is vulnerable to MAI impairments. As a consequence, this near-far effect gives rise to an irreducible error floor for 10 and 20 user communication systems.

### 4.5.3.2 Selective Interference Cancellation

As stated previously, proposed detector basic approach, is to divide replicas into two different groups according to the received signals power level.

Matched filter outputs $y\bar{l},\bar{k}$ $(i)$, obtained in (13), are compared with a proper threshold value $SSD$ in order to define the reliability term reliability term $\tilde{R}_{l,k}(i)$ as:

$$\tilde{R}_{l,k}(i) = \begin{cases} -1 & -S_{SD} \geq y_{l,k}(i); \\ 0 & -S_{SD} < y_{l,k}(i) < S_{SD}; \\ 1 & y_{l,k}(i) \geq S_{SD}. \end{cases} \tag{4.57}$$

If $\tilde{R}_{l,k}(i) = 1$, we have a reliable signal replica. For all the reliable replicas, MAI, self-noise and AWGN influences are assumed negligible and their effects on all the other signals are cancelled on the second step. Let replica $\bar{l}$ of user $\bar{k}$ be unreliable: for this signal the cancellation term due to the other interfering replicas is equal to:

$$C_{l,k}(j) = \sum_{k=1}^{K}\sum_{l=1}^{L}\tilde{R}_{l,k}(i)y_{k,l}(i)F_k(t - jT - \tau_{k,l})e^{j\vartheta_{k,l}(j)} \quad \text{for } j = i-1, i, i+1 \tag{4.58}$$

where choice of index $j$ depends on user relative time offset. It is worth emphasising that in (17), by (16), only reliable replicas cause a non-zero contribution to $C\bar{l}, \bar{k}$ $(j)$. Therefore cancellation operation gives:

$$\hat{r}(t) = r(t) - \sum_{j=i-1}^{i+1} C_{l,k}(j) \tag{4.59}$$

For reliable replicas cancellation is not performed, i.e. $C\bar{l}, \bar{k}$ $(j)$ is set to zero, so that $\hat{r}(t) = r(t)$.

Cancelled signal $\hat{r}(t)$ is input to correlator, combined with the other replicas, detected through a zero threshold. The flow chart of cancellation process in a multipath fading environment is shown in Fig. 4.57. It is worth stressing that in single rate asynchronous system a two bits buffer is required to accomplish a correct parallel interference cancellation: in particular, by this two bits implementation (see Fig. 4.68), MAI produced by bits transmitted by interfering users on the successive (or previous) signalling interval could be cancelled.

Anyway, perfect MAI cancellation cannot be implemented since lack of a good received signals parameters estimates leads to approximate operations. From (16–18) it is evident that the effectiveness of the proposed interference cancellation methods are dependent on threshold $SSD$, so that a suitable values can be chosen also for this system.

In performing our simulations the following conditions have been assumed:

- Satellite up-link channel for 2 GHz band;
- Symbol rate for the PSK modulation equal to 78.740 ksymbols/sec;
- Spreading obtained through Gold sequences with processing gain equal to 127;
- Doppler spread equal to 8 kHz (No AFC is supposed);
- User $k$ power $Pk$ equal to 1 $\forall k$ (i.e. free space attenuation is not considered).

The channel model considered is derived from (101). In Figs 4.64 and 4.65 multiuser canceller performance is compared with conventional rake receiver results, for 10 and 20 user systems respectively, in satellite channel.

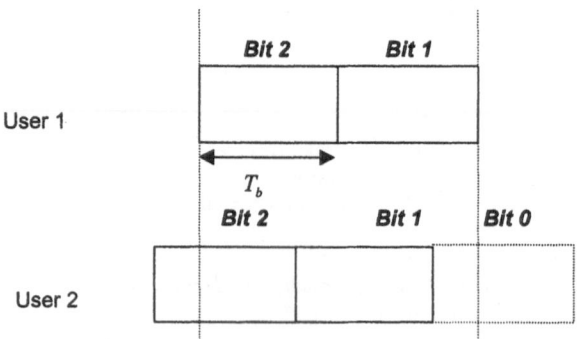

**Figure 4.63.** Two-bit buffer for asynchronous communications

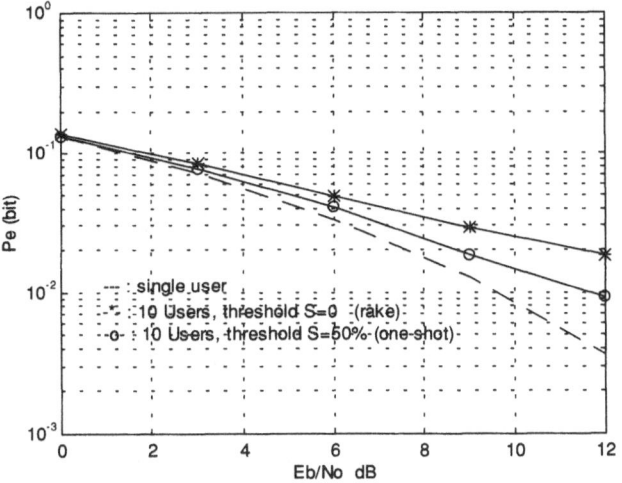

**Figure 4.64.** BER comparison.

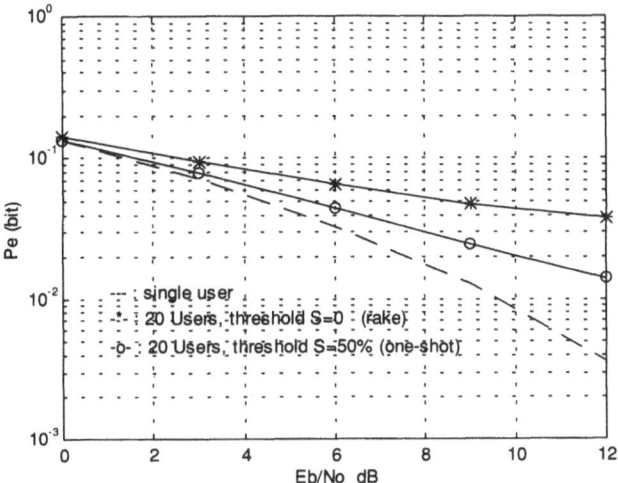

**Figure 4.65.** BER comparison.

In both figures the threshold is fixed at 50 per cent of the single user power in the considered channel and the conventional single user rake receiver performance is reported for a comparison. Figures 4.66 to 4.67 shows how the One-shot receiver performance is influenced by the threshold value. Three curves are shown for different values of SNR.

We can observe that the optimum threshold value is almost the same for all SNR considered. This feature is indeed positive towards design and development of the proposed receiver. We can fix the optimum threshold at 40–60 per cent of the single user power in the considered channel. Anyway we can deduce that the One-shot detector shows robustness with respect to eventual estimation error in the determination of the optimum threshold value.

### 4.5.3.3 Blind Adaptive Detector for Up-Link DS-CDMA Communications

For what Blind Adaptive multiuser detectors are concerned (102), we consider different schemes for DS/CDMA satellite communication systems in which the BPSK modulation technique is used in transmission: the proposed receivers are supposed to be used at the end of up-link satellite channel in the base station, so that the considered communication system is characterised by lack of synchronism between the users and multipath fading satellite channel. The proposed multiuser detectors,

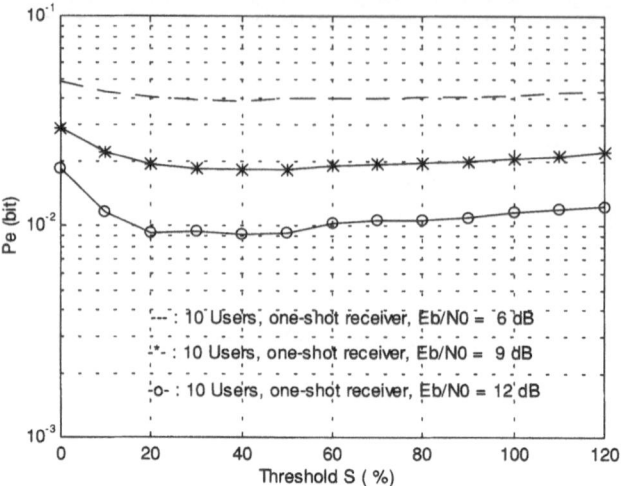

**Figure 4.66.** BER comparison.

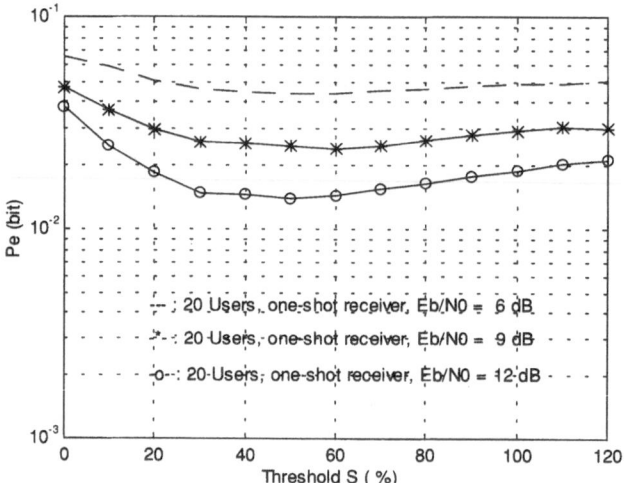

**Figure 4.67.** BER comparison.

described in Fig. 4.68 to 4.69, show remarkable near-far resistance and requires knowledge of the same parameters needed by the conventional single-user receiver, i.e. desired user's signature waveform and its timing, while no training sequence is needed, but only essential information about all the user, i.e. the number of active users and the processing gain. As for IC scheme considered before, proposed receiver real time implementation is allowed by the light computational complexity.

Blind adaptive MultiUser Detector (BA-MUD), proposed by Verdù, Honig and Madhow (103), achieves very good performance in synchronous AWGN environment. In order to consider a more realistic communication environment, satellite channel (104) is considered for our simulations.

Due to bandwidth specifications, for each user, only direct path and one attenuated echo is discriminated at receiving end: besides, reflected echo is nearly negligible.

The received signal is composed as in the following:

$$r(t) = \sum_{k=1}^{K} \left( \sum_{l=1}^{L} A_{k,l} S_k(t - \tau_{k,l}) \, b_k \cos{(\omega_0 t + \phi_{k,l})} \right) + n(t) \qquad (4.60)$$

where $L$ is the number of replicas, assumed equal to two in our simulations. User of interest is assumed to be the first. With no loss of generality we suppose that each replica is perfectly tracked,

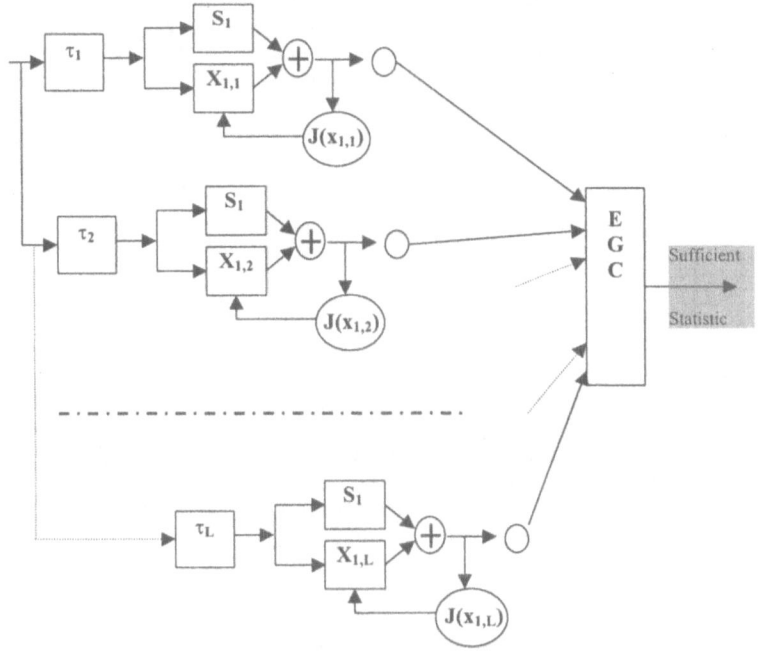

**Figure 4.68.** Scalar blind adaptive MUD structure.

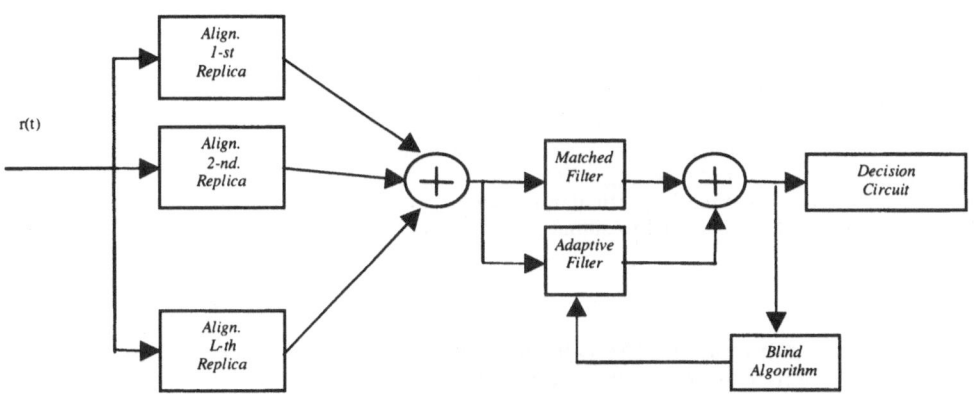

**Figure 4.69.** Vectorial blind adaptive MUD structure.

i.e. $\phi_{k,l}$ is known and perfectly compensated $\forall\, k, l$. Baseband equivalent sampled signals are considered in the following.

For this satellite a generalisation of the blind receiver (BA-MUD) is needed (105): in particular, a single blind receiver for each replica is introduced as it is shown in Fig. 4.69. With this structure each user replica faces the other replicas of the same user as they were pure interference; this receiver modification doesn't affect convergence procedure since each delayed signal is effectively seen as a further interfering user and its perturbation contribution is determined by signature waveform autocorrelation correspondent to the replicas relative delays. This solution can be considered as a generalisation of the solution presented in (106) to completely asynchronous system.

As stated in (107)(108), adaptive filter impulse response $c_{1,l}$ is equal to sum of spreading sequence $s_1$ and orthogonally anchored sequence. For what concerns convergence procedure, in (109) orthogonal sequence $x_{1,l}$, i.e. filter impulse response adaptive component, is derived as in the following:

$$x_{1,j}^{i+1} = x_{1,j}^{i} - \mu \langle y^i, s_1 + x_{1,l}\rangle(y^i - \langle y^i, s_1\rangle s_1) \tag{4.61}$$

where $i$ means adaptive algorithm index, i.e. considered bits position from the start of the procedure. This algorithm converges regardless of the initial condition to the MMSE detector if the step size decreases as $1/i$ (110). As it is described in (111), in blind adaptive detection definition of an energy threshold is mandatory to prevent from the desired user signal cancellation, eventually caused by signature waveforms mismatch.

Receiver impulse response energy have to be constrained according to following relation

$$||c_1||^2 = ||s_1||^2 + ||x_{1,l}||^2 \leq 1 + \chi_{l,SC} \tag{4.62}$$

Starting from considerations presented in (112, 113), suitable threshold value $\chi_{l,SC}$ can be deduced as

$$\chi_{l,SC} = \frac{\alpha(6K-1)}{N - \alpha(6K-1)} \tag{4.63}$$

where, since considered system is completely asynchronous, parameter is defined equal to two.

Parameter $\chi_{l,SC}$ setting is the prelude of the introduction of a proper adaptive algorithm that takes into account this energy constraint. A new embedded adaptive procedure has been recently proposed (114): in particular, starting from (20), adaptive impulse response $x_{1,l}$ is supposed to be derived for each single replica; before applying it to the received signal, a control is implemented to verify whether impulse response energy is greater than threshold, i.e. inside the detector, for each replica, the following comparison is made:

$$||x_{1,l}^i||^2 \gtrless \chi_{l,SC} \tag{4.64}$$

If impulse response energy is less or equal than $\chi_{l,SC}$, the adaptive algorithm is carried on; otherwise a new adaptive impulse response is defined and is assumed equal to:

$$\tilde{x}_{1,l}^i = \frac{x_{1,l}^i}{\sqrt{||x_{1,l}^i||^2}} \cdot \chi_{l,SC} \tag{4.65}$$

After normalisation, impulse response obtained $\tilde{x}_{1,l}^i$ is used as receiver adaptive component and the recursive relation (20) is carried on with $x_{1,l}^i = \tilde{x}_{1,l}^i$.

Since the adaptive algorithm chosen is derived from (20) through the introduction of the successive normalisation operation, the definition of the proper step size $\mu$ is a key point in receiver design: in particular, taking into account suggestion of Miller and Smith (115), a new step size has been proposed for satellite channel: in particular, the step proposed is dependent from mean power $\overline{E_{1,l}}$ that is output of the filter matched to signature waveform $s_1$:

$$\mu = \frac{0.1}{ind(i) \cdot \overline{E_{1,l}}} \tag{4.66}$$

where index $ind(i)$ goes from 1 to $IND_{MAX}$ in monotonously increasing way and, successively, linearly increases from $IND_{MAX}/2$ to $IND_{MAX}$ as it represented a saw blade, $i$ is the algoritm index defined in (20) and

$$\overline{E_{1,j}} = \frac{1}{T} \int_0^{T_b} r(t) \, s_1 \, (t - \tau_1) \, dt \tag{4.67}$$

is the output energy of the considered user $l$th replica matched filter. The same step size updating general rule is assumed for each path, i.e. for each BA-MUD.

In this approach, each detector presents its convergence procedure related to a specific signal, considering other replicas as pure interference: weaker replicas could not reach stability region causing impairments because of replicas combining, so causing remarkable performance loss.

Another approach has been considered (116): adaptive sequence $x_1$ orthogonal not to the nominal spreading vector $s_1$, but to the space spanned by all replicas. Since received information signal vector must lie in this space, desired signal cancellation risk is reduced.

In this detection scheme an unique adaptive vector $x_1$ is generated, aiming to suppress MAI from all pre-combined replicas of desired signal at the same time, instead of a set of $L$ independent adaptive vectors, as described in Fig. 4.68. Moreover, a single receiver after RAKE recombination structure has been implemented, so yielding a remarkable reduction of implementation complexity.

Pre-combined approach main features help obtaining convergence situation for the whole received signal of the user of interest: in particular, the weaker echoes are considered as a part of the overall signal and not as an individual informative signal, so yielding a more reliable decision variable and a more robust convergence procedure.

The proposed scheme performance is evaluated in BER terms by means of simulations and the following conditions have been assumed:

- Satellite up-link channel for 2 GHz band.
- Symbol rate for the PSK modulation equal to 31.496 ksymbols/sec.
- Spreading obtained through Gold sequences with processing gain equal to 127.
- Doppler spread equal to 100 kHz (ideal AFC is supposed).
- User $k$ power $P_k$ equal to 1 $\forall k$ (i.e. free space attenuation is not considered).

In Figs 4.70 to 4.71 blind adaptive multiuser detectors performance is compared with conventional rake receiver results for 4 users systems. As it can be deduced from Fig. 4.68–4.69, considered detectors main difference lie in the way each replica energy contribution is used: in particular, while scalar blind receiver consists of six different detector, one for each replica, in the vectorial detector the received signal is considered as a whole, with a better complexity-performance trade-off.

### 4.5.3.4 OFDM-CDMA for Multi-Bit-Rate Radio Interface on LEO Satellites

Orthogonal Frequency Division Multiplexing (OFDM) together with a Code Division Multiple Access could be an interesting solution in a LEO satellite link when different service classes are present on the same radio channel (117). By using a special arrangement of the carriers we succeed in preserving the orthogonality between signals bearing different classes of information.

Bit-rate classes are deduced from ETSI UTRA proposal and can be summarised by the following formula:

$$Bit\_rate = 32*2k$$

$$Spreading\_Factor = 256/2k \qquad\qquad (4.68)$$

$$k = 0 \dots 6$$

This classification of bit-rates leads to a choice of services characterised by low bit-rates associated with an high spreading gain and high bit-rates with low spreading gains. This structure perfectly fits the carrier assignment in a OFDM-CDMA transmission system as in Fig. 4.72.

The main advantage of the proposed system is the conjunction of two properties, namely OFDM system intrinsic rejection of delayed version of the transmitted signal, always present in mobile

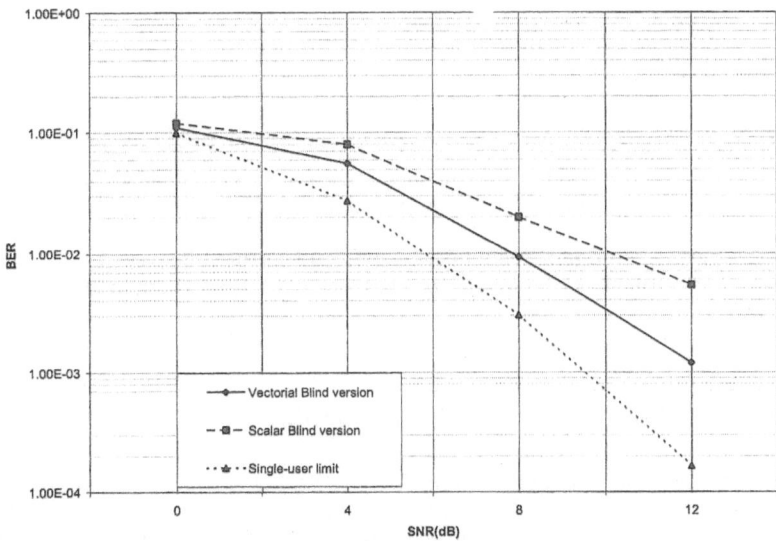

**Figure 4.70.** BER comparison.

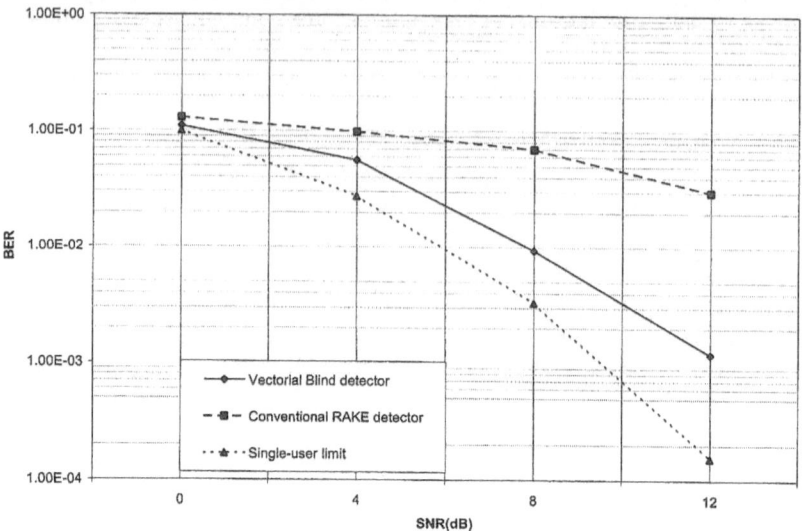

**Figure 4.71.** BER comparison

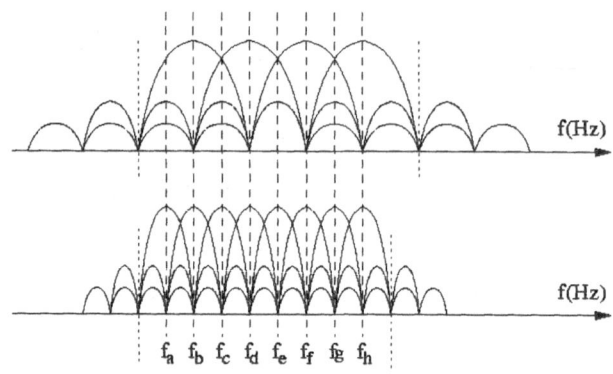

**Figure 4.72.** OFDM-CDMA carrier assignment.

asynchronous LEO channels and the CDMA soft-capacity concept, stating that the number of users may increase at the expense of a smooth degradation of the service quality, starting from the users which employ a small spreading gain.

The above properties result in a quality profile perfectly balanced between bit-rate and bit-error-rate. This thesis is confirmed by simulations on a LEO satellite channel both in line-of-sight and non-line- of-sight conditions.

The coexistence of multiple bit-rate spread signals is shown in Fig. 4.73, where the BER/bit-rate relationship is highlighted.

### 4.5.3.5 Simplified Receiver Structure with New Sequences for CDMA using FFT Implementation

It is well known that Code Division Multiple Access (CDMA) systems need sequences with good cross-correlation properties in order to minimise inter-user interference but it also desirable that their auto-correlation be equally good because of synchronisation. On the other hand, if the system requires coding, i.e. FEC, than a suitable code should be chosen with good performance and

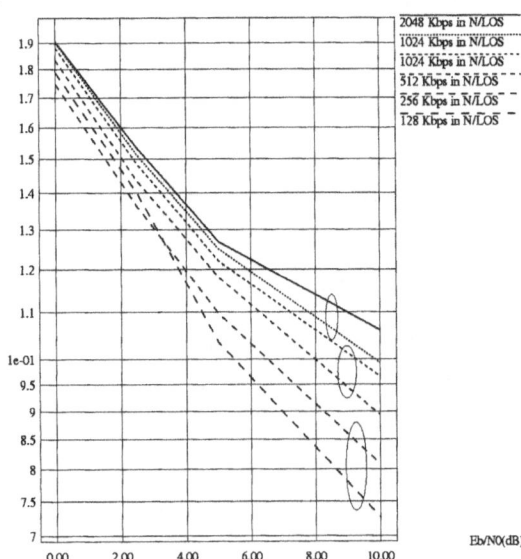

Figure 4.73. OFDM-CDMA BER performance.

relatively simple implementation. These and other decorrelating functions of the receiver can be very simplified by the use of DSP techniques, namely using an FFT. Two new types of sequences have been proposed (118, 119, 120) in order to satisfy the above requirements: one of this is synthesised in the time domain, the other in the frequency domain, both can be easily demodulated with an FFT and applied to a CDMA receiver with all the benefits of a spread spectrum system.

The following class of signals has been considered:

$$s_i(t) = \sqrt{\frac{2}{T_b}} \sum_{k=1}^{K} S_i(k) \, e^{j(2\pi kt)/T_b} \tag{4.69}$$

where $s_i(t)$ is the signal assigned to user $i$ in a communication system where the channel is shared among $u$ users, $1 \leq i \leq U$, $T_b$ is the time length of the bit, $S_i(k)$ is a complex-valued spectral coefficient.

In order to generate signals with flat amplitude spectrum and unitary energy, the values of $S_i(k)$ are chosen with constant modulus and variable phase, as in the following:

$$S_i(k) = \frac{1}{\sqrt{K}} e^{j\vartheta_i(k)} \tag{4.70}$$

A poliphase or N-Phase sequence (121, 122, 123, 124, 125, 126) is a sequence whose elements are of the form $\exp(j2\pi n/N)$, with $0 \leq n \leq N$.

Attention has been focused on so called perfect arrays, whose definition is the following:

$$s(x, y) = \exp\left\{ j \frac{2\pi}{N} (x \cdot y) \right\} \tag{4.71}$$

In this case $N$ different sequences of length $N$ can be generated. It can be easily shown that each sequence is perfectly orthogonal to any other. The number of codes generated is equal to spreading factor, as in the case of direct sequences derived from the Walsh's matrix. Let's assume that same information symbol is continuously repeated by both the desired and the interfering user. Moreover, no phase distortion effects are taken into account and a spreading factor K of 32 is considered. Fig. 4.73 shows the cross-correlation profile (continuous line) and the auto-correlation profile (dotted line) of two spreading signals.

In the upper part of Fig. 4.74, two signals generated by the proposed PS technique are compared. The upper-left part shows the effect of delay on two signals derived by Chu Perfect Sequences. Chu

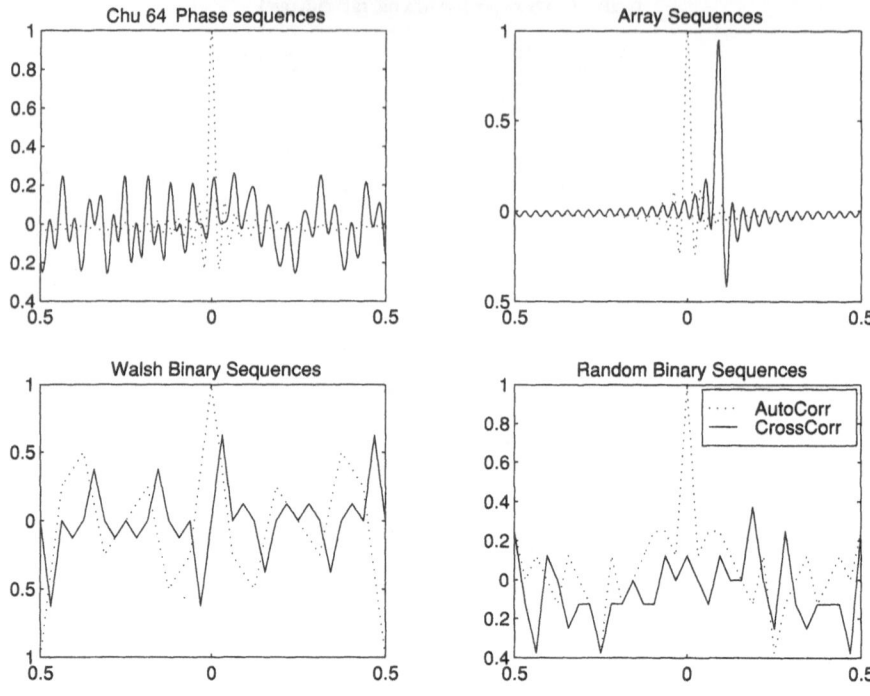

**Figure 4.74.** Correlation profiles.

Sequences do not guarantee the perfect orthogonality with r = 0, but cross-correlation is bounded in the whole range of delay.

On the upper-right part profiles of two signals derived by Perfect Arrays are drawn: the orthogonality is perfect for the synchronous case ($\tau = 0$) and the delayed profile is extremely low, with the exception of a narrow pulse. Basically the spreading signals derived from a Perfect Array are composed by a discrete cyclic delayed copy of the first sequence.

This may appear as an undesired property, but the narrowness of the impulse is the key issue here. As confirmed by computer simulations, later described, the cross-correlation impulse is so narrow that only a few times a multipath replica is likely to fall inside it. This results in a global averaged good rejection of the delayed replicas of the interference signals.

As the lower half of Fig. 4.74 is concerned, the same cross/auto-correlation profile is computed for an ordinary Direct Sequence spreading signal. Each chip modulates a square pulse and, both Walsh and random binary codes, are represented. Orthogonality holds in the synchronous case and also for Walsh DS codes but, as soon as the signal is delayed, DS signals manifest poor cross and auto-correlation profiles.

DS-CDMA and PS-CDMA have been tested for different values of received interfering users power. In Fig. 4.75 values of the vectors involved.

In Fig. 4.76 a comparison between DS-CDMA and PS-CDMA systems is performed in a severe multipath environment. The performance of the proposed technique is clearly better than that of direct sequences. As expected this gap increases as the AWGN noise is reduced. The cross-correlation properties of Phase sequences are useless when the dominant cause of error is the AWGN noise.

### 4.5.3.6 Code Spreading CDMA with TCH Codes for Simplified Radio Interface

#### 4.5.3.6.1 Introduction
In this section a Code Spread CDMA system is proposed which is based on TCH codes. System performance is evaluated and compared with other previously proposed solutions.

In a conventional DS-CDMA system spreading is achieved by multiplication (or modulo 2 addition) of the signal by a multiple access (pseudo-random) spreading sequence, such as the Gold or

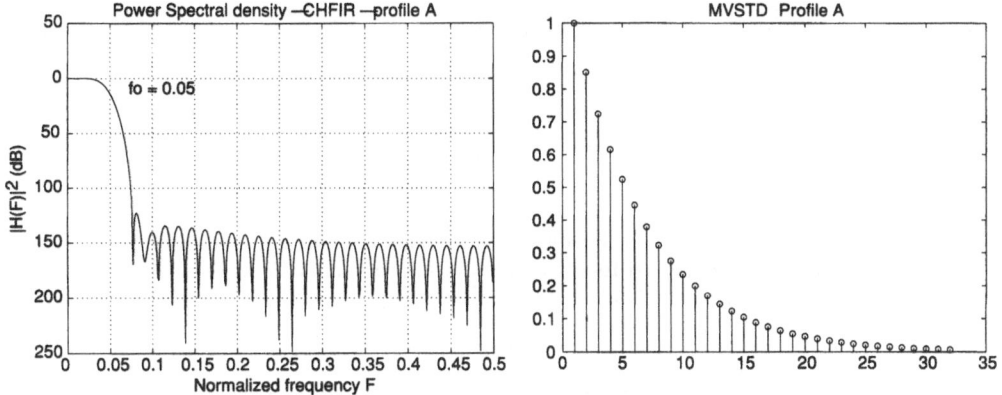

**Figure 4.75.** Phase-sequence profiles.

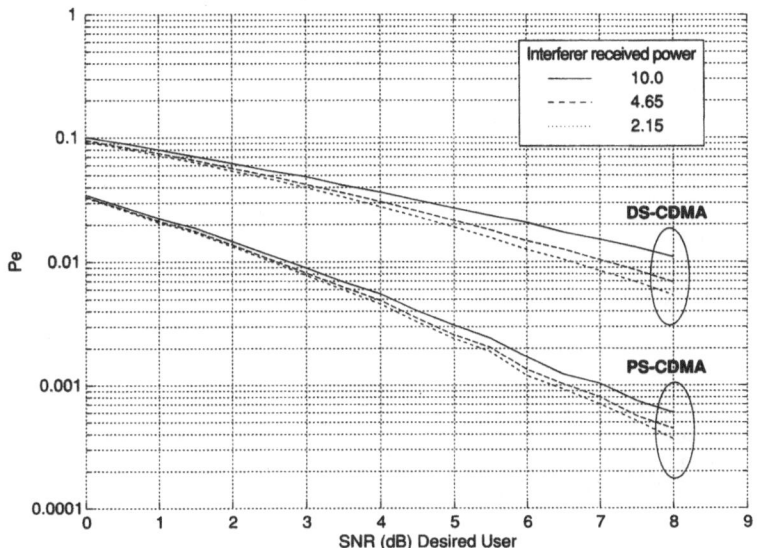

**Figure 4.76.** BER comparisons.

Kasami family sets. In order to achieve sufficient low error rates (for a given number of active users), some kind of forward error correction (FEC) must be applied. These FEC codes are responsible for some spreading over the available bandwidth, while the reminder is used by the multiple access (MA) sequence.

From Information Theory we know that the maximum theoretical DS-CDMA system capacity can only be achieved by employing very low rate FEC codes using the entire available bandwidth (127, 128), without further spreading by the MA sequence. In this case the MA sequence and the coded symbol stream are added modulo 2 at the same rate. The MA sequence continues to assure the multiple access to the communication channel, though is only responsible for a "scrambling" of the signal of the active users. Such a system is often referred as Code Spreading CDMA. A limiting factor however, has been the lack of adequate very low rate codes. In (129) is proposed the use of Orthogonal Convolutional codes (130). In this paper we consider the application of (block) TCH codes for the same purpose.

### 4.5.3.6.2 System Analysis
The theoretical performance of TCH codes in a AWGN (Additive White Gaussian Noise) channel can be determined via the Union Bound, as the upper bound

$$P_e \leq \frac{2^{m-1}}{2^m - 1} \left[ \sum_{\substack{d=d_{min} \\ d=d+2}}^{(n/2)-2} 2\overline{M}(d) \left( Q(\sqrt{G_e}d) + Q(\sqrt{\Gamma_c n} - d) \right) + 2\,\overline{M}\!\left(\frac{n}{2}\right) Q(\sqrt{\Gamma_c\,(n/2)}) \right] \tag{4.72}$$

where

$$Q(x) = \frac{1}{\sqrt{2\pi}} \int_x^{+\infty} e^{-t^2/2}\,dt\,, \tag{4.73}$$

$n$ is the number of coded symbols, $m$ is the number of information bits, $d_{min}$ is the minimum distance of the code, $\Gamma_c = R_c E_b/N_0$ is the signal-to-noise ratio per coded symbol, $Rc$ is the code rate and

$$\overline{M}(d) = \sum_{p=1}^h M_p(d) \tag{4.74}$$

is the mean value, over $h$ TCH polynomials, of the number $M_d$ of codewords at a distance $d$ between themselves.

An upper bound for the BEP (Bit Error Probability) of convolutional codes in an AWGN channel, considering soft decisions, is given by (131),

$$P_e \leq \frac{1}{2} \frac{\partial T(D, I)}{\partial I}\Bigg|_{I=1,D=\exp(-\Gamma_c)} \tag{4.75}$$

where $T(D, I)$ is the Transfer Function of the convolutional code. For Orthogonal Convolutional codes with constraint length $L$, expression (4.75) can be approximated by (132)

$$\frac{\partial T(D, I)}{\partial I}\Bigg|_{I=1} \approx D^{L2^{L-1}} \tag{4.76}$$

and for Biorthogonal Convolutional codes by

$$\frac{\partial T(D, I)}{\partial I}\Bigg|_{I=1} \approx D^{(L+1)s^{L-2}} \tag{4.77}$$

disregarding second order terms.

The upper bounds for the BEP of the Orthogonal Convolutional and Biorthogonal Convolutional codes can then be stated, respectively, as

$$P_e \leq \frac{1}{2} \exp\left(-\Gamma_c L 2^{L-1}\right) \tag{4.78}$$

$$P_e \leq \frac{1}{2} \exp\left[-\Gamma_c (L + 1) 2^{L-2}\right] \tag{4.79}$$

where $L$ is the constraint length, $\Gamma_c = R_c E_b/N_0$ is the signal-to-noise ratio per coded symbol and $R_c$ is the code rate.

In order to analyse the performance of our CDMA system, we consider the self-user interference as additive white gaussian noise and apply the Gaussian Approximation.

Let us consider a system with $K$ simultaneous active users with perfect power control (we mean that all users reach the receiving station with the same power level), perfect word and bit synchronisation and let us treat the MA sequences as pure random sequences (133). In this context, it is possible to evaluate the probability of error $Pe$ of an uncoded (without FEC) asynchronous CDMA system with BPSK (Binary Shift Keying) modulation in an AWGN channel as

$$P_e = Q(\sqrt{2\Gamma}) \tag{4.80}$$

where $\Gamma$ is the "efficient" signal-to-noise ratio given by (134, 135),

$$\Gamma = \left(\frac{N_0}{E_b} + \frac{2(K-1)}{3S}\right)^{-1}, \tag{4.81}$$

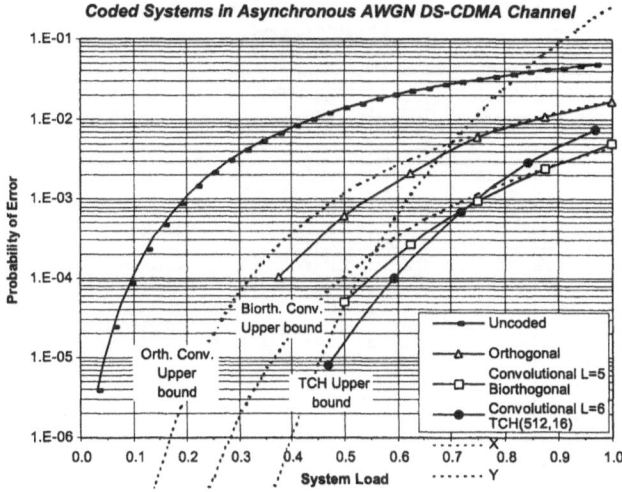

**Figure 4.77.** Upper bounds and simulation results on performance of different codes in an asynchronous AWGN DS-CDMA channel, $E_b/N_0 = 10$dB, $S_{tot} = 32$.

where $E_b/N_0$ is the signal-to-noise (thermal) ratio, $K$ is the number of active users and $S$ is the spreading factor caused by the MA sequence.

In this context, the probability of error of the coded CDMA systems can then be determined modifying expression (4.81) to

$$\Gamma_c = \left( R_c^{-1} \frac{N_0}{E_b} + \frac{2(K-1)}{3S} \right)^{-1} \tag{4.82}$$

and applying (4.82) to the expressions (4.72), (4.78) or (4.79).

*4.5.3.6.3 Results*

Let us define by system load the quantity $U = K/S_{tot}$, where $S_{tot} = S/R_c$, $S$ is the spreading factor caused by the MA sequence and $R_c$ is the code rate. The system load is then the number of active user normalised to the overall spreading factor. When FEC is used alone for bandwidth expansion, as in code spreading CDMA, the spreading factor caused by the MA sequence is thus $S = 1$.

In Fig. 4.77 we show the evaluation of the upper bounds on the BEP for the three types of codes considered, considering an AWGN DS-CDMA channel, with $E_b/N_0 = 10$ dB and an overall spreading factor $S_{tot} = 32$. The BEP for an uncoded (without FEC) system is also known. We have assumed pure random MA sequences for both the coded and uncoded systems. The actual Bit Error Rate (BER) was estimated by computer simulation and the obtained results are also plotted for comparison. It was assumed in all computer simulations that the FEC process is accomplished with soft decision decoding and that the overall spreading factor of the codes is the same for a fair comparison between the different systems.

The analysis of the performance of these systems in a more realistic channel is also evaluated via computer simulations. We have considered a fully interleaved Rayleigh fading channel with perfect channel estimates. We mean that each period of duration $T_c$ (one coded symbol duration) suffers a constant Rayleigh fading event and that this fading is uncorrelated along the time and between the active users. This model is adequate if we consider a fully interleaved system.

In Fig. 4.78 we show some simulation results of the BER versus the system load, with $E_b/N_0 = 10$ dB, for the considered code spreading CDMA schemes and also for a conventional half rate convolutional $L = 7$ code plus a MA sequence spreading of $S = 16$. In Fig. 4.79 the BER is plotted versus the signal-to-noise (thermal) ratio for a system load $U = 0.625$, corresponding to $K = 20$ users, since we considered $S_{tot} = 32$.

*4.5.3.6.4 Conclusion*

As we can verify, TCH codes perform better than the other schemes simulated considering the same conditions. The maximum allowable number of simultaneous users is slightly better when we use TCH codes, namely for lower values of $Pe$, which is a desirable feature.

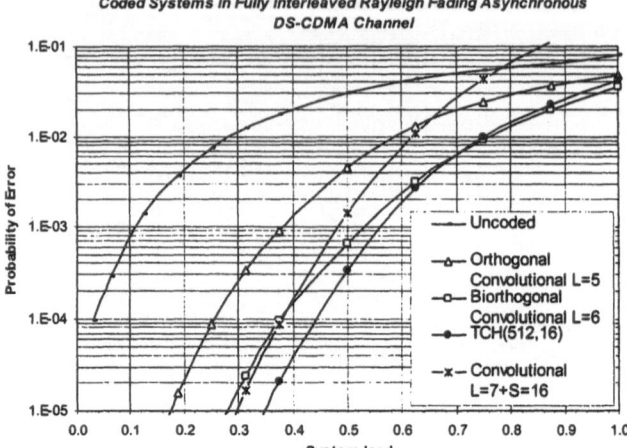

**Figure 4.78.** Simulation results on performance of different codes in a Rayleigh fading asynchronous DS-CDMA channel, $E_b/N_0 = 10$dB, $S_{tot} = 32$.

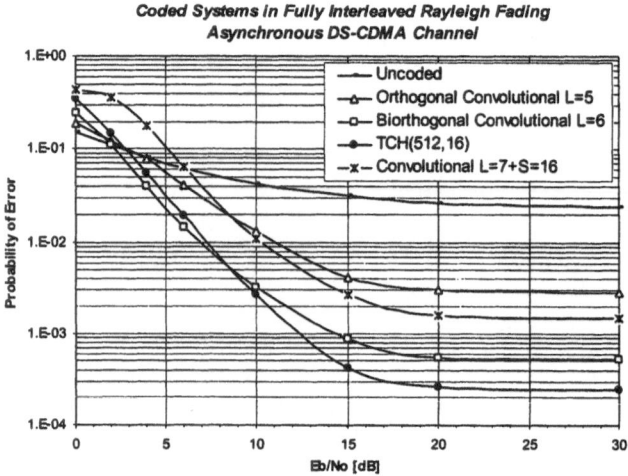

**Figure 4.79.** Simulation results on performance of different codes in a Rayleigh fading asynchronous DS-CDMA channel, $S_{tot} = 32$, $U = 0.625$.

The type of receiver studied can be used to build a complete CDMA system including appropriate channel models and all the required parameters in order to determine its overall capacity both in terms of maximum number of simultaneous users and their useful data rates.

## 4.6 Conclusions

The Air Interface aspects for future generations of satellite communications, dealled with in WG3, achieved many results thanks to the cooperation of the several institutions involved in this COST 252 action and also to the cooperation of other similar Actions, as for example, COST 253 and COST 255 and according to the work plan agreed by all.

As previously mentioned, particular attention was taken with WB-CDMA due not only to its standardization for the third generation of mobile but also because of its possible use in global communication systems, using necessarily satellites, as IMT2000 proposed by ESA.

The major contributions on WG3100, dealling with channel characteristics, propagation, measurements and modelling, are presented in Section 4.2. Results for the propagation characteristics at

L-Band and EHF-Band are presented, both for narrow and wideband which can be very useful for the new generations. This section also presents results that can be used as inputs in statistical models to evaluate the system availability. These results can be used for testing new proposed constellations for third generation mobile communications and is based on studies of shadowing effects and their correlation. Another study included in this section, based on non-GEO satellite constellations, shows that dynamic coverage brings a significant increase in the capacity for systems with high satellite diversity probability, it is a feature, therefore, desirable in future system design.

Multiple access techniques (WG3200) were also mentioned but they were integrated in the studies of WG3300 concerning receivers. So, Section 4.3 presents the results of new modulation techniques concerning variable rate CPFSK modulation that outperform constant rate 4-CPFSK modulation and is suitable for mobiles.

Section 4.4 presents new coding techniques using TCH codes. The main feature of these codes is the simplification of the receiver design, while keeping a similar performance to other codes, which also is a desirable feature for low-cost satellite receivers. Their performance was estimated both for AWGN and bursty channels.

The use of CDMA receivers was also addressed and two different sudies have shown that the use of certain sequences can also simplify its design. It was shown that TCH codes can not only simplify the receiver design in a FEC application but also in a CDMA receiver where its codewords are used as spreading sequences with its correlation done in the frequency domain by an FFT. Another type of sequences, denominated Phase Sequences, was also proposed for this same purpose with good results.

Receivers for wideband DS-CDMA communications were addressed in different aspects and a new technique, "One Shot Receiver" was presented for multi-rate communications. Other related studies addressed, namely with results for LEO satellites, were OFDM-CDMA, one-shot and selective interference cancelation and a blind adaptive detector for up-link DS-CDMA communications.

## 4.7 Appendix

### 4.7.1 Correlation Modelling Using a Physical Statistical Approach

The parameters of the physical-statistical model are defined as shown in Fig. 4.21 and in Table A4.1.

Simple trigonometry yields the following relationships

$$h_r = \begin{cases} h_m + \dfrac{d_m \tan \phi}{\sin \theta} & 0 < \theta \leq \pi \\[2ex] h_m - \dfrac{(w - d_m) \tan \phi}{\sin \theta} & -\pi \leq \theta \leq 0 \end{cases} \tag{A.1}$$

Under the assumptions made in the model (see 4.2.4.6), the mean value for the shadowed state is given by

$$\bar{S}_i = \int_0^{2\pi} S_i(\theta_s) p(\theta_s) \, d\theta_s = \frac{1}{2\pi} \int_0^{2\pi} u_t(h_r - h_b) \, d\theta \tag{A.2}$$

where $S_i(\theta s)$ denotes the shadowing state for satellite-$i$, with azimuth angle $\theta_s$. Where, $u_i$ is the unit step function

$$u_i(h_r - h_b) = \begin{cases} 1 & h_r - h_b \geq 0 \\ 0 & \text{otherwise} \end{cases} \tag{A.3}$$

**Table A4.1.** Physical-statistical model parameters

| | |
|---|---|
| $\phi$ | Elevation angle of the satellite from the mobile |
| $\theta$ | Azimuth angle of the satellite from the mobile relative to the axis of the street |
| $w$ | Street width (m) |
| $d_m$ | Perpendicular distance of the mobile from the building face (m) |
| $h$ | The height of the phase centre of the mobile antenna above local ground level (m) |
| $h_b$ | The height of the building immediately below the direct ray relative to local ground level (m) |
| $h_r$ | The height of the direct ray above the building face relative to local ground level (m) |
| $d_r$ | The distance along the direct ray from the mobile to the point on the ray immediately above the building (m) |

The value 1 corresponds to the line-of-sight state and the value 0 to the non-line-of-sight state. For simplicity, the following notation will be used

$$a_i \equiv \sin^{-1}\left(\frac{d_m \tan \phi_i}{h_b - h_m}\right) \quad b_i \equiv \sin^{-1}\left(\frac{(w - d_m) \tan \phi}{h_b - h_m}\right) \tag{A.4}$$

The terms inside the integral (Eq. A.2) are equal to unity when, for $0 < \theta s < \pi$,

$$h_r \geqslant h_b \Rightarrow \sin(\theta_s) \leqslant \frac{d_m \tan \phi_i}{h_b - h_m} \tag{A.5}$$

for $0 < \theta_s < a_i$ and $\pi - a_i < \theta_s < \pi$ and also when, for $-\pi < \theta_s < 0$,

$$h_r \geqslant h_b \Rightarrow \sin(\theta_s) \geqslant -\frac{(w - d_m) \tan \phi}{h_b - h_m} \tag{A.6}$$

for $-b_i \leqslant \theta_s < 0$ and $-\pi \leqslant \theta_s \leqslant -\pi + b_i$.

Therefore, the mean state is given by

$$\bar{S}_i = \frac{1}{2\pi}\left\{\int_0^{a_i} d\theta + \int_{\pi - a_i}^{\pi} d\theta_s + \int_{\pi}^{\pi + b_i} d\theta_s\right\} = \frac{1}{\pi}(a_i + b_i) \tag{A.7}$$

The term $E(S_1 S_2)$ is given by

$$E(S_1 S_2) = \int_0^{2\pi} S_1(\theta_s) S_2(\theta_s) p(\theta_s) d\theta_s = \frac{1}{2\pi} \int_0^{2\pi} S_1(\theta_s) S_2(\theta_s) d\theta_s \tag{A.8}$$

By substituting the signal amplitudes it becomes

$$E(S_1 S_2) = \frac{1}{2\pi}[I_1 + I_2] \tag{A.9}$$

where

$$I_1 = \int_0^{\pi} u_1(h_r - h_b) u_2(h_r - h_b) d\theta_s \tag{A.10}$$

$$I_2 = \int_{\pi}^{2\pi} u_1(h_r - h_b) u_2(h_r - h_b) d\theta_s \tag{A.11}$$

Each integral is split into two integrals, since $u_2$ changes value within the integration limits, so by setting $\theta_1 = 0$, without any loss of generality,

$$I_1 = I_{11} + I_{12} \tag{A.12}$$

where

$$I_{11} = \int_0^{\pi - \Delta\theta} u_1\left(h_m + \frac{d_m \tan \phi_1}{\sin(\theta_s)} - h_b\right) \cdot u_2\left(h_m + \frac{d_m \tan \phi_2}{\sin(\Delta\theta + \theta_s)} - h_b\right)$$

$$I_{12} = \int_{\pi - \Delta\theta}^{\pi} u_1\left(h_m + \frac{d_m \tan \phi_1}{\sin(\theta_s)} - h_b\right) \cdot u_2\left(h_m - \frac{w - d_m \tan \phi_2}{\sin(\Delta\theta + \theta_s)} - h_b\right) d\theta_s \tag{A.13}$$

and $\Delta\theta = |\theta_1 - \theta_2|$.

As regards the term $I_{11}$, the final result depends on the values of the physical parameters, so many different cases exist.

In order to show how these terms can be solved, the expression of $I_{11}$ is illustrated for the case when $\Delta\theta > a_1$. In this case, the integral expression reduces to

$$\int_0^{z_1} u_2\left(h_m + \frac{d_m \tan \phi_2}{\sin(\theta_s + \Delta\theta)}\right) d\theta_s = e_1 + e_2 \tag{A.14}$$

**Table A4.2.** WG3 participants

| Institution | Country | Abbreviation |
|---|---|---|
| University of Aveiro | Portugal | UAV |
| Bradford University | United Kingdom | BRA |
| DLR, Institut fuer Nachrichtentechnik | Germany | DLR |
| Ecole Polytechnic Federal de Lausanne | Switzerland | EPF |
| Institute Josef Stefan | Slovenia | IJS |
| Telecom Paris | France | ENST |
| University of Florence | Italy | UFI |
| University Rome Tor Vergata | Italy | TOR |
| University of Surrey | United Kingdom | SRU |
| Instituto Superior Técnico | Portugal | IST |

where

$$e_1 = \begin{cases} a_2 - \Delta\theta & \Delta\theta < a_2 < z_1 + \Delta\theta \\ z_1 & z_1 + \Delta\theta \leq a_2 \\ 0 & a_2 < \Delta\theta \end{cases} \tag{A.15}$$

and

$$e_2 = \begin{cases} z_1 + \Delta\theta - \pi + a_2 & \pi - \alpha_2 \geq \Delta\theta \\ z_1 & z_1 + \Delta\theta \leq a_2 \end{cases} \tag{A.16}$$

and the term $e_2$ is evaluated only when $z_1 + \Delta\theta > \pi - a_2$. The term $z_1$ represents the minimum of $a_1$ and $\pi - \Delta\theta$. Similar expressions exist for the other cases resulting in a complicated solution.

Following the same approach for $I_2$,

$$I_2 = I_{21} + I_{22} \tag{A.17}$$

where

$$I_{21} = \int_{\pi}^{2\pi - \Delta\theta} u_1 \left( h_m - \frac{(w - d_m)\tan\phi_1}{\sin(\theta_s)} - h_b \right) \cdot u_2 \left( h_m - \frac{(w - d_m)\tan\phi_2}{\sin(\Delta\theta + \theta_s)} - h_b \right) d\theta_s$$

$$I_{22} = \int_{2\pi - \Delta\theta}^{2\pi} u_1 \left( h_m - \frac{(w - d_m)\tan\phi_1}{\sin(\theta_s)} - h_b \right) \cdot u_2 \left( h_m - \frac{d_m \tan\phi_2}{\sin(\Delta\theta + \theta_s)} - h_b \right) d\theta_s \tag{A.18}$$

Again, the final result for $I_2$ depends on the values for the physical parameters.

A common conclusion is that the final value of $E(S_1 S_2)$ and hence, $\rho$, depends only on the difference of the azimuth angles of the two satellites and not their absolute values. The standard deviations are determined by:

$$\sigma_i = \sqrt{E[S_i^2] - E^2[S_i]} \tag{A.19}$$

## 4.7.2 COST 252-WG3 Participants

The table below lists the participants in WG2, their country of origin and the abbreviation used to refer to them within this document.

## References

(1) Davarian F (1994) Earth-satellite propagation research. IEEE Comm Mag 74–79, April 1994.
(2) Goldhirsch J, Vogel WJ (1994) Mobile satellite propagation measurements from UHF to K band. Proceedings 15th AIAA International Communications Satellite Systems Conference, pp 913–920
(3) Lutz E et al (1991) The land mobile satellite communication channel – recording, statistics and channel model. IEEE Trans Veh Tech, 40:375–386.
(4) Butt G, Evans BG, Richharia M (1992) Narrowband channel statistics from multiband propagation measurements applicable to high elevation angle land-mobile satellite systems. IEEE J Sel Areas in Comm 10:1219–1226.
(5) Kleiner N, Vogel WJ (1992) Impact of propagation impairments on optimal personal mobile SATCOM system design. Conference Proceedings Mobile/Personal Communications Systems.

(6) Parks M, Saunders S, Evans B (1996) A wideband channel model applicable to mobil satellite systems at L- and S-Band. IEE Colloquium on propagation aspects of future mobile systems.

(7) Jahn A, Lutz E (1994) DLR channel measurement programme for low earth orbit satellite systems. Proceedings International Conference on Universal Personal Communications ICUPC'94, pp 423.429.

(8) Ernst H, Jahn A (1998) Channel models for land mobile satellite systems: a survey. ITG- Fachtagung Wellenausbreitung bei Funksystemen und Mikrowellensystemen, this issue.

(9) Lutz E et al (1991) The land mobile satellite communication channel – recording, statistics and channel model. IEEE Trans Veh Tech 40:375–386.

(10) Jahn A et al (1995) A wideband channel model for land mobile satellite systems. Proc Fourth Int Mobile Satellite Conf IMSC'95, pp 122–127.

(11) Jahn A, Bischl H, Heiß (1996) Channel characterisation for spread spectrum satellite communications. Proc IEEE Fourth International Symposium on Spread Spectrum Techniques and Applications (ISSSTA'96), pp 1221–1226.

(12) H. Ernst, A. Jahn (1998) Channel models for land mobile satellite systems: a survey. ITG- Fachtagung Wellenausbreitung bei Funksystemen und Mikrowellensystemen, this issue.

(13) Höher P, et al (1997) A suitability study of satellite emulation by airborne platforms. Int J Sat Comm 15:51–64.

(14) Jahn A (1994) Propagation data and channel model for LMS systems. Final Report, ESA Purchase Order 141742, DLR, Institut für Nachrichtentechnik.

(15) Lutz E et al (1991) The land mobile satellite communication channel – recording, statistics and channel model. IEEE Trans Veh Tech 40:375–386.

(16) Ernst H, Jahn A (1998) Channel models for land mobile satellite systems: a survey. ITG- Fachtagung Wellenausbreitung bei Funksystemen und Mikrowellensystemen, this issue.

(17) Jahn A, Bischl H, Heiß G (1996) Channel characterisation for spread spectrum satellite communications. Proc. IEEE Fourth International Symposium on Spread Spectrum Techniques and Applications (ISSSTA'96), pp 1221–1226.

(18) Jahn A, Bischl H, Lutz E (1996) Wideband channel model for UMTS satellite communications – detailed model. ETSI SMG5 (96) TD 006/96.

(19) Jahn A, Bischl H, Lutz E (1996) Wideband channel model for UMTS satellite communications – tapped delay model. ETSI SMG5 (96) TD 007/96.

(20) Jahn A, Bischl H, Lutz E (1996) Wideband channel model for UMTS satellite communications. ITU, REVAL, ITU-TG81, Meeting Mainz, Germany, 15–26 April.

(21) Lutz E et al (1991) The land mobile satellite communication channel – recording, statistics and channel model. IEEE Trans Vehicular Technology, 40:375–386.

(22) Jahn A et al (1995) A wideband channel model for land mobile satellite systems, Proc Fourth Int. Mobile Satellite Conf IMSC'95, pp 122–127.

(23) Lutz E (1996) A Markov model for correlated land mobile satellite channels, International Journal of Satellite Communications, 14, 333.339.

(24) Robet P.P., Evans B.G. and Ekman A (1992) Land mobile satellite communications channel model for simultaneous transmission from a land mobile terminal via two separate satellites, Int J Sat Comm 10:139–154.

(25) Lutz E et al (1991) The land mobile satellite channel – recording, statistics and channel model. IEEE Trans of Veh Tech, VT-40: 375–386.

(26) Lutz E, Cygan D, Dippold M, Dolainsky F, Papke W, (1991) The land mobile satellite channel – recording, statistics and channel model IEEE Trans of Veh Tech, VT-40, pp 375–386.

(27) Lutz E (1996) A Markov model for correlated land mobile satellite channels, Int J Sat Comm 14:333–339.

(28) Jahn A, Bischl H, Heiß G (1996) Channel charaterisation for spread spectrum satellite communications, Proc IEEE Fourth Symposium on Spread Spectrum Techniques and Appl, ISSSTA.

(29) Akturan R, Vogel W, (1997) Path diversity for LEO-satellite-PCS in the urban environment, IEEE on Ant and Prop 45(7) July 1997.

(30) Vogel WJ (1997) Satellite diversity for personal satellite communications – modelling and measurements, Tenth Int Conf on Ant and Prop, ICAP'97, pp 1269–1272.

(31) Akturan R, Penwarde K (1997) Satellite diversity as a propagation impairment mitigation technique for non-GSO MSS systems. International Mobile Satellite Conference, Pasadena.

(32) Karasawa Y, Kimura K, Minamisono K (1997) A propagation channel model for personal mobile-satellite services, IEEE Trans on Veh. Tech, VT-46:4.

(33) Meenan C et al (1998) Availability of First Generation Satellite Personal Communications Network Service in Urban Environments, IEEE VTC'98, pp 1471- 1475.

(34) Lutz E (1996) A Markov model for correlated land mobile satellite channels, Int J Sat Comm 14:333–339.

(35) Robet PP, Evans BG, Ekman A (1992) Land mobile satellite communication channel model for simultaneous transmission from a land mobile terminal via two separate satellites. Int J Sat Comm 10:139–154.

(36) Akturan R, Vogel WJ (1997) Path Diversity for LEO Satellite-PCS in the Urban Environment, IEEE Trans Veh Tech 45:1107–1116.

(37) Lutz E (1996) A Markov model for correlated land mobile satellite channels. Int J Sat Comm 14:333–339.

(38) Jahn A, Bischl H, Heiß G, (1996) Channel Charaterisation for Spread Spectrum Satellite Communications, Proc IEEE Fourth Symposium on Spread Spectrum Techniques and Appl, ISSSTA.

(39) Tzaras C., Evans B.G. and Saunders R.S. (1998) A physical-statistical analysis of the land mobile satellite channel, IEEE Electronics Letters 34(13):1355–1357.

(40) Fontan FP, et al (1998) A methodology for the characterisation of environmental effects on global navigation satellite system (GNSS) Propagation, Int Sat Comm 16:1–22.

(41) Tzaras C, Saunders SR, Evans BG (1998) A Physical-Statistical Propagation model for diversity in mobile Satellite PCN, IEEE VTC'98.

(42) Saunders SR et al (1997) A physical statistical model for land mobile satellite propagation in built-up areas. Tenth International Conference on Antennas and Propagation, ICAP'97, pp 2.44–2.47.

(43) Tzaras C, Evans BG, Saunders RS (1998) A physical-statistical analysis of the land mobile satellite channel. IEEE Electronics Letters 34(13):1355–1357.

(44) Fontan FP et al (1998) A methodology for the characterisation of environmental effects on global navigation satellite system (GNSS) propagation. Int J Sat Comm 16:1–22.

(45) Vazquez MA, (1998) Land mobile satellite channel modelling by means of statistical and deterministic methods, PhD thesis, University of Vigo, Spain (in Spanish).

(46) Jahn A, Bischl H, Heiß G (1996) Channel charaterisation for spread spectrum satellite communications, Proc IEEE Fourth Symposium on Spread Spectrum Techniques and Appl, ISSSTA.

(47) Tzaras C, Saunders SR, Evans BG (1998) A physical-statistical propagation model for diversity in mobile satellite PCN, IEEE VTC'98.

(48) Krewel W, Maral G (1998) Single and multiple satellite visibility statistics of first-generation non-GEO constellations for personal communications. Int J Sat Comm 16:105–125.

(49) Akturan R, Vogel W (1997) Path Diversity for LEO-Satellite-PCS in the Urban Environment, IEEE on Ant and Prop 45(7), July (1997).

(50) Bosch J (1996) Impact of diversity reseption on fading channels with coded modulation', PhD thesis, Politecnico di Torino.

(51) Akturan R, Vogel W (1997) Path diversity for LEO-satellite-PCS in the urban environment. IEEE on Ant and Prop 45(7), July 1997.

(52) Bosch J (1996) Impact of diversity reseption on fading channels with coded modulation. PhD thesis, Politecnico di Torino.

(53) Goldhirsh J, Vogel W (1992) Propagation effects for land mobile satellite systems: overview of experimental and modelin results, NASA Reference Puiblication 1274.

(54) Davidoff M (1994) The Satellite Experimenter's Handbook. The American Radio Relay League.

(55) Krewel W, Maral G (1998) Single and multiple satellite visibility statistics of first-generation non-GEO constellations for personal communications. Int J Sat Comm 16;105–125.

(56) Zaghloul A (1990) Advances in multibeam communications satellite antennas. Proc. IEEE 78(7):1214–1232.

(57) Rao, KS et al (1992) Reconfigurable L-Band active array antennas for satellite communications, Can J Elect and Comp Eng 17(3).

(58) Cances J-P, et al (1994) Coverage reconfiguration for dynamicallocation in a multibeam satellite system, Fifteenth AIAA International Communications Satellite Systems Conference (ICSSC-15), pp 1032–1041, San Diego, 28 February–3 March 1994.

(59) Luglio M, Forcella A, Vatalaro F (1998) Mitigation of Interference Impairments due to Multibeam Coverage for Geostationary Satellite Systems in case of Site and Time Dependent Traffic Distribution for MF_TDMA Access, Technical Document TD(98)25, COST252.

(60) Del Re E, Fantacci R, Giambene G (1995) Efficient Dynamic Channel Allocation Techniques with Hand over Queuing for Mobile Satellite Networks', IEEE J Sel Areas Comm 13(2):397–404.

(61) Lüders R (1961) Satellite Networks for continous zonal coverage. Am Rocket Soc J.

(62) Cercas FAB (1996) A new family of codes for simple receiver implementation, PhD Thesis, Technical University of Lisbon, Instituto Superior Técnico, March 1996.

(63) Tomlinson M, Cercas FAB, Hughes EC (1991) Aspects of coding for power efficient satellite VSAT systems, ESA Journal 15:165–185.

(64) Cercas FAB (1996) A new family of codes for simple receiver implementation, PhD thesis, Technical University of Lisbon, Instituto Superior Técnico, March 1996.

(65) Rice M et al (1996) K-band land-mobile satellite channel characterisation using ACTS. Intern Journ of Sat Comm 14:283–296.

(66) Hans Riesel (1985) Prime numbers and computer methods for factorization, Progress in Mathes, vol. 57.

(67) Cercas FAB (1996) A new family of codes for simple receiver implementation, PhD thesis, Technical University of Lisbon, Instituto Superior Técnico, March 1996.

(68) Cercas FAB, Tomlinson M, Albuquerque AA (1993) TCH: A new family of cyclic codes length 2m, 1993 IEEE International Symposium on Information Theory, Hilton Palacio del Rio Hotel, San Antonio, Texas, USA, 17–22 January 1993, p 198.

(69) TD(98) 15 Simplified Receiver Structure with new Sequences for CDMA using FFT Implementation, Cercas F, Del Re E, Fantacci R, Ronga LS.

(70) Cercas FAB (1996) A new family of codes for simple receiver implementation, PhD Thesis, Technical University of Lisbon, Instituto Superior Técnico, March 1996.

(71) Gilbert E (1960) Capacity of a Burst-Noise Channel, The Bell System Tech J, September 1960, pp 1253.

(72) Jeruchim M, Balaban P, Shanmugan K (1994) Simulation of Communications Systems, 2nd edn, Plenum Press, New York, 1994, p. 386.

(73) Proakis JG (1995) Digital Communications, McGraw-Hill, 3rd edn.

(74) Cercas FAB (1996) A new family of codes for simple receiver implementation, PhD Thesis, Technical University of Lisbon, Instituto Superior Técnico, March 1996.

(75) Bian Y, Popplewell A, O'Reilly J (1994) Novel Simulation Technique for Assessing Coding System Performance, Electronics Letters, 10 November 1994, 30(23).

(76) Bian Y, Popplewell A, O'Reilly J (1994) Novel Simulation Technique for Assessing Coding System Performance, Electronics Letters, 10 November 1994, 30(23).

(77) Jeruchim M, Balaban P, Shanmugan K (1994) Simulation of Communications Systems, 2nd edn, Plenum Press, New York, p. 393.

(78) P. Sebastião, Efficient Simulation of the Performance of TCH Codes Using Stochastic Models (in Portuguese), M.Sc. Thesis, Instituto Superior Técnico, Lisbon, October 1998.

(79) Jeruchim M, Balaban P, Shanmugan K (1994) Simulation of Communications Systems, 2nd edn, Plenum Press, New York, p. 394.

(80) Sebastião P (1998) Efficient Simulation of the Performance of TCH Codes Using Stochastic Models (in Portuguese), M.Sc. Thesis, Instituto Superior Técnico, Lisbon, October 1998.

(81) Jeruchim M, Balaban P, Shanmugan K (1994) Simulation of Communications Systems, 2nd edn, Plenum Press, New York, p. 394.

(82) Jeruchim M, Balaban P, Shanmugan K (1994) Simulation of Communications Systems, 2nd edn, Plenum Press, New York, p. 395.

(83) Cercas FAB (1996) A new family of codes for simple receiver implementation, PhD Thesis, Technical University of Lisbon, Instituto Superior Técnico, March 1996.

(84) ITU WWW site on Radio Transmission Technology proposals for IMT-2000, http://www.itu.int/imt/2-radio-dev/proposals/index.html

(85) Adachi, Sawahashi F, Okawa M (1997) Tree-structured Generation of Orthogonal Spreading Codes with Different Lengths for Forward Link of DS-CDMA Mobile Radio, Electronics Letters, 33:27–28, January 1997.

(86) Verdù S (1998) Multiuser Detection, Cambridge University Press, Cambridge UK.

(87) Del Re E et al (1999) Multi-user Cancellation Detector for S-UMTS Multirate Communications, Proc. Of the Fifth European Conference on Satellite Communications, November 1999, Toulouse, France.

(88) Patel P, Holtzman J (1994) Analysis of a simple successive interference cancellation scheme in DS-CDMA system, IEEE J Sel Areas Comm., 12:796–807, June 1994.
(89) Viterbi A (1990) Very low rate convolutional codes for maximum theoretical performance, of spread spectrum multiple access channels, IEEE J on Sel Areas Comm 8(4), May 1990.
(90) Patel P, Holtzman J (1995) Performance comparison of a DS-CDMA system using a successive interference cancellation scheme and a parallel IC scheme under fading, IEEE Comm. Mag 33:58–67, January 1995.
(91) Yoon Y, Kohno R, Imai H (1996) A spread-spectrum multi-access system with co-channel interference cancellation over multipath fading channels, IEEE J Sel Areas Comm 11:1519–1521, 1996.
(92) Del Re E et al (1998) One-shot multiuser cancellation receiver for UMTS satellite CDMA systems, Proc of the sixth International Workshop on Digital Signal Processing Techniques for Space Applications, September 1998, ESTEC, NordWijk, The Netherlands.
(93) Del Re E et al (1998) Multi-user cancellation detector for UMTS CDMA satellite communications, Proc of the Third European Workshop on Mobile/Personal Satcoms, EMPS98, November 1998, Venice, Italy.
(94) Natali F (1984) AFC tracking algorithms. IEEE Trans Comm, 32:935–947, Aug. 1984.
(95) Bischl H, Jahn A, Lutz E (1998) Wideband channel model for UMTS satellite communications, TD(98)05 Temporary Document inside COSTaction252.
(96) Bischl H, Jahn A, Lutz E (1998) Wideband channel model for UMTS satellite communications, TD(98)05 Temporary Document inside COSTaction252.
(97) Fantacci R, Morosi S, Panchetti F (1999) One-Shot Multiuser Cancellation Receiver for Wireless CDMA Communication Systems, Proc of Vehicular Technology Conference (VTC99-Fall), Amsterdam, Holland, September 1999.
(98) Bischl H, Jahn A, Lutz E (1998) Wideband Channel Model for UMTS Satellite Communications, TD(98)05 temporary document inside COSTaction252.
(99) TD(98) 28. One-Shot Multi-User Cancellation Receiver for UMTS Satellite CDMA Systems, E. Del Re, R. Fantacci, S. Morosi, F. Panchetti, P. Bagnoli.
(100) Bischl H, Jahn A, Lutz E (1998) Wideband Channel Model for UMTS Satellite Communications. TD(98)05 temporary document inside COSTaction252.
(101) Bischl H, Jahn A, Lutz E (1998) Wideband channel model for UMTS satellite communications. TD(98)05 temporary document inside COSTaction252.
(102) Proposta di un Ricevitore Multiutente Vettoriale per Sistemi di Comunicazione Radiomobile DS/CDMA, L. Mucchi, Telecommunications Engineering Degree Thesis.
(103) Verdú S, Honig ML, Madhow U (1995) Blind Adaptive Multiuser Detection, IEEE Transactions on Information Theory 41(4), July 1995.
(104) Bischl H, Jahn A, Lutz E (1998) Wideband Channel Model for UMTS Satellite Communications, TD(98)05 Temporary Document inside COSTaction252.
(105) Del Re E et al (1999) Advanced Blind Adaptive Multiuser Detector for Communications in Non Stationary Multipath Fading Channel, Proc of the Fifth Bayona Workshop on Emerging Technologies in Telecommunications, September 1999, Bayona, Spain.
(106) De Gaudenzi R, Giannetti F, Luise M (1998) Design of a low-complexity adaptive interference-mitigating detector for DS/SS receivers in CDMA Radio Networks. IEEE Trans Comm, January 1998.
(107) Verdú S, Honig ML, Madhow U (1995) Blind adaptive multiuser detection, IEEE Trans on Inf Theory 41(4), July 1995.
(108) De Gaudenzi R, Giannetti F, Luise M (1998) Design of a Low-Complexity Adaptive Interference-Mitigating Detector for DS/SS Receivers in CDMA Radio Networks, IEEE Trans Comm, January 1998.
(109) Verdú S, Honig ML, Madhow U (1995) Blind adaptive multiuser detection, IEEE Trans on Inf Theory 41(4), July 1995.
(110) Madhow U, Honig ML (1993) MMSE detection of CDMA signals: analysis for random signature sequences, Proc 1993 IEEE Int Symp on Information Theory, January 1993.
(111) Verdú S, Honig ML, Madhow U (1995) Blind Adaptive Multiuser Detection. IEEE Trans on Inf Theory 41(4), July 1995.
(112) Verdú S, Honig ML, Madhow U (1995) Blind Adaptive Multiuser Detection. IEEE Trans on Inf Theory 41(4), July 1995.
(113) De Gaudenzi R, Giannetti F, Luise M (1998) Design of a Low-Complexity Adaptive Interference-Mitigating Detector for DS/SS Receivers in CDMA Radio Networks. IEEE Trans Comm, January 1998.
(114) Del Re E et al (1999) Advanced Blind Adaptive Multiuser Detector for Communications in Non Stationary Multipath Fading Channel. Proc of the Fifth Bayona Workshop on Emerging Technologies in Telecommunications, September 1999, Bayona, Spain.
(115) Smith RF, Miller SL, Acquisition Performance of an MMSE Receiver for DS-CDMA, submitted to IEEE Transactions on Vehicular Technology.
(116) Proposta di un Ricevitore Multiutente Vettoriale per Sistemi di Comunicazione Radiomobile DS/CDMA, Mucchi L, Telecommunications Engineering degree thesis.
(117) TD(98) 27 Application of the OFDM-CDMA Technique in a LEO satellite system for communications with multiple bit-rate services, L. Branchetti, E. Del Re, R. Fantacci, L. Ronga.
(118) TD(98) 27 Application of the OFDM-CDMA Technique in a LEO Satellite System for Communications with Multiple Bit-Rate Services, Branchetti L, Del Re E, Fantacci R, Ronga L.
(119) TD(98) 08 FFT Implementation of OFDM-CDMA using Perfect Sequences in Multipath Environment, Del Re E, Fantacci R, Ronga L.
(120) TD(98) 08 FFT Implementation of OFDM-CDMA using Perfect Sequences in Multipath Environment, Del Re E, Fantacci R, Ronga L.
(121) Pursley MB, Performance evaluation for phased-coded spread-spectrum multiple-access communication – Part I: System analysis, IEEE Trans Commun, COM-25(8): pp 795–799, August 1977.
(122) Pursley MB, Sarwate DV (1977) Evaluation of correlation parameters for periodic sequences", IEEE Trans Inform Theory, IT-23(4): pp 508–513, July 1977.
(123) Bomer L, Antweiler M (1992) Perfect n-phase sequences, IEEE J Sel Areas Comm 10(4):782–789, May 1992.
(124) Chu DC (1972) Polyphase codes with good periodic correlation properties, IEEE Trans Inform Theory, IT-18:531–532.
(125) Frank RL, Zado S (1962) Phase shift pulse code with good periodic correlation properties, IEEE Trans Inform Theory, IT-8:381–382.
(126) Milewski A (1983) Periodic sequences with optimal properties for channel estimation and fast start-up equalisation, IBM J Res Develop 27:426–431, 1983.

(127) Viterbi A (1990) Very low rate convolutional codes for maximum theoretical performance of spread spectrum multiple access channels, IEEE J Sel Areas Comm, 8(4), May 1990.
(128) Frenger P, Orten P, Ottosson T (1998) Combined coding and spreading in CDMA systems using maximum free distance convolutional codes, Forty-eighth Annual Vehicular Tech Conf, Ottawa, Canada, 18–21 May 1998.
(129) Viterbi A (1990) Very low rate convolutional codes for maximum theoretical performance of spread spectrum multiple access channels, IEEE J Sel Areas Comm 8(4), May 1990.
(130) Viterbi A (1967) Orthogonal tree codes for communications in the presence of white gaussian noise, IEEE Trans on Comm Tech, COM-15, April 1967.
(131) Viterbi AJ, Omura JK (1979) Principles of digital communication and coding. McGraw-Hill.
(132) Gilbert E (1960) Capacity of a Burst-Noise Channel, The Bell System Tech J, September 1960, p 1253.
(133) Frenger P, Orten P, Ottosson T (1998) Combined coding and spreading in CDMA systems using maximum free distance convolutional codes, Forty-eighth Annual Vehicular Tech Conf, Ottawa, Canada, 18–21 May, 1998.
(134) Pursley MB (1977) Performance evaluation for phased-coded spread-spectrum multiple-access communication – Part I: System analysis, IEEE Trans Comm COM-25(8):795–799 August 1977.
(135) Prasad R (1996) CDMA for wireless personal communications. Artech House Publishers, 1996.
(136) Andrson J, Aulin T, Sundberg C-E (1986) Digital phase modulation. Plenum Press Company.
(137) Blahut RE (1983) Theory and practice of error control codes. Addison Wesley.
(138) Del Re E et al (1999) Multi-user cancellation detector for S-UMTS multirate communications. Proc of the Fifth European Conference on Satellite Communications, November 1999, Toulouse, France.
(139) Gilhousen et al (1991) On the capacity of a cellular CDMA System, IEEE Trans Veh Tech pp 303–312, May 1991.
(140) Gold R (1968) Maximal recursive sequences with 3-valued recursive cross-correlation functions, IEEE Trans Inform Theory IT-14:154–156, Jan. 1968.
(141) Joseph Hui (1984) Throughput analysis for code division multiple accessing of the spread spectrum channel, IEEE J Sel Areas in Comm SAC-2(4) July 1984.
(142) Losquadro G, Luglio M, Vatalaro F (1997) A geostationary satellite system for mobile multimedia applications using portable, aeronautical and mobile terminals. Proceedings Fifth International Mobile Satellite Conference (IMSC'97), pp 427–432, 1997.
(143) Sarwate, Pursley (1980) Crosscorrelation Properties of Pseudo-random and Related Sequences, Proceedings of the IEEE, May 1980.
(144) Steele (1992) Mobile Radio Communications, Pentech Press, London.
(145) Webb W, Hanzo L (1994) Modern Quadrature Amplitude Modulation, IEEE Press, Pentech Press.

# CHAPTER 5

# Conclusions and Future Actions

**E. Del Re and L. Pierucci** University of Florence, Italy

New generation satellites will provide a complementary service to the future fixed or terrestrial mobile networks, since they can offer coverage to remote users. It is expected that the International Mobile Telecommunications after the year 2000 (IMT-2000) and the Universal Mobile Telecommunication System (UMTS) standards will provide seamless mobile multimedia services to users throughout the world, at rates up to 2 Mb/s on the terrestrial segment and slightly lower on the satellite segment. Current work in the European Advanced Communications Technologies and Services (ACTS) projects such as (SECOMS, ASSET, WISDOM) under the Mobile Broadband Services (MBS) and from the Ka-band (SPACEWAY, GALAXY, EUROSKYWAY, SKYBRIDGE and TELEDESIC) commercial systems propose a possible portable implementation up to 2 Mb/s. As far as time scales are concerned UMTS will be operational by 2002 and the second generation of LEO and GEO satellites will address to the mobile/portable multimedia market in the 2002–2010.

According to the points mentioned in the above paragraph, the kind of work that has been carried out in the COST 252 programme is completely justified. The COST 252 work provide many innovative results which can be the basis for new global systems generation with services at high rates.

The main activities of COST 252 have been developed considering the following identified scenarios in the evaluation of the satellite component in an integration system with terrestrial networks:

- S-UMTS scenario: current guidelines suggest that rates of up to 144 kbit/s are possible for S-UMTS. This scenario necessitates some form of adaptation to allow S-UMTS network to interface with the B-ISDN network.
- S-ATM scenario: transmission of ATM-like cells via satellite, assuming a B-ISDN backbone network.
- S-IP scenario: internet protocol transmissions allowing direct connection to an IP backbone network.

As a result of this project a new COST Action, in conjunction to the COST253 Action named COST 272 "Packet-oriented Service Delivery via Satellite" started in June 2001.

COST 272 objectives can be found on the site http://www.tesa.pzd.fc/cost272